QH	80433
546	Dormancy and
.D67	developmental
	arrest :
	experimental
	analysis in
	plants and

DATE			

Contributors

R. J. Berger

G. L. Florant

S. F. Glotzbach

H. C. Heller

Arthur M. Jungreis

Larry D. Nooden

Marilyn B. Renfree

Graham M. Simpson

Ian M. Sussex

Virginia Walbot

J. M. Walker

James A. Weber

Dormancy
and
Developmental Arrest

Experimental Analysis
in Plants and Animals

Edited by

Mary E. Clutter

Department of Biology
Yale University
New Haven, Connecticut

ACADEMIC PRESS   New York San Francisco London 1978
A Subsidiary of Harcourt Brace Jovanovich, Publishers

ACADEMIC PRESS, INC.
111 Fifth Avenue, New York, New York 10003

United Kingdom Edition published by
ACADEMIC PRESS, INC. (LONDON) LTD.
24/28 Oval Road, London NW1 7DX

Library of Congress Cataloging in Publication Data

Main entry under title:

Dormancy and developmental arrest.

Includes bibliographies and index.
1. Dormancy (Biology). I. Clutter, Mary E.
QH546.D67   574.5'4   78−18482
ISBN 0−12−177050−8

Contents

5. Environmental and Hormonal Control of Dormancy in Terminal Buds of Plants
Larry D. Nooden and James A. Weber

6. Sleep and Torpor—Homologous Adaptations for Energy Conservation
H. C. Heller, G. L. Florant, S. F. Glotzbach, J. M. Walker, and R. J. Berger

7. Dormancy and Development
Ian M. Sussex

List of Contributors

Numbers in parentheses indicate the pages on which the authors' contributions begin.

R. J. Berger (269), Thimann Laboratories, University of California, Santa Cruz, California 95064

G. L. Florant* (269), Department of Biological Sciences, Stanford University, Stanford, California 94305

S. F. Glotzbach (269), Department of Biological Sciences, Stanford University, Stanford, California 94305

H. C. Heller (269), Department of Biological Sciences, Stanford University, Stanford, California 94305

Arthur M. Jungreis (47), Department of Zoology, University of Tennessee, Knoxville, Tennessee 37916

Larry D. Nooden (221), Department of Botany, University of Michigan, Ann Arbor, Michigan 48109

Marilyn B. Renfree (1), School of Environmental and Life Sciences, Murdoch University, Murdoch, Western Australia 6153

Graham M. Simpson (167), Crop Science Department, University of Saskatchewan, Saskatoon, Saskatchewan, Canada

Ian M. Sussex (297), Department of Biology, Yale University, New Haven, Connecticut 06520

Virginia Walbot (113), Department of Biology, Washington University, St. Louis, Missouri 63130

J. M. Walker (269), Thimann Laboratories, University of California, Santa Cruz, California 95064

James A. Weber (221), Biological Station, University of Michigan, Ann Arbor, Michigan 48109

* Present address: Albert Einstein College of Medicine and Mortefiore Hospital Medical Center, Department of Neurology, 111 East 210th Street, Bronx, New York 10467.

Preface

"Eels in blighted Wheat"

In a month of August 1743, a small Parcel of blighted Wheat was sent by Mr. Needham to Martin Folkes, Esq; President of the Royal Society (with an Account of his new Discovery) which Parcel the President was pleased to give to me, desiring I would examine it carefully. With this material he was very soon "entertained with the pleasing sight of this wonderful Phenomenon". He repeated these experiments at different times and even with grains kept dry for 4 years, always with the same success and demonstrated it to Mr. Folkes and some other friends. "We find [he writes] an Instance here, that Life may be suspended and seemingly destroyed; that by an Exhalation of the Fluids necessary to a living Animal, the Circulations may cease, all the Organs and Vessels of the Body may be shrunk up, dried, and hardened; and yet, after a long while, Life may begin anew to actuate the same Body; and all the animal Motions and Faculties may be restored, merely by replenishing the Organs and Vessels with a fresh supply of Fluid. Here is, I say, a Proof, that the Animalcules in the Grains of blighted Wheat can endure having their Bodies quite dried up for the Space of four Years together, without being thereby deprived of their living Power: and since, after they are become thus perfectly dry and hard, there seems little room for farther Alteration, unless their Organs should be broken or torn asunder; may they not possibly be restored to Life again, by the same Means, even after twenty, forty, an hundred, or any other Number of Years provided their Organs are preserved intire? This Question future Experiments alone can answer.

> **Henry Baker** (1698–1774) biologist, microscopist, Fellow of The Royal Society, in *Employment for the microscope*, Chapter IV as reported by D. Keilin in The Leeuwenhoek Lecture: The problem of anabiosis or latent life: history and current concept delivered at the University of Cambridge, June 19, 1958 [*Proc. R. Soc. London B* **150**, 1959].

This report by Henry Baker in 1743 provides the question of the present volume.

The possibilities and potentialities of the condition variously known as dormancy, hibernation, diapause, cryptobiosis, anabiosis, anhydrobiosis, abiosis, quiescence, aestivation, latent life, and other synonyms of suspended animation have long held a compelling fascination not only for scientists and students of biology and exobiology, but also for religious thinkers, philosophers, writers of science fiction, tellers of tales, such as "Sleeping Beauty" and other fairy stories, and cryonics entrepreneurs. In short, any mechanism which might serve to extend life, especially human life, has been of longstanding interest.

There are a variety of states of developmental or metabolic arrest which occur in microorganisms, plants and animals. Keilin (1958) placed them on a scale of metabolic activity ranging from cryptobiosis (0%) to active life (100%). The following scale was adapted from Keilin's:

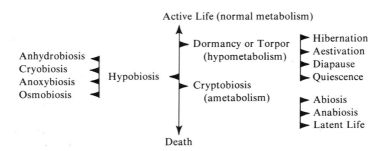

Dormant states can occur over time spans as brief as minutes as sometimes occurs in sleep or torpor or as long as hundreds of years in the case of spores or seeds. An exciting idea advanced in the last century was the doctrine of panspermy. It is no longer widely believed but it is postulated that life may have been introduced on earth from an origin in outer space via dormant spores released from fallen meteorites. Intriguing as this idea is there is no direct supporting evidence, except that spores are known to survive extremes in environmental conditions.

Some seeds are also known to remain viable for many years. The famous case of the viable lotus seeds found in peat of an ancient lake bed in Manchuria brought forth speculations that the seeds may have been deposited in the Pleistocene, about 50,000 years ago. Other evidence suggests that they may have been only a few hundred years old. Whether the seeds were hundreds or thousands of years old is immaterial. The importance is that development was discontinued or arrested for a period of time, then resumed without damage to the organism.

Experimentation on states of suspended animation did not begin or were not recorded until Henry Power, a physician and Fellow of the Royal Society, reported on his experiments on freezing and thawing eel-worms (*Turbatrix aceti*) in 1663 and Leeuwenhoek reported his discovery in 1702 that certain "Animalcules" (rotifers) could be resuscitated after having been evaporated to dryness. As a result experimentation for the next two

centuries was largely centered on microorganisms which could be placed in
a cryptobiotic or hypobiotic state by physical or chemical means and then
revived. Similar experiments are being done today, albeit on a more
sophisticated level. However, the state of the field is probably best indicated
in the title of a 1971 paper by J. H. Crowe, "Anhydrobiosis: an unsolved
problem."

Contemporary research on dormancy encompasses the whole range of
hypo and ametabolic states in all forms of life and has been extended to
include metabolic arrest of whole organisms as well as parts of organisms,
organs, tissues, and cells, either as a natural or developmental phenomenon
or as a controlled artificial arrest.

The present book grew out of a AAAS symposium entitled "Mechanisms
of Dormancy and Developmental Arrest" held in December 1972. Since all
living organisms exhibit some form of dormancy, the objective of the book
is to examine experimental evidence that may shed light on some of the
common underlying mechanisms.

The plant and animal examples were selected to show that dormancy may
occur at a particular stage of development, or that it may be a recurrent
phenomenon occurring in relation to changes in the environment. The
chapters illustrate the range and complexity of the phenomenon. It is not
simply a cessation of development, but is an interconnected process involving structural, physiological, and molecular alterations in the organism
which causes it to enter the arrested state, to maintain that state for a
period of time, then to exit from it into another period of activity.

The timing of events of these phenomena is critical. It may be controlled
by environmental stimuli, but the primary mechanism appears to be an
endogenous rhythm which is phased by photoperiod or other environmental
factors.

Recent efforts to discover the connection between physiological responses
to environmental signals and cellular metabolism have been increasingly
directed to the level of gene activity. In many examples of dormancy, nucleic acid and protein synthesis cease prior to entry into the dormant state
and resume rapidly when this state is broken. Experimental evidence favors
a transcriptional mode of regulation, but there is increasing evidence for
regulation at the translational level as well. The examples presented here
examine the phenomenon at all levels. It is hoped that they will provide
insight into the complex problem and stimulate further research in this field
so that at last Baker's questions can be answered.

I should like to thank the authors of the chapters for their patience and
cooperation.

1

Embryonic Diapause in Mammals—A Developmental Strategy

MARILYN B. RENFREE

I. INTRODUCTION: PATTERNS OF DELAY—OBLIGATE AND FACULTATIVE

Early studies of reproduction in mammals led to the finding that each species had a characteristic period of gestation with a fairly narrow range.

1

DORMANCY AND DEVELOPMENTAL ARREST

In the roe deer, however, although the rut occurred in the beginning of August, there was no embryo visible in the uterus until January. It was assumed that mating must have occurred during a period of "silent heat," for the observed matings seemed well separated in time from the first obvious signs of pregnancy.

Ziegler in 1843 and Bischoff in 1854 made the discovery that in the roe deer, *Capreolus capreolus*, mating was followed by a period of quiescence and an almost complete cessation of growth of the embryo (Short and Hay, 1966). This led to a new concept that certain species manifest a delay in implantation, quite distinct from a delay in fertilization following sperm storage, which came to be known as "delayed implantation" or "embryonic diapause." The roe deer is the only ungulate whose embryo has so far been found to show a delayed period of growth. Several other workers at the turn of the century subsequently described the occurrence of embryonic diapause in a variety of species, for example, badgers, rodents, and armadillos (reviewed by Wimsatt, 1975). However, most of the impetus to study diapause has occurred in the last two decades; diapause was discovered in marsupials only relatively recently (Sharman, 1954, 1955a, 1955b).

Reproduction and development generally occur when environmental conditions are favorable for a species. Unfavorable conditions, especially when they recur periodically, may be avoided by migration or by the intervention of a resting or dormant stage in the animal's life cycle. Dormancy can involve the suspension of some physiological activity in the organism, or it may comprise the suspension of reproductive growth and development. Seasonal reproductive cycles are distinct from suspension of reproductive activity. In insects, for example, where adults can suspend metabolism, the condition is usually considered true dormancy or diapause. Among mammals in contrast, there are seasonally breeding species that have developed such adaptations to take advantage of favorable conditions, whereas species with a reproductive dormancy have often evolved mechanisms to avoid unfavorable periods in what would otherwise be a continuous reproductive life.

There are many different forms of dormancy among mammals. There is differentiation among the various kinds of quiescence, torpor, estivation, and hibernation, on the one hand, which are all clearly related to environmental conditions, and "embryonic diapause," on the other, which is probably more closely related to the physiology of early development.

This account is restricted to the latter type of dormancy in mammals, that is, "delayed implantation" or "embryonic diapause," and emphasizes the characteristics of the phenomenon in three disparate wild species that have recently been studied in depth: the roe deer, *Capreolus capreolus*; the western spotted skunk, *Spilogale putorius latifrons*; and the tammar wallaby, *Macropus eugenii*.

The usual mammalian development pattern begins with fertilization of the ovum. The resultant zygote spends a predictable period of time in the oviduct and eventually reaches the uterus. During the passage down the oviduct, numerous cleavage divisions occur, resulting in the formation of a hollow ball of cells known as the blastocyst. The blastocyst continues to expand by cell division and by accumulation of fluids from the uterine secretions. Eventually it attaches to the uterine wall via its chorion or trophoblast (Eckstein *et al.*, 1959). The duration of the preimplantation stage is determined by the species, but it ranges from about 4 to 6 days in rabbits and mice, through nearly 20 days in some marsupials, to 40–60 days in ungulates (Wimsatt, 1975). Implantation and the formation of the yolk sac or allantoic placenta facilitate maternal–fetal exchanges; embryogenesis and organogenesis generally proceed without interruption.

In some mammals, a period of delay occurs during this process. Implantation itself may be delayed or, alternatively, development before implantation may be delayed. The term "delayed implantation" is used to denote a prolonged period in which the blastocyst is not directly attached to the uterus although it remains within the uterine lumen. In many but not all species showing delay, implantation follows immediately after the resumption of development. For those animals whose blastocysts do not implant immediately, the term, strictly speaking, is incorrect, since the delay occurs before implantation and the process of implantation itself is not inhibited. Several authors have suggested instead that "embryonic diapause" better describes this condition in mammals (Tyndale-Biscoe, 1963a; Short and Hay, 1966; Baevsky, 1963).

Insect dormancy is also known as diapause, a name used originally in insect embryology to describe a stage just before the reversal of movement of the embryo around the posterior pole of the egg. The meaning in insect biology has now changed and refers to a "condition" of arrested growth, whether embryonic or postembryonic, rather than to a stage of morphogenesis. In these metazoans, diapause can occur at any stage of the life cycle—in the fertilized egg, in larval instars, in the prepupal or pupal instars, and in the adult instar. The critical point to note is that the diapausing insect requires an environmental cue coupled to a physiological trigger to release it from diapause, just as mammals do.

Both "delayed implantation" and "embryonic diapause" are used more or less interchangeably in mammalian studies. In this review, "embryonic diapause" is used. Two types of delay occur, and, as in insects, distinction is usually made between "obligatory" and "facultative" diapause. "Obligate

TABLE I

Species of Marsupial Mammals Showing Embryonic Diapause

Taxon	Common name	Reference
Subclass Marsupialia		
Order Diprotondonta		
Family Macropodidae		
Subfamily Potoroinae		
Potorous tridactylus	Potoroo	Hughes (1962)
Bettongia lesueuri	Boodie	Tyndale-Biscoe (1968)
B. penicillata	Woylie (brush-tailed bettong)	Samson (1971)
B. gaimardii (cuniculus)	Rat kangaroo (eastern bettong)	Hill (1900), Flynn (1930)
Subfamily Macropodinae		
Macropus eugenii	Tammar wallaby	Berger (1966), Renfree and Tyndale-Biscoe (1973a)
M. rufogriseus	Red-neck or Bennetts wallaby	Sharman *et al.* (1966)
M. giganteus	Eastern gray kangaroo	Poole and Pilton (1964), Kirkpatrick (1965), Clark and Poole (1967), Poole (1975)
M. parma	Parma wallaby	Maynes (1973)
M. robustus	Euro	Ealey (1963), Sadlier and Shield (1960)
M. irma	Western black-gloved wallaby	Ride and Tyndale-Biscoe (1962)
M. agilis	Agile wallaby	Merchant (1976)
Megaleia rufa	Red kangaroo	Sharman and Calaby (1964), Clark (1966)
Wallabia bicolor	Swamp wallaby	Sharman *et al.* (1966)
Lagorochestes hirsutus	Western hare wallaby	Ride and Tyndale-Biscoe (1962)
Lagostrophus fasciatus	Banded hare wallaby (munning)	Tyndale-Biscoe (1965)
Thylogale thetis	Red-necked pademelon	Sharman and Berger (1969)
T. billarderii	Tasmanian pademelon	Sharman and Berger (1969)
Setonix brachyurus	Quokka	Sharman (1955a,b)
Family Burramyidae		
Cercatetus conninnus	Pigmy possum (mundarda)	Clark (1967), Hill (1900)
Acrobates pygameus	Pigmy glider (feathertail glider)	Clark (1967), Hill (1900)

TABLE II

Species of Eutherian Mammals Showing Embryonic Diapause

Taxon	Common name	Reference
Subclass Eutheria		
Order Insectivora		
Family Soricidae		
Sorex araneus	Common shrew	Brambell (1935)
S. minutus	Lesser shrew	Brambell (1937), Brambell and Hall (1936)
Order Chiroptera		
Family Pteropidae		
Eidolon helvum	Fruit bat	Mutere (1965, 1967)
Family Vespertilionidae		
Miniopterus australis	Little bent-winged bat	Dwyer (1963)
M. schreibersii	Bent-winged bat	Courrier (1927), Dwyer (1968), Peyre and Herlant (1967)
Order Edentata		
Family Dasypodidae		
Dasypus novemcinctus	Nine-banded armidillo	Enders (1966)
D. hybridus	Mulita armidillo	Hamlett (1932)
Order Rodentia		
Family Muridae		
Peromyscus sp.	Deer mouse	Svihla (1932)
Mus musculus	Common mouse	Reviewed by Psychoyos (1973a and b)
Rattus norvegicus	Laboratory rat	Reviewed by Psychoyos (1973 a and b)
Phenacomys longicaudus	Red tree mouse	Hamilton (1962)
Family Cricetidae		
Meriones unguiculatus	Mongolian gerbil	
Clethrionomys glaneolus	Vole	Brambell and Rowlands (1936)
Order Artiodactyla		
Family Cervidae		
Capreolus capreolus	Roe deer	Short and Hay (1966), Aitken (1974a)
Order Carnivora		
Family Usidae		
Ursus americanus	American black bear	Wimsatt (1963)
U. arctos arctos	European brown bear	Hamlett (1935)
U. arctos horribilus	Grizzly bear	Craighead *et al.* (1969)
U. maritimus	Polar bear	Hamlett (1935), Asdell (1964)
Selenarctos tibetanus	Himalayan black bear	Dittrich and Kronberger (1962)

(Continued)

TABLE II   *(Continued)*

Taxon	Common name	Reference
Family Mustelidae		
Martes americana	American marten	Wright (1963)
M. pennanti	Fisher	Wright (1963)
M. zibellina	Sable	Baevsky (1963)
Gulo gulo	Wolverine	Wright (1963)
Lutra canadensis	River otter	Wright (1963)
Enhydra lutrus	Sea otter	Sinha *et al.* (1966)
Mustela fenata	Long-tailed weasel	Wright (1963)
M. vison	Mink	Pearson and Enders (1944), Dukelow (1966)
M. erminea	Stoat (short-tailed weasel)	Deanesly (1943), Wright (1963)
Spilogale putorius latrifons	Western|spotted skunk	Mead (1968a,b)
Taxidea taxus	American badger	Wright (1966)
Meles meles	European badger	Canivenc and Bonnin-Laffargue (1963), Canivenc (1966)
Order Pinnipeda		
Family Otariidae		
Eumetopias jubata	Steller sea lion	Harrison)1969)
Arctocephalus pusillus	Cape fur seal	Rand (1954)
Callorhinus ursinus	Northern fur seal	Pearson and Enders (1951)
Family Phocidae		
Erignathus barbatus	Bearded seal	Harrison (1969)
Mirounga leonina	Southern elephant seal	Harrison *et al.* (1952), Laws (1956)
Phoca vitulina	Common or harbor seal	Fisher (1954), Harrison (1963, 1969)
Halichoerus grypus	Gray seal	Harrison (1963), Backhouse and Hewer (1956)
Pagophilus groenlandicus	Harp seal	Harrison (1963)
Pusa hispida	Ringed seal	Harrison (1969)
Leptonychotes weddelli	Weddell seal	Harrison (1969)
Lobodon carcinophagus	Crabeater seal	Harrison (1969)
Cystophora cristata	Hooded seal	Harrison (1969)

diapause" describes the condition in which every embryo enters a period of dormancy before successfully completing its development. "Facultative diapause" describes those species in which the embryo *may* enter a period of delay but in which the delay is not essential for the successful development of the embryo. This condition most commonly occurs during lactation; if lactation ceases, the dormant embryo resumes development. In some species, such as some marsupials, both obligatory and facultative diapause may occur. There is no obvious ecological or evolutionary relationship among the various species with delay, but there are numerous common factors.

Embryonic diapause is known to occur in representatives of both eutherian and marsupial mammals (Tables I and II), but the greatest number of species are found in the order Carnivora (Eutheria) and the family Macropodidae (Marsupialia). Indeed, there is only one species among the Macropodidae in which diapause has not been observed: *Macropus fuliginosus* (Western gray kangaroo) (Poole, 1973).

A large number of studies on delay have been carried out on laboratory rodents because of their availability. These studies will be discussed by way of comparison to the wild species in the final section. Several recent reviews (e.g., McLaren, 1971, 1973; Psychoyos, 1973a,b; Weitlauf, 1974) summarize present knowledge of diapause in rodents.

II. BRIEF SURVEY OF THE NATURE AND OCCURRENCE OF DIAPAUSE

A. Characteristics of Diapause

Embryonic diapause occurs during the blastocyst stage of development. It is significant that this is the point in embryonic development when the growing embryo is no longer self-sufficient. During the cleavage divisions, little expansion has occurred; the endogenous material in the ovum has not increased significantly by the accumulation of substances from the uterus. From the detailed studies made in the mouse, it is clear that fertilized eggs are able to progress to the blastocyst stage *in vitro* in very simple culture media; they apparently require relatively little in the way of exogenous resources. However, the transition from blastocyst to implanted embryo occurs *in vivo* in a secreting, luteal uterus and has often been difficult to achieve *in vitro* because of the more complex requirements during this phase of rapid expansion. It appears that blastocysts may also require a physiological trigger to release them from diapause.

During diapause, blastocysts either remain totally quiescent or expand at a very slow rate. For example, a few mitotic figures or only an occasional one can be observed in marsupial, rat, and armadillo blastocysts (Clark, 1967; Schlafke and Enders, 1963; Enders, 1962), but the blastocyst of the roe deer grows during the period of delay (Short and Hay, 1965).

Sadleir (1969) suggested that, in addition to recognizing obligate and facultative as broad subdivisions within those species with delay, it may be useful to distinguish between seasonal (divided into "synchronous" and "asynchronous" species) and aseasonal delay. In synchronous seasonal embryonic diapause, mating, implantation, and birth occur at fixed times of the year, presumably related to environmental influences. In asynchronous seasonal delay, mating occurs over fixed, although long, periods of the year, but the time of implantation is not fixed relative to the annual cycle. This type occurs in some rodents in which lactation may result in a delay or in a few of the seasonally breeding macropodid marsupials such as the tammar (*Macropus eugenii*), Bennett's wallaby in Tasmania (*Macropus rufogriseus*), and the quokka (*Setonix brachyurus*), in which loss of the pouch young will allow implantation sooner than usual (Berger, 1966; Sharman and Berger, 1969; Renfree and Tyndale-Biscoe, 1973a; Sharman *et al.*, 1966; Tyndale-Biscoe *et al.*, 1974).

Aseasonal embryonic diapause occurs in species that breed continuously so that the time of implantation is not fixed according to the time of the year but rather is related to other factors. It may be related to the birth of the previous offspring or to nutritional or ecological factors.

In bats, apart from those species that store sperm, two types of dormancy can be recognized: those in which delay occurs before implantation and those in which implantation is followed by a period of delayed development. In the latter group, development is not arrested but is slow in comparison with that of related species (Bradshaw, 1962; Bodley, 1974; Wimsatt, 1975). In one species, *Artibeus jamaicensis*, blastocysts conceived after the summer parturition implant in the uterus but then remain dormant between September and November. Blastocysts conceived after the spring parturition, however, undergo no such delayed development; so their gestation periods are much shorter than those of their earlier siblings (Fleming, 1971).

Blastocyst growth in mammals that do not have embryonic diapause occurs by rapid cell division and fluid accumulation. Although diapausing blastocysts can also accumulate fluid, there appears to be little or no mitotic activity and therefore little or no new cell proliferation (Daniel, 1970). Dormancy occurs when the embryo is 200–400 cells with the exception of the badger (900–2000) and marsupials and rodents (80–100 cells) (Figs. 1 and 2).

The characteristics of the delay in marsupials differ in several ways from those observed in eutherians. One important feature is the long separation in time between resumption of development and implantation. In the mouse, for example, the embryo implants almost immediately after the delay is broken, whereas there is a period of some 18 days before this event occurs in the tammar wallaby. In the mouse, estrogen is the key hormone in resumption of development, whereas progesterone acts to maintain the delay (McLaren, 1973). In marsupials, progesterone alone is required to initiate development after diapause (Renfree and Tyndale-Biscoe, 1973a). In the marsupial, as in the roe deer (Short and Hay, 1966), there is no

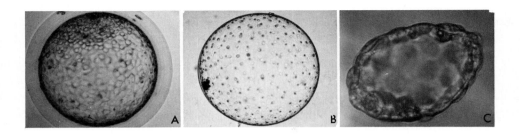

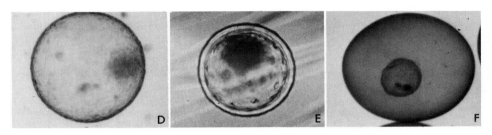

Fig. 1. Photographs of the blastocysts of (A) rabbit, (B) mink, (C) rat, (D) armadillo, (E)fur seal, (F) black bear. The bear blastocyst is fixed and stained; the others are fresh specimens. (Courtesy Professor J. C. Daniel.)

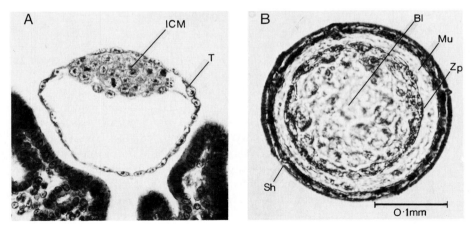

Fig. 2. Photographs of the diapause blastocysts of (A) bent-winged bat (courtesy Mr. G. I. Wallace), (B) tammar wallaby (courtesy Dr. C. H. Tyndale-Biscoe). Abbreviations: Bl, blastocyst; ICM, inner cell mass; Mu, mucolemma; Sh, shell membrane; T, trophoblast; Zp, zona pellucida. Note the lack of zona pellucida in the bat blastocyst in delay. The magnification of the bat blastocyst is twice that of the wallaby blastocyst.

measurable change in either corpus luteum size during the entire year-long delay (Berger and Sharman, 1969). The zona pellucida, however, is present during the delay in marsupials as it is in carnivores (Wright, 1963), but it is absent in the armadillo (Enders, 1966) roe deer (Short and Hay, 1966) and bat (Wallace, 1978). Finally, the diapause in some marsupials may occur during the breeding season, in which case it is lactation controlled, or during the nonbreeding season, when it is controlled by photoperiod.

B. Occurrence of Diapause in Mammals

The phenomenon of diapause occurs in a wide number of mammalian species (Tables I and II). Comparisons of individual species in general indicates that delay is species specific and each species has its own particular characteristics. However, there are a number of common phenomena that regulate the delay, particularly environmental and nutritional factors. Nevertheless, the ultimate control in all cases is certainly an endocrine one. These common features are summarized in the final section; a brief survey of the occurrence of delay within different mammalian groups follows.

1. GROUPS IN WHICH REPRESENTATIVES OF MOST GENERA ARE KNOWN TO EXHIBIT EMBRYONIC DIAPAUSE

a. Mustelidae.   There are 25 mustelid genera, and members of at least 8 have embryonic diapause. Within the mustelids, the nature of the diapause varies quite considerably, although the duration of the postimplantation period is quite similar: 28–31 days in *Spilogale putorius*, 25–28 days in *Martes americanus*, 23–24 days in *Mustela frenata*, and 28–30 days in *Mustela vison* (Enders, 1952; Wright, 1963; Mead, 1973). In some, such as the ferret (*Mustela furo*), no delay occurs at all (Buchanan, 1966, 1969).

The mink (*Mustela vison*) shows variable, but short, periods of delay. Gestation ranges from 42 to 79 days (mean 51 days), and the length of delay (up to 45 days) is dependent on how early in the breeding season (which extends from late February to early April) the female mates (Hansson, 1949; Enders and Enders, 1963; Wright, 1966). The mink is unusual in that both copulation and fertilization may occur during the delay period. The last fertilized blastocysts are usually the ones that carry to term (C. E. Adams, personal communication). The mink blastocyst increases in size during the delay (mostly due to an increase in the cavity of the embryo), and there is also an increase in cell nuclear size, although there is low mitotic activity (Baevsky, 1963). Sable (*Martes zibellina*) blastocysts do likewise. In mink there is a hypophyseal luteinizing hormone (LH) deficiency during delay and a storage phase during late delay. At implantation there is synthesis and secretion of LH, a change that is not mirrored in cells that secrete follicle-stimulating hormone (Murphy, 1972). There is also a rise in plasma progesterone concentration in twice-mated females, which correlates with resumption of corpus luteum activity and the resumption of blastocyst development (Møller, 1973). It is interesting, however, that all attempts to induce implantation in mink with exogenous progesterone alone or with estrogen failed (Cochrane and Shackelford, 1962).

Most mustelids, however, have a long period of delay. Generally, the longest delay periods are shown by those in which breeding occurs in the postpartum estrus. These are the fisher, *Martes pennanti* (11–12 months); the otter, *Lutra canadensis* (9½–12 months); and the European badger, *Meles meles* (12 months) (Wright, 1963; Canivenc and Bonnin-Laffargue, 1963; Hamilton and Eadie, 1964; Neal and Harrison, 1958; Harrison, 1963). Mating may also occur some time after parturition but while the animal is lactating, such as occurs in the stoat, *Mustela erminea* (Deanesley, 1943; Wright, 1963). In most other mustelids studied, copulation occurs between July and September after cessation of lactation. Delay in these species (long-tailed weasel, *Mustela frenata*; marten, *Martes americana*; wolverine, *Gulo gulo*; American badger, *Taxidea taxus*; and western spotted skunk, *Spilogale putorius latifrons*) lasts until the following spring (Wright, 1963; Mead, 1968b).

b. Macropodidae. The macropodid marsupials have a restricted distribution and are found only in the Australasian region. It is probable that members of all of the 16 macropodid genera have embryonic diapause since all except one species so far studied exhibit the phenomenon. The exception is *Macropus fuliginosus*, the Western gray kangaroo, which, despite much detailed work, has never been shown to have diapause (Poole, 1975; Poole and Catling, 1974).

All species of Macropodidae are monovular and polyestrous, the cycle length varying from 22 to 46 days. Gestation is longer than the estrous cycle only in *Wallabia bicolor*, the swamp wallaby, so that in most macropodids there is a postpartum estrus. Some species have long estrous cycles, so that birth occurs before the proestrous phase is initiated, and suckling prevents further development. In these species (*Macropus giganteus*, *Macropus parma*, *Macropus parryi*, and *Aepyprymnus rufescens*) diapause usually occurs after fertilization in late stages of pouch suckling (Calaby and Poole, 1971; Maynes, 1973; Tyndale-Biscoe, 1973; Poole and Catling, 1974; Moors, 1975). *Macropus fuliginosus*, as discussed above, does not have diapause since although the proestrous phase of the cycle is suppressed, the lactational inhibition of ovarian activity is complete.

In the remaining macropodids, gestation extends into the postluteal and proestrous phases without suppressing follicle growth, so that ovulation occurs a few days after parturition in these species (*Setonix brachyurus*, *Macropus eugenii*, *Macropus rufogriseus*, *Megaleia rufa*, *Bettongia lesueur*, and *Potorous tridactylus*). Suckling prevents the full development of the corpus luteum, and if fertilization has occurred postpartum the embryo is retained, as a unilaminar blastocyst in diapause (Sharman, 1955a, b; Hughes, 1962; Sharman, 1965b; Tyndale-Biscoe, 1968; Calaby and Poole, 1970).

The delay depends on the presence of the suckling pouch young, but in at least two species, *Macropus eugenii* and *Macropus rufogriseus*, the quiescent blastocyst and corpus luteum persist for several months after weaning (Sharman and Berger 1969), and resume development after the summer solstice (Berger, 1966; Renfree and Tyndale-Biscoe, 1973a; Tyndale-Biscoe *et al.*, 1974).

2. GROUPS IN WHICH SEVERAL GENERA ARE KNOWN TO EXHIBIT EMBRYONIC DIAPAUSE BUT IN WHICH MOST PROBABLY POSSESS IT

a. Ursidae. Embryonic diapause has been studied in detail in the American black bear, *Ursus americanus* (Wimsatt, 1963), and has been demonstrated in the grizzly bear, *Ursus arctos horribilis* (Craighead, *et al.*, 1969, but not in the remaining species. Hamlett (1935) considered that it also occurs in the European brown bear, *Ursus arctos arctos*, and polar bear, *Ursus maritimus*.

All bears studied so far have essentially similar breeding cycles. In all of them, estrus and copulation occur in early summer, and parturition occurs in the winter. Pregnancy is characterized by a period of developmental arrest of the embryos for about 5 months before implantation in the autumn. It is not yet clear, however, whether growth and expansion of the blastocyst are fully suppressed or simply slowed down during the delay (Wimsatt, 1963). However, as in macropodid marsupials, the corpus luteum remains quiescent during the delay period.

b. Pinnipeda. Delay of implantation appears to be a common feature in seals and may even be the rule among the Otariidae (6 genera) and Phocidae (13 genera) (Harrison and Kooyman, 1968). The diapause results in birth and mating occurring very close together, although the period of time spent out of the water may be quite extended. Mating may occur during lactation, toward the end, or after it has ceased, and there seems to be no firm relation between the cessation of lactation and the start of the delayed period of blastocyst growth (Harrison, 1969). It has been reported in the common or harbor seal, *Phoca vitulina* (Fisher, 1954; Harrison, 1960); the gray seal, *Halichoerus grypus* (Harrison, 1963); the cape fur seal,

Arctocephalus pusillus (Rand, 1954); the elephant seal, *Mirounga leonina* (Harrison *et al.*, 1952); and the northern fur seal, *Callorhinus ursinus* (Pearson and Enders, 1951). In the gray fur seal, the interval between birth and mating is about 16 days whereas, in the common seal, birth occurs in June and mating and ovulation occur in late July/early August (Harrison, 1963, 1969). The embryo of the northern fur seal remains dormant for 4 months, resulting in a gestation length of about a year (Harrison and Kooyman, 1968). Mitoses in the blastocyst cells occur at a very low rate, and it takes 50–60 days for the cells to double in number (Daniel, 1971). Activation occurs 1–2 weeks before implantation in November and is characterized by a dramatic increase in cell numbers and blastocyst size.

c. Muridae and Cricetidae. The classic work of Cochrane and Meyer (1957) in the rat demonstrated for the first time that embryonic diapause could be induced by progesterone. Although more is probably known about embryonic diapause in laboratory mice and rats than in most other mammals, little work has been done on the other 100-odd murine genera. In the rat and mouse embryonic diapause usually occurs during lactation in a progesterone-dominated uterus, and implantation can be induced by estrogen (Whitten, 1958; Mayer, 1963; Nutting and Meyer, 1963; Mantalenakis and Ketchel, 1966).

The length of the delay is proportional to the number of young nursing. Delay can also occur after early ovariectomy or hypophysectomy, and implantation can be stimulated under these circumstances only when both progesterone and estrogen are administered (reviewed by Psychoyos, 1973 a and b). In intact animals it is generally accepted that both hormones are present, although estrogen is the "trigger." Further details on the characteristics of diapause in the mouse and rat are given. In Mongolian gerbils (*Meriones unguiculatus*) blastocysts enter delay (still in their zona pellucidae) if conceived postpartum and implant after weaning (Meckley and Ginther, 1971, 1972; Norris and Adams, 1971).

Embryonic diapause has been recorded in a few wild rodents: the vole, *Clethrionomys glareolus* (Brambell and Rowlands, 1936); the deer mouse, *Peromyscus* sp. (Svihla, 1932); and the red tree mouse, *Phenacomys longicaudus* (Hamilton, 1962). It is probable that many other species of rodents will show delay in implantation when studied in the wild.

3. GROUPS IN WHICH ONLY A FEW REPRESENTATIVE GENERA SHOW DELAYED IMPLANTATION

a. Chiroptera: Pteropidae and Vespertilionidae. Embryonic diapause is probably not widespread among bats, although this group of mammals shows a number of interesting reproductive strategies that directly or indirectly delay the birth of the young (Table III). This includes a large group in the Vespertilionids that store sperm, those with "true" embryonic diapause, and those, such as *Artibeus* and *Macrotus* (Phyllostomatidae) and *Natalus* (Natalidae), in which implantation occurs but development is delayed (Wimsatt, 1975). A few species with delay have been studied in detail—*Eidolon helvum* (Mutere, 1967), *Miniopterus australis* (Dwyer, 1963), and *Miniopterus schreibersii* (Peyre and Herlant, 1967; Wallace, 1975, 1978). Two parameters seem to be involved in the termination of delay in these species—seasonal rainfall and temperature rhythms (Wimsatt, 1975)—although there are no obvious reasons why the tropical species have embryonic diapause. In *Miniopterus* the delay appears to be regulated by body temperature and metabolism changes as they relate the ambient temperature (Wimsatt, 1975). In the pipistrelle bat, *Pipistrellus pipistrellus*, pregnancy may be extended in response to adverse conditions in the natural environment, as the mean length of gestation has been extended by the induction of torpor at different stages during pregnancy (Racey, 1969, 1973).

b. Dasypodidae. The armadillos have a restricted range from the southern United States southward to Argentina, and of the nine genera, only one species of one genus has been studied thoroughly—*Dasypus novemcinctus* (e.g., Enders and Buchanan, 1959; Enders, 1963, 1966, 1967)—although the mulita, *Dasypus hybridus*, probably also has delay (Hamlett, 1932). The nine-banded armadillo is a monestrous, monovular, polyembryonic animal. Most females ovulate in July and August, implantation not occurring until November/December. Birth takes place in March/April. During the delay period the corpus luteum is apparently active, while the blastocyst remains in the fundic portion of the uterus (Enders, 1967; Enders and Given, 1977).

Bilateral ovariectomy during delay results in implantation 18–20 days later, but unilateral ovariectomy does not (Enders, 1966). The corpus luteum grows rapidly during the passage of the egg down the oviduct, and it continues its growth to become fully developed during the delay. No obvious changes occur in the corpus luteum at implantation, but appreciable changes occur in the endometrium. Enders (1961) concludes that the retarded development of the blastocyst in delay occurs during a prolonged preimplantation phase rather than during one of low progesterone levels.

c. Cervidae. Embryonic diapause was first discovered in the deer, *Capreolus capreolus*, and this species is the only ungulate proved to show delay (Short and Hay, 1966). Fertilization occurs in late July/early August,

TABLE III

Patterns of Reproduction in Bats Showing Embryonic Diapause and Delayed Development[a]

Estrous frequency	Seasonality	Timing of copulation, fertilization, and implantation	Taxon		Source
			Family	Genus	
Monestrous	Seasonal	Copulation, ovulation, and fertilization in autumn with a period of embryonic diapause	Vespertilionidae	*Miniopterus australis*[b]	Dwyer (1963)
				M. schreibersii	Courrier (1927)
				M. schreibersii	Dwyer (1963)
			Pteropodidae	*Eidolon helvum*	Mutere (1967)
			Natalidae	*Natalus stramieneus*	Mitchell (1965)
		Copulation, ovulation, fertilization, and implantation occurring as a direct sequence but followed by a period of delayed development	Phyllostomidae	*Macrotus californicus*	Bradshaw (1962)
			Vespertilionidae	*Miniopterus australis*[b]	Medway (1971)

[a] Adapted from Wallace (1975, 1978).
[b] *Miniopterus australis* shows variation in reproductive cycles.

and the embryo grows only slowly, until December (Short and Hay, 1966; Aitken, 1974a,b). After the delay period, but immediately before implantation, there is a period of rapid elongation. One unusual feature of the species, compared to most other mammals with delay, is that the corpus luteum is fully active during the delay. The corpus luteum is also active in the armadillo and in mice (Enders, 1966; Whitten, 1958). The roe deer is treated in detail in Section III,B.

4. GROUPS OF UNCERTAIN STATUS

a. Soricidae. The European common shrew, *Sorex araneus*, has a postpartum mating, and the resulting pregnancy is thought to be prolonged by a delay in implantation (Brambell, 1935). Delay may also exist in the lesser shrew, *Sorex minutes* (Brambell, 1937; Brambell and Hall, 1936).

b. Burramyidae. The possible occurrence of embryonic diapause in the pygmy possum, *Cercatetus concinna*, was pointed out by Hartman (1940) commenting on Bowley's (1939) observation that a second litter was produced in the absence of a male. Although Sharman (1963) and Tyndale-Biscoe (1973) both list *Cercatetus* as having delay, Clark (1967) found no evidence of a stay in development during lactation. Sharman and Berger (1969) subsequently suggested that it is likely that growth of embryo is continuous through lactation and that embryonic diapause is confined to the one family of marsupials, the Macropodidae. However, both *Cercatetus* and *Acrobates* carry uterine blastocysts during the suckling of a previous litter (Clark, 1967; Hill, 1900). Tyndale-Biscoe (1973) suggests either that there is a lack of total cessation of growth, as in macropodids, but a continued slow growth, as in the roe deer and European badger, or that there are periodic halts in embryonic growth during the seasonal pattern of torpor. He considers the second alternative most unlikely.

Clearly, both the Soricidae and the Burramyidae must be studied further before any firm statements can be made about the nature of their delay, if any.

III. EMBRYONIC DIAPAUSE IN THREE WILD SPECIES

The brief general survey of the last section does not provide a sufficient basis for speculation about the mechanisms that control diapause. Studies on laboratory rodents indicate that delay generally occurs as a facultative event during lactation. However, in the last 5 years or so, there have been more data gathered on the reproductive biology of wild species so that a better understanding of the phenomenon can be attained. In particular, efforts have centered on the blastocyst–uterine and pituitary–ovarian interactions. In this section detailed data are given for three species to bring out a number of unique features as well as some of the common ones controlling embryonic diapause in mammals.

A. Western Spotted Skunk, *Spilogale putorius latifrons*

1. NATURAL HISTORY AND REPRODUCTIVE BIOLOGY

There are two distinct populations of the spotted skunk, *Spilogale putorius*, in North America, comprising several subspecies. The eastern population (*S. p. interrupta*, *S. p. ambarvalis*, and probably *S. p. putorius*), found east of the Continental Divide of the Rocky Mountains, breeds in April and produces two litters per year after a short gestation period of 55–65 days (Mead, 1968a). The western population (*S. p. latifrons*, *S. p. leucoparia*, *S. p. gracilis*, and *S. p. phenax*) breeds in late September and has a long gestation period (230–250 days). Implantation usually occurs in April (at a similar time to the eastern form) after a period of dormancy of 210–230 days (Mead, 1968b; Foresman and Mead, 1973).

The reproductive organs of both male and female spotted skunks from the two populations cannot be distinguished histologically except during the period of delayed implantation, when western forms have small, inactive corpora lutea.

Males from the eastern population become fertile by March, when the subadults are 9–10 months old, whereas adults from the western population become fertile in May. Some juveniles are fertile in September, when they are only 4–5 months old. The estrous cycles of the females of the two populations do not overlap, with the possible exception of the southern subspecies. Females of the eastern population breed at 9–10 months of age, whereas the western females do so at 4–5 months. Embryonic diapause is the rule in the western forms; it is not known to occur in eastern skunks.

At the time the embryos enter the uterus, the corpora lutea become involuted; plasma progesterone levels remain low, and embryonic development is arrested at the blastocyst stage (Mead and Eik-Nes, 1969a). Implantation occurs in late April, and parturition takes place 28–31 days later (Foresman and Mead, 1973; Mead, 1973).

2. THE OVARY AND STEROID HORMONES

Luteal cells of recently formed corpora lutea assume an inactive appearance about the time the morulae enter the uterus. The histological appearance of these cells remains unchanged until implantation takes place (Mead and Eik-Nes, 1969a). The diameters of the corpora lutea remain reasonably constant until nidation, when they enlarge, the individual luteal cells nearly doubling in size.

Ultrastructural changes in pre- and postimplantation corpora lutea of mustelids such as the mink, *Mustela vison* (Enders and Enders, 1963), the European badger, *Meles meles* (Canivenc *et al.*, 1967), and spotted skunks (Sinha and Mead, 1975) have been studied in detail. In the skunk during the preimplantation period, the corpus luteum contains undifferentiated, small granulosa cells and differentiated, large granulosa lutein cells (Sinha and Mead, 1975). About 2 days before implantation the ratio of large to small cells is about 1, but 8 to 12 hr before nidation the corpora lutea contain mostly large cells.

The process of luteinization is completed within 72 hr of nidation. The changes in cell size before and during implantation are due mostly to increased synthesis of smooth endoplasmic reticulum, differentiation of mitochondria, plications of plasma membranes, and accumulation of lipid material (Sinha and Mead, 1975).

Peripheral blood plasma levels of progesterone in nonpregnant females show little variation through the year and are similar to the concentrations found in pregnant animals during the first part of the delay (Mead and Eik-Nes, 1969a). Plasma progesterone levels are variable late in the preimplantation period and during nidation and appear to be correlated with the differentiation of the lutein cells (Sinha and Mead, 1975). In the mink, however, progesterone concentrations show a gradual rise beginning about 40 days before parturition (around the time of blastocyst resumption) to reach a peak 10–25 days later (Møller, 1973).

Peripheral plasma levels of estrone could not be quantitated throughout the year, and levels found in pregnant and nonpregnant skunks did not differ significantly (Mead and Eik-Nes, 1969b). Estradiol-17β could be detected only during estrus (3.75 ng/ml) and around the time of implantation (0.1–0.2 ng/ml). The gradual rise in plasma estradiol-17β levels parallels the rise in plasma progesterone in the late stage of delay described by Mead and Eik-Nes (1969a), but they point out that, since plasma levels of neither estrone nor estradiol-17β were elevated during the early preimplantation period, it is unlikely that these steroids participate in preventing nidation by a negative feedback system (Mead and Eik-Nes, 1969b).

3. EFFECTS OF LIGHT AND THE PINEAL GLAND

Premature nidation has been induced in mustelids by increasing the length of the daily photoperiod (e.g., Wright, 1963; Mead, 1971; Canivenc *et al.* 1971), although attempts to induce implantation by injections of steroids, gonadotropins, or ovariectomy have failed (Canivenc and Bonnin-Laffargue, 1963).

The spotted skunk is more strictly nocturnal than other mustelids. Exposure to 14 hr of light per day causes premature implantation but has no effect on blinded skunks. Blinding prolongs the delay period and prevents the preimplantation rise in plasma progesterone (Mead, 1971). It is possible that light, acting via the eyes and hypothalamus, is the proximate factor in timing of implantation after delay in the spotted skunk.

Synthesis of melatonin by the pineal increases in darkness; melatonin also has antigonadotropic properties (Wurtman *et al.*, 1968). However, pinealectomy is ineffective in reducing the preimplantation period, and on the present evidence it must be concluded that the observed effects of light are not mediated through the pineal gland and that the pineal does not produce a factor inhibiting implantation (Mead, 1972).

4. THE PITUITARY AND GONADOTROPINS

The plasma LH level in pregnant skunks gradually rises from 4.9 ng/ml at 120 days preimplantation to 7.9 ng/ml at implantation (Foresman and Mead, 1974). No surge in LH is observed before nidation, but the gradual rise in LH is paralleled by the rise in progesterone (Mead and Eik-Nes 1969a). This observation suggested that LH might have a direct or indirect role in initiating resumption of luteal function and embryonic development at implantation, i.e., that LH in the skunk is luteotropic. However, the pituitary responsiveness to synthetic LH-releasing hormone (LH-RH) did not change during any of the last months of the preimplantation period (February, March, or April) either in the amount of LH released or in the time interval over which such release occurred, nor did the embryos of treated animals implant earlier than those of control animals (Foresman *et al.*, 1974). This indicates that the rise in plasma gonadotropins and change in pituitary activity during delay is not due to any inability of the pituitary to respond to hypothalamic control (Foresman *et al.*, 1974).

Hypophysectomy resulted in a decrease in cell numbers in the inner cell mass of the blastocyst and in a reduced peripheral plasma LH concentration (Mead, 1975). However, luteal histology was unaffected. The only effect on ovarian histology was the loss of all antral follicles, although implantation was inhibited by hypophysectomy. The pituitary is apparently needed for implantation but slightly less for blastocyst survival, since some blastocysts remained alive although with fewer cells in the inner cell mass. The interactions between pituitary and ovary are not clear from these experiments; however, the histological data suggest that corpora lutea of the spotted skunk do not require pituitary hormones for maintenance during delay but probably require luteotropic hormone(s) for reactivation (Mead, 1975).

5. THE UTERUS

Although it has been reported that the uterine glands show signs of increased activity during the free vesicle stage (Mead, 1968b), there are unfortunately no data as yet about the role of the uterine secretions and proteins. It would be interesting to study the role of the uterus during delay and implantation and the responsiveness of the uterine secretions to the various changing hormonal levels in this species. It would be surprising if no changes occurred in the uterine secretions at reactivation. In most mustelids, the endometrium contains several distinct cell types, which alter quite significantly at the time of blastocyst activation, with a release of contained materials (Enders and Given, 1977). In the northern fur seal and in the mink corresponding changes in the uterine fluid composition correlate with implantation after diapause (Daniel and Krishnan, 1969; Daniel, 1970, 1971).

B. Roe Deer, *Capreolus capreolus*

1. NATURAL HISTORY AND REPRODUCTIVE BIOLOGY

The roe deer (*Capreolus capreolus*) is a spontaneously ovulating species, which appears to be monoestrous and polyovular, with an obligatory period of delay (or "pseudopregnancy") during which the corpus luteum is fully functional over the five month phase of delay (Aitken *et al.* 1973, Aitken 1977c).

The roe deer is the only artiodactyl known to exhibit embryonic diapause, but Short and Hay (1966) suggest that the two closely related species of roe deer, the Siberian deer (*Capreolus pygarus*) and the Manchurian deer (*Capreolus bedfordi*) are also likely to have embryonic diapause. In the roe deer, the gestation period lasts about 10 months, with mating occurring during the rut in late July/early August (Aitken, 1974a). The blastocyst loses its zona pellucida a few days after ovulation and enters the delay period, which continues for 5 months until late December/early January. The roe deer blastocyst is not attached to the uterine epithelium and lies free in the uterine lumen (Aitken, 1974a), but during the delay period the blastocyst increases in diameter from 1 to 5 mm. Embryonic diapause in roe deer differs from that in most other mammals in that the corpus luteum is fully active during the delay period and the zona pellucida is absent (Short and Hay, 1966). Delay is unaffected by lactation. However, as in the rabbit and pig, sex chromatin formation occurs well before implantation (Aitken, 1974b).

Implantation follows a rapid elongation of the blastocyst in early January (Fig. 1) and, after a further 5 months of "normal" gestation, between one and three young are born in May (Aitken *et al.*, 1971; Aitken, 1974a,b,c).

2. BLASTOCYST, OVARY, AND STEROID HORMONES

During delay, roe deer ovaries have active corpora lutea and show signs of follicular growth and atresia. There is no change in either the weight of corpora lutea or the luteal progesterone concentrations from September to February (Short and Hay, 1966). The cytoplasm of the luteal cells during this time is hypertrophied and contains aggregations of small osmophilic granules (Aitken et al., 1973). Follicles showing all stages of active development occur, whereas atresia is observed in secondary, tertiary, and mature follicles (Aitken, et al. 1973). However, during elongation of the blastocyst there is a marked increase of plasma estrogen levels (Aitken 1974a). No changes are observed in the ovary at this time, indicating an extraovarian source of the hormone. Aitken (1974a) suggested that the estrogen is embryonic in origin. During the delay period the trophoblast cells have numerous lipid droplets, but during rapid embryonic growth the content of cytoplasmic organelles increases in trophoblast cells and there is a decrease in the number of lipid droplets and granular inclusions. There is some evidence that estrogen can accelerate blastocyst development in roe deer. Subcutaneous implantation of crystalline estradiol 17-β (100μg to 1 mg) appeared to precociously stimulate resumption of embryonic growth (Short, 1967). However, one control animal also had a prematurely developed embryo.

3. THE UTERUS AND UTERINE SECRETIONS

In mid-October, uterine gland cells are of moderate height, but by early November there is extensive proliferation of the Golgi apparatus and an accumulation of lipid droplets (Aitken, et al., 1971). The rapid elongation of the blastocyst is therefore preceded by secretory activity in the cells of the endometrial glands. Throughout the delay the endometrium appears to be progestational (Short and Hay, 1966). The arrest, with respect to the endometrium, occurs when the uterus is in an early pregnancy or preimplantation condition and in this resembles that of the armadillo (Enders, 1967). Aitken (1974a, 1975) suggests that the endometrial glands play a major role in the initial restraint and subsequent stimulation of embryonic growth. In the blastocyst during delay there is a lack of cytoplasmic organelles, indicating a minimum of metabolic activity and cell growth (Aitken, 1975). The intense endometrial secretory activity during the resumption of rapid embryonic growth is followed by secretory activity in the ductal epithelium by the hypertrophied granular endoplasmic reticulum of the glands (Aitken 1975). The latter type of secretion reaches maximal development at the time of placental attachment, and may be stimulated by the embryonic tissue itself. An important component of the endometrial secretion is free fructose, which increases dramatically in January when blastocysts resume development (from 2 to 31 μgm/ml) (Aitken, 1976). The fetal fluids also contain fructose both before and after placental attachment. Zinc in the endometrium, on the other hand, is maintained at a constant high level, although it is very low or not detectable in the uterine flushings (Aitken, 1974c). There is a progressive rise in calcium in the endometrium during delay, reaching maximal levels on resumption of embryonic growth, and a lower but significant increase in calcium in the uterine flushings (Aitken, 1974c). The role of these elements is, however, unclear.

Uterine flushings also contain a number of proteins, both uterine-specific and serum proteins, at the time of embryonic elongation (Aitken, 1974a). During delay, only low concentrations of protein are found, but a major component after resumption of development is a uterine-specific protein. Likewise, the level of α-amino nitrogen (comprising 20 amino acids) rises dramatically after delay, whereas during delay only 8 amino acids are found.

It seems likely that the uterus is able to restrain the growth of the roe deer blastocyst during embryonic diapause by withholding the release of an endometrial secretion capable of inducing and supporting the elongation of the embryo (Aitken, 1974a). The endometrial glands may, in turn, be under the control of estrogen, and the elevated levels observed at resumption of development may stimulate the release of secretion from the glands (Aitken, 1974a). However, it remains possible that the estrogen is embryonic in origin since no obvious ultrastructural changes are seen in the ovaries throughout this period (Aitken *et al.*, 1973).

C. Tammar Wallaby, *Macropus eugenii*

1. NATURAL HISTORY AND REPRODUCTIVE BIOLOGY

The tammar wallaby, *Macropus eugenii*, is a small (3–6 kg) member of the kangaroo family Macropodidae. It occurs on islands in southwestern Australia, as far north as the Abrolhos Islands (29° S), and in South Australia, where it is restricted to Kangaroo Island (35° S). It also has a limited distribution on the mainland in southwestern Australia.

The tammar wallaby is polyestrous and monovular. Females retain a blastocyst in the uterus in a state of diapause throughout the 6-month breeding season and the 6-month quiescent period of the year (Berger, 1966). Blastocysts normally remain in diapause for the whole year. The breeding season extends from midsummer in January to June. The normal sequence of events begins with the birth of the pouch young in late January/early February. Two days later, the female returns to estrus and mates, and the resulting fertilized egg grows during the next 2 days to become a unilaminar blastocyst of about 100 cells. The presence of a suckling pouch young (by now about 4–5 days old) prevents further development of the uterine blastocyst throughout the remainder of the breeding season.

Should the suckling pouch young be lost, the blastocyst resumes its development, a new pouch young is born, and the cycle begins again. However, should this occur after June, no initiation of blastocyst development occurs and the female remains reproductively quiescent, although carrying a viable, dormant blastocyst. Around the time of the longest day (December 22) the blastocyst resumes development. Twenty-six days later it is born, and the cycle resumes (Renfree and Tyndale-Biscoe, 1973a).

2. THE BLASTOCYST, OVARY, AND STEROID HORMONES

Pregnancy is accommodated within the length of an estrous cycle, estrus occurring 18–24 hr after birth. At estrus a single Graafian follicle completes maturation in one ovary. After spontaneous ovulation a corpus luteum forms from the follicle in about 2 days. A luteal or secretory phase develops in the endometrium of the two separate uteri (Tyndale-Biscoe, 1973). The corpus luteum inhibits further follicular growth during the first half of the estrous cycle, but its effect declines during the second half of the estrous cycle (or the second half of pregnancy). In the presence of a suckling pouch young the corpus luteum grows for only a few days; it enters a quiescent phase, and further development is inhibited. If conception has occurred during the postpartum estrus, the embryo also enters a quiescent phase as a unilaminar blastocyst (Tyndale-Biscoe *et al.*, 1974). The quiescent corpus luteum can, however, inhibit follicular development on the contralateral ovary throughout the period of delay. During both lactational and seasonal quiescence, the corpus luteum does not (or cannot) induce the luteal phase in the uterus.

The blastocyst consists of a unilaminar layer of embryonic cells, lacking the familiar inner cell mass of eutherian blastocysts. It is surrounded in turn by the zona pellucida, an acid mucopolysaccharide ("albumin") layer, and finally the keratinous "shell" membrane—an egg membrane that no eutherian mammal possesses (Hill, 1910; Hughes, 1969a,b; Tyndale-Biscoe, 1973).

The blastocyst, conceived in late January of one year, is carried for 11 months in diapause (Berger, 1966). Lactation inhibits both corpus luteum and blastocyst growth until June, after which time the withdrawal of the suckling stimulus does not result in development of the delayed embryo (Renfree and Tyndale-Biscoe, 1973a). The tammar thus has a period of facultative delay (from January to June during "lactational quiescence") and then passes through a period of obligate delay (from June to January during "seasonal quiescence"). Young animals maturing and entering estrus for the first time after June ovulate and mate; the resultant blastocyst immediately enters delay (Berger and Sharman, 1969; Yadav, 1973). These blastocysts are then born a month (one gestation period) after the summer solstice, at the same time as the remainder of the population.

Before resumption of development, blastocyst cells show relatively low levels of nuclear RNA polymerase activity, but at resumption nucleolus-associated DNA polymerase activity of embryos is elevated (Moore, 1975). However, no mitoses are observed at this time.

Steroid hormone levels during the estrous cycle, pregnancy, and the period of embryonic diapause are largely unknown. The corpus luteum of pregnancy increases in size, reaching a peak at day 17 of the 26-day gestation after removal of the pouch young (Renfree and Tyndale-Biscoe, 1973a). The luteal cells hypertrophy and reach a peak size between day 13 and 17. Hyperplasia and hypertrophy of luteal cells of the corpus luteum coincide with the onset of cell division in the blastocyst between days 4 and 6, and the luteal condition is well developed in the endometrium by day 6. Ovariectomy between days 2 and 8 results in the blastocyst remaining in diapause, but if ovariectomy is performed after day 8 the blastocyst continues development to full term (Tyndale-Biscoe, 1963a,b, 1970; Sharman and Berger, 1969).

During diapause, the blastocyst can be stimulated to resume development with injections of progesterone alone (10 mg/day for 10 days) (Renfree and Tyndale-Biscoe, 1973a), but under these conditions the corpus luteum remains quiescent; the progesterone apparently stimulates the uterus and/or blastocyst directly and bypasses the ovary. Estrogen is apparently not needed for the resumption of development. In animals given estrogen the majority of the blastocysts recovered after treatment had highly abnormal cells and none developed to full term as they do after progesterone treatment (Smith and Sharman, 1969; Berger and Sharman, 1969; Clark, 1968). The corpus luteum in estradiol-treated animals on day 4 was more vascularized than the quiescent corpus luteum; so the estrogen may have acted by stimulating it to produce progesterone (Lemon, 1972).

Only one study has been made of circulating levels of plasma progestins and progesterone (Lemon, 1972), although preliminary work is in progress (Renfree and Heap, 1977). The steroid concentrations are low compared with those of other mammalian species (1 ng/ml). An increase in plasma progesterone concentration, apparently of ovarian origin, occurs before birth and/or estrus. The plasma progestin concentration during pregnancy is greater than during the estrous cycle (Lemon 1972).

In the tammar the critical stage for successful maintenance of development to full term occurs after the resumption of development and just before the blastocyst undergoes a period of rapid expansion prior to attachment (or implantation) (Tyndale-Biscoe et al., 1974). The role of the uterus is of prime importance at this stage (Tyndale-Biscoe, 1970; Renfree and Tyndale-Biscoe, 1973a).

3. THE UTERUS, UTERINE SECRETIONS, AND FETAL FLUIDS

The tammar has two separate uteri, but only one becomes gravid during each pregnancy. The luteal phase appears in both uteri by day 8 and persists until about day 20, when degenerative changes occur (Berger and Sharman, 1969; Renfree and Tyndale-Biscoe, 1973a). The endometrium is highly glandular with loose, fibrous connective tissue stroma. Even during lactation the endometrium is quite highly developed compared to that of some eutherian mammals during suckling. The luteal phase is characterized by elongated columnar epithelial cells, basal round nuclei, and chromophobic cytoplasm, and the stroma becomes highly edematous (Enders, 1967). During lactation, however, the endometrium is distinctly less hypertrophied and has an unchanging, "quiescent" appearance. Its cells do not atrophy and are probably capable of secretory activity during quiescene (Tyndale-Biscoe, 1974; Enders and Given, 1977).

During lactational quiescence both uteri are small and of a similar size, but after removal of the pouch young the pregnant uterus enlarges from 10 × 8 mm to a maximum of 30 × 23 mm (Renfree and Tyndale-Biscoe, 1973a). The endometrium likewise increases in size and amount in both uteri, but only the endometrium of the pregnant uterus remains enlarged and actively secreting. The increase in endometrial size and secretory activity appears to be due to two factors: the stimulation by the secretions of the corpus luteum (presumably progesterone) during the early phases of pregnancy and the presence of the embryo or, more precisely, the presence of the yolk sac placenta (Renfree, 1972b; 1973a, Renfree and Tyndale-Biscoe, 1973a). The corpus luteum effect can be duplicated by injections of progesterone (Renfree and Tyndale-Biscoe, 1973a), and the embryo effect can be induced in the absence of an active corpus luteum either by blastocyst transfer or after development has been induced by progesterone. Ovariectomy after day 6 has no effect on the development of the secretory endometrium (Tyndale-Biscoe, 1970).

Concomitant with the weight changes in the endometria are changes in the amount and quality of the proteins in the uterine secretion (Renfree, 1972b, 1973a). During quiescence uterine fluids are almost undetectable, but after the resumption of development the amount increases three fold. The uterine secretion contains uterine-specific proteins as well as serum proteins. The main source of the uterine-specific proteins appears to be the actively secreting uterine glands. Of particular interest are the uterine prealbumins, which may be of importance as stimulators of the early stages of embryonic growth (Renfree, 1973a,b; Renfree and Tyndale-Biscoe, 1973b). The uterine fluids are also three times more concentrated in all amino acids than in maternal plasma, including high amounts of glutamic acid, which is 10–15 times more concentrated than glutamate in fetal fluids and 6 times more concentrated than glutamate in serum.

The prealbumins as well as a number of other proteins are found in the fetal fluids and particularly in the yolk sac fluids (Renfree, 1973b). During the free vesicle stage and before implantation, although protein concentrations are quite low, concentrations of glucose and urea in the yolk sac fluid resemble those found in maternal serum; levels of free amino acids are 10 times higher than in serum and 3 times higher than in uterine fluid (Renfree, 1973b). As in the roe deer, fructose is also found in fetal and uterine fluids, but little change is observed at implantation (Renfree, 1972a). After attachment, the composition of the yolk sac fluid is noticeably changed, particularly in the protein components and in glucose (Renfree, 1970, 1973b). Urea accumulates in the allantoic fluid. The uterine secretions do not reflect such changes; their greatest contribution is probably at the blastocyst stage and up to the time of attachment. The importance of the uterine environment has been demonstrated by Tyndale-Biscoe (1970) using blastocyst transfer techniques. The development of transferred blastocysts depends on the secretory activity of the uterus and does not apparently require direct stimulation by the corpus luteum secretions. The uterus and its secretions closely interact, then, both with the corpus luteum and its secretions and with the embryo and its placenta. Although there seems to be little doubt that the blastocyst is held in diapause by the lack of an appropriate uterine environment and the absence of a stimulator to "switch" it on, it is still far from clear whether the signal derives directly from a progestagen acting on the blastocyst as well as on the uterus, or whether the blastocyst simply waits for the uterus to become secretory before it is released from the quiescent phase (Fig. 3).

4. THE PITUITARY AND GONADOTROPINS

In the tammar, the critical factor in diapause is apparently the arrested development of the corpus luteum. However, although hypophysectomy causes lactation to cease, as expected, the diapause blastocyst resumes development (Hearn, 1973, 1975). The resumption of development presumably results from the reactivation of the corpus luteum, which also ceases to be quiescent after hypophysectomy (Hearn, 1973, 1974). The conclusion is that the corpus luteum of the tammar does not require a gonadotropic stimulus from the pituitary but is tonically suppressed by it. The inhibition is probably not luteolytic, since the quiescent corpus luteum is viable, and it is not due to the absence of a luteotropic factor, since

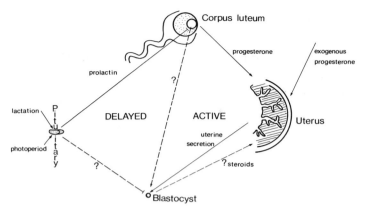

Fig. 3. Summary of possible pituitary–ovary–uterus interactions during delay in the tammar, *Macropus eugenii*. Diapause is initiated in the early part of the year by the suckling stimulus, which acts via neural pathways to signal the pituitary. The pituitary tonically inhibits further development of the corpus luteum. It is likely that the inhibitory factor is prolactin. It is possible that the pituitary also directly inhibits the blastocyst but more likely that the blastocyst is held in delay by the lack of nutritive uterine secretions (a function dependent on progesterone or on an active corpus luteum). During the latter part of the year, removal of the suckling stimulus does not cause reactivation of the blastocyst; the pituitary may be producing the same inhibitors as in suckling or different one(s). Reactivation occurs at the summer solstice and presumably involves suppression of neuroendocrine inhibitory mechanisms, allowing the corpus luteum to become active, the uterus to become secretory, and the blastocyst to resume development. It remains possible that the blastocyst is stimulated directly by pituitary and/or corpus luteum secretions, but secretory activity of the uterus is essential for blastocyst reactivation.

hypophysectomy allows the quiescent corpus luteum to grow. It seems likely, however, that the factors responsible are prolactin and/or oxytocin (Tyndale-Biscoe *et al.*, 1974). Recent experiments seem to show conclusively that prolactin is the inhibitor of the corpus luteum activity (Tyndale-Biscoe and Hawkins, 1977a,b).

In adult females, peripheral plasma gonadotropins remain at a steady level throughout the year, at 2–5 ng/ml, and rise only during a transient preovulatory peak to 10–18 ng/ml (Hearn, 1972b, 1974, 1975). Gonadotropin is undetectable in plasma of hypophysectomized animals, but in ovariectomized animals gonadotropin levels are elevated above basal levels. It is possible that the steady levels of gonadotropin are needed to maintain the corpus luteum in its quiescent condition. It is postulated that in the absence of this low (2.5 ng/ml) level of gonadotropin the corpus luteum would eventually degenerate and the uterus would assume the anestrous condition observed in the quokka, *Setonix brachyurus* (Tyndale-Biscoe *et al.*, 1974).

5. EFFECTS OF LIGHT ON THE PINEAL GLAND

The strict seasonality of the resumption of blastocyst development at the summer solstice in December suggests that photoperiod is critical in the initiation of growth after diapause. In addition, tammars moved to the Northern Hemisphere adjusted their breeding season to the July summer solstice (Berger, 1970). One attempt to manipulate photoperiod by advancing it one month did not give conclusive results, since the experimental group gave birth subsequent to the control group (Hearn, 1972c). Recently, Sadleir and Tyndale-Biscoe (1977) repeated and enlarged this experiment, and the results, although also somewhat difficult to interpret, show that photoperiodic manipulation can effect the termination of embryonic diapause in the tammar. A sudden increase in day length is not stimulatory, but a sudden decrease is.

The pineal may be responsible for the recording of photoperiod changes. Melatonin inhibits gonadotropin activity in mammals. This can be most readily measured by assaying for one of the enzymes involved in melatonin synthesis. In the tammar, preliminary experiments have shown that blastocyst activation occurs when pineal HIOMT (hydoxyindole-O-methyltransferase) activity falls to a low level (Kennaway and Seamark, 1975), and this coincides with the longest day.

However, as yet HIOMT activity has been measured only around the time of the summer solstice. It would be interesting to know more of the role of the pineal during the 6-month quiescent period (a period of slowly increasing day length) and whether pinealectomy would induce blastocyst development.

IV. DIAPAUSE AS DEVELOPMENTAL STRATEGY

Diapause, as we have seen, is widespread, and yet within each taxonomic group there are few indications of common factors (Fig. 4). Why, then, do the species concerned mate up to a year in advance of birth, although their "true" gestation periods may be quite short? One must assume that there was some evolutionary advantage in this pattern of development to a species that adopted it and survived. The fact that representatives of a number of unrelated groups exhibit this characteristic confuses any attempt to define a single advantage in embryonic dormancy that is common to all mammals. The occurrence of periods of delay in embryonic growth in diverse groups is more likely to indicate that the evolutionary advantage is real, although the way in which it has been achieved may differ. The diversity of groups showing delay and the diversity of mechanisms used result from convergent, not

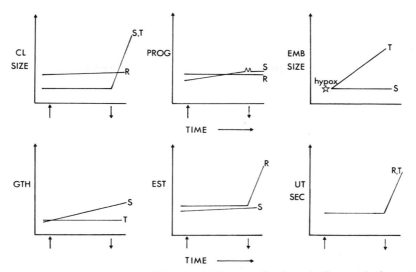

Fig. 4. Schematic summary of the essential features of embryonic diapause in the spotted skunk (S), the roe deer (R), and the tammar wallaby (T). Arrows show initiation of diapause (↑) and resumption of development (↓). Abbreviations: CL, corpus luteum; GTH, gonadotropin levels; PROG, progesterone levels; EST, estrogen levels; EMB, embryo; UT SEC, uterine secretion; hypox, hypophysectomy.

parallel, evolution. The answer presumably lies in the advantages associated with the evolution of viviparity; diapause is simply one strategy for achieving a lengthened intrauterine development. To this end, a variety of mechanisms have been brought into play to make it physiologically possible. The mechanisms involve environmental factors (especially nutrition and photoperiod), endocrinological factors (particularly in the translation of environmental factors), and those associated with the offspring themselves such as occur during lactation or in the uterus before implantation. In this section, I attempt briefly to synthesize some of these features and finish by speculating on the role of diapause in developmental processes.

A. Physiological Basis and Regulatory Mechanisms

1. ENVIRONMENTAL CUES

The precise timing of the sequence of events associated with resumption of development after delay in some species strongly implicates photoperiodic influences. The availability of nutritional resources, the environmental temperature, and rainfall can also affect the regulation of the patterns of delay.

Experimental demonstration of the role of photoperiod has been attempted in the mustelids, the roe deer, and the tammar wallaby. Premature implantation has been induced in these species by increasing the length of the daily photoperiod (Pearson and Enders, 1944; Wright, 1963). Embryos of mink, long-tailed weasel, western spotted skunk, and pine marten implant as day length increases, whereas embryos of the European badger and the tammar wallaby (Wright, 1963; Hearn, 1972c; Sadleir and Tyndale-Biscoe, 1977) implant as day length decreases. In the spotted skunk, photoperiod manipulations can influence the timing of implantation but have no effect on the *initiation* of implantation (Mead, 1971); blinding prolongs the preimplantation period in this species. In the tammar wallaby, blastocysts resume development on or about the summer solstice day, and the tammar gives birth a month later. The photoperiod response in the tammar correlates well with the drop in pineal HIOMT activity (Kennaway and Seamark, 1975), possibly implicating the pineal as a mediating gland. However, in the skunk although blinding prevented implantation, pinealectomy had little or no effect (Mead, 1971, 1972).

In the badger, temperature as well as photoperiod seems to be involved (Canivenc *et al.*, 1971), but in the experiment described it is difficult to separate the individual effects. However, photoperiod manipulations are notoriously difficult to conduct in a way that produces reliable results. Thus, for example, the suggestion that the roe deer does not respond to increased or decreased photoperiod should be treated as preliminary information, because so few animals were available (Lincoln and Guinness, 1972).

In many species the synchrony that diapause attains is related to such factors as seasonal rains, e.g., the bats *Eidolon helvum* (Mutere, 1965, 1967) and *Artibeus* (Fleming, 1971), or rain-induced periods of good nutrition, as is observed in the red kangaroo, *Megaleia rufa*, and the euro, *Macropus robustus* (Ealey, 1963; Frith and Sharman, 1964; Sadleir, 1965; Newsome, 1965, 1971). The red kangaroo is especially interesting because it is a species with delay which lives in a habitat where erratic rainfall can occur at any time of the year. It is an opportunistic breeder, and in favorable conditions it suckles a young at foot and a smaller young in the pouch and carries a diapausing blastocyst (Ealey, 1963; Frith and Sharman, 1964; Newsome, 1965, 1966). If the quantity of available food declines, reproductive activity declines. There is a heavy mortality of the young. If poor conditions continue, older pouch young are lost. The delayed embryo then resumes development and enters the pouch but remains alive for only about 2 months if the drought prevails. This cycle of replacement of 6- to 8-week-old pouch young continues until conditions improve after rain. If the drought persists for 3 months in summer or 5 months in winter, 50% of female kangaroos in central Australia enter anestrus but resume cycling activity immediately following rain.

Photoperiod is probably the synchronizing factor in seals, too, since many give birth within a narrow time range. In most of these species a relatively short period of time is spent out of water, and for mating to occur both sexes must be responsive to some common environmental feature.

Thus, photoperiod and nutrition are quite closely linked. The photoperiod synchronizes internal events related to external, seasonal ones (such as seasonal deficit or abundance of food). Recently, it was demonstrated that in lactating rats implantation can be further delayed by reducing the food intake (Shapira *et al.*, 1974).

In the quokka, *Setonix brachyurus*, on Rottnest Island, the viability of the diapausing embryo decreases at the end of the winter, and in this species, too, seasonal nutritional stress is postulated to be the cause of blastocyst loss (Shield and Woolley, 1963). Sadleir (1969) suggests that "the phenomenon of delayed implantation always acts to enable the female mammal to undergo those periods of nutritional stress, late pregnancy, and the whole of lactation at periods when environmental nutrition is maximal."

2. ENDOCRINE MEDIATORS

The endocrine requirements for the implantation of delayed blastocysts are not well understood for most of the species listed in Tables I and II. In addition to studies of the three species discussed in Section III, a few other detailed studies have been carried out, notably on the European badger by Canivenc and colleagues and the armadillo by Enders and associates.

Bilateral ovariectomy in the badger during quiescence has no effect on the blastocyst and implantation does not occur, probably due to the lack of secretory activity in the uterus which would normally be stimulated by renewal of corpus luteum activity (Canivenc and Bonnin-Laffargue, 1963; Wright, 1966). In the nine-banded armadillo, *Dasypus novemcinctus*, bilateral ovariectomy early in the quiescent period causes implantation in 18–20 days (Enders, 1966). In this species, ovariectomy after implantation causes loss of embryos; the ovary is essential for maintenance of pregnancy during the first third of the postimplantation period (Buchanan *et al.*, 1956).

This is in direct contrast to the situation found in marsupials. Ovariectomy alone has no effect on the delayed embryos; they remain quiescent. In the quokka and the tammar, ovariectomy after resumption of development, if carried out between days 0 and 6, causes failure to the embryos to implant. If ovariectomy is carried out after this time, the embryos implant and go to full term, although parturition is inhibited (Tyndale-Biscoe, 1963a,b, 1970; Tyndale-Biscoe *et al.*, 1974). Hypophysectomy in the tammar causes resumption of blastocyst growth (Hearn

1972a,b, 1973) but not in the western spotted skunk, where hypophysectomy prolongs the delay (Mead, 1975).

Attempts to terminate embryonic diapause in mustelids by means of exogenous gonadotropins or steroids have all been unsuccessful (Cochrane and Shackelford, 1962; Wright, 1963; Canivenc et al., 1972; Mead, 1972; Sheldon, 1972, 1973). Mead and Eik-Nes (1969a,b) suggest that although both estrogen and progesterone play an important role in preparing the uterine environment for implantation they probably do not induce implantation, at least in the spotted skunk. However, the search for other controlling hormones has not yielded any clear-cut candidates (Mead, 1971, 1972, 1975; Foresman and Mead, 1974), although LH may have a direct or indirect role in the initiation of development of the corpus luteum and embryo.

A totally different situation obtains in mice, rats (Psychoyos, 1973a,b), and marsupials (Tyndale-Biscoe et al., 1974; Sharman and Berger, 1969). Lactational delay of implantation in mice can be terminated by injecting estrogen (Whitten, 1955, 1958; McLaren, 1968). Mice ovariectomized early in pregnancy and given maintenance doses of progesterone can be induced to implant with a single dose of estrogen (Yoshinaga and Adams, 1966; Smith and Biggers, 1968) but not without estrogen (Bloch, 1958). In the red kangaroo and tammar, either estrogen and progesterone allows blastocyst resumption, although estrogen probably causes later loss of the embryos (Clark, 1968; Smith and Sharman, 1969). In the tammar, progesterone alone can induce development to full term, although there is a significant proportion of intrauterine mortality, generally in the first 7–11 days (Renfree and Tyndale-Biscoe, 1973a). McLaren (1971), however, reported an injection schedule of progesterone alone that consistently resulted in implantation in mice both after ovariectomy and during lactation. It should be noted that relatively high doses of progesterone were given to obtain this result.

Other hormones may also be important in delay. For example, prolactin and oxytocin are implicated in the initiation and maintenance of delay in the tammar wallaby (Hearn, 1973, 1974; Tyndale-Biscoe et al., 1974). Prolactin produced by the pituitary inhibits the activity or further development of the corpus luteum, which in turn prevents the progesterone-dependent development of the secretory uterus (Tyndale-Biscoe and Hawkins, 1977a,b). In the armadillo, the pituitary has a high level of LH during delay, which is depleted by 17 days after bilateral ovariectomy. Implantation occurs 18–20 days after bilateral ovariectomy (Labhsetwar and Enders, 1969; Enders, 1966).

There seem to be no common pathways for switching the blastocysts "on" or "off," an observation supporting the idea stated above that the phenomenon of delay has arisen independently in the various groups; the mechanisms for obtaining it can be different in each group. It is clear from the studies briefly reviewed above that in some mammals the embryos may be held in abeyance by the presence of an inhibitor, whereas in others they may be delayed due to the lack of a stimulator. The hormones that obviously control the production of such factors may not necessarily themselves *be* the factors. In several species the more likely candidates are the secretory products of the uterus.

3. UTERUS–EMBRYO RESPONSES

Blastocysts in delay lie free and are not attached to the uterine epithelium. The uterus, which provides the immediate environment of the blastocyst, must respond to the endocrines by changing that environment in such a way that the blastocyst will respond by resuming its development.

Enders (1967) and Enders and Given (1977) made a detailed study of the endometrium of a wide variety of mammals during the delay period. They found that the majority, with the exception of the rat and mouse, appeared to show some signs of glandular secretion, so that the endometrium itself is well maintained during the delay. Tyndale-Biscoe *et al.* (1974) distinguished between the various breeding patterns in macropodid marsupials, pointing out that the anestrous condition of the uterus is not analogous to quiescence and that diapausing blastocysts are found only in quiescent uteri. Enders' (1967) results suggest that this difference probably holds for all mammals. However, as Enders points out, the ultrastructural or histological condition of the uterus is of limited value in determining the importance of secretion to the uterine milieu, and a change in the composition may be more important than a change in the rate of secretion, although undoubtedly both influence blastocyst growth. On the other hand, other factors also influence the uterine environment—oxygen tension, organic and inorganic materials, etc.

Nevertheless, the condition of the uterus has been clearly established by blastocyst transfer experiments in several species as the controlling influence. Dickmann and De Feo (1967) transferred dormant rat blastocysts to "active" uteri (hormonally stimulated) and found that the blastocysts resumed development. Conversely, "active" blastocysts became dormant when placed in dormant uteri. Similarly, quiescent quokka and tammar wallaby embryos developed in "active" uteri after transfer to either intact animals or animals ovariectomized at the time of transfer (Tyndale-Biscoe, 1963c, 1970).

Chang (1968), in a slightly more elaborate experiment, showed that delayed mink blastocysts grew when placed in the uteri of ferrets (which have no delay of implantation), whereas the reciprocal transfers were unsuccessful, presumably because the ferret blastocysts became "dormant" and died.

a. Uterine Secretions.   Recently there have been several studies designed to elucidate the role of the uterine secretions and also the responses of the blastocysts during and after periods of delay.

The production of endometrial secretion is influenced by the level of ovarian steroids. Adminstration of such hormones changes the relative proportions of the intrauterine protein fractions (Beier, 1970; 1974, Arthur and Daniel, 1972; Renfree, 1973a), and such changes have been correlated with embryonic growth (Daniel, 1970, 1972; Beier, 1970; Daniel and Krishnan, 1969; Aitken, 1974a).

In the roe deer and the tammar wallaby embryonic diapause is associated with a deficiency of secretory material, whereas the rapid elongation of the blastocyst is accompanied by the production of several types of secretion (Renfree, 1972b, 1973a; Aitken, 1974c, 1975). Specific uterine proteins are produced in these species at resumption of development. In the northern fur seal, uterine protein concentrations are very low during the delay period, with a specific protein fraction similar to the uteroglobin (blastokinin) fraction found in rabbits (Daniel, 1967, 1971). The diapausing blastocysts of a variety of mammals with embryonic diapause (armadillo, black bear, fur seal, mink, and rat) became activated when cultured in a medium supplemented with blastokinin from the rabbit (Daniel and Krishnan, 1969). These authors concluded that embryonic diapause results from the failure of the mother to provide sufficient protein and/or certain specific proteins that are needed for active growth in the blastocysts.

In mice, implantation is associated with an increased uterine protein content, but during delay protein levels are low (Aitken, 1977a). Estradiol-17β induced a biphasic increase in uterine luminal protein, whether or not blastocysts were present. Nilsson (1974) showed ultrastructural changes 8 hr after estrogen injection in the uterine glands and blastocyst. These results suggest that as in the fur seal, roe deer, and wallaby, diapause in the mouse is associated with low protein. No evidence could be found for the existence of inhibitory proteins in the uterine fluids of mice in diapause using protein electrophoresis (Aitken, 1977a). Surani (1975) also observed an increase in high molecular weight proteins in the uterine lumen of the rat 13–20 hr after estradiol injection.

These results are difficult to equate with the ability of actinomycin D, a protein synthesis inhibitor, to terminate embryonic diapause in the mouse within 48–72 h (Finn, 1974). It was postulated that the actinomycin D reduced the transcription of a protein inhibitor and released blastocysts from diapause. In repeating Finn's experiments, Aitken (1977b) observed that induction of implantation with actinomycin D was associated with a fall in the protein content of the uterine lumen. Aitken concluded that the highly significant increase in luminal protein observed at implantation during normal pregnancy or following estradiol-17β administration is not an absolute requirement for implantation. He further suggested that actinomycin D might reduce the level of a small molecular weight inhibitor (R. J. Aitken, personal communication), a suggestion consistent with the observation by Weitlauf (1976) that the *in vitro* inhibitor is dialyzable.

Once blastocysts have reactivated, numerous changes occur in the uterus. In the roe deer and the tammar wallaby, endometrial secretion reaches highest levels at the time of implantation (Aitken, 1975; Renfree, 1973b), and it is possible that these secretions are stimulated by embryonic tissue or membranes (Aitken, 1974a; Renfree, 1972b). Considerable attention has been given to the ability to the blastocyst to synthesize steroids. Dickmann and Dey (1973) suggested that the preimplantation embryo is a source of steroid hormones influencing blastocyst growth and implantation. Rabbit blastocysts contain progesterone and 20α-hydroxyprogesterone (Seamark and Lutwak-Mann, 1972). Pig blastocysts synthesize estrogen and progesterone (Perry and Heap, 1973; Perry *et al.*, 1973), and in coculture experiments endometrial protein synthetic activity is greater when the tissue is cultured with blastocyst tissue (Heap *et al.*, 1977). Such steroidogenic activity in blastocysts was first reported by Huff and Eik-Nes (1966) in the rabbit. These experiments suggest that the blastocyst may influence metabolic activity of the endometrium as well as being dependent on it. Diapause may thus be caused by interference with either blastocyst or uterine metabolism.

b. Metabolic Changes in the Blastocyst During Dormancy. Mouse blastocysts, during embryonic diapause, are dormant metabolically as well as developmentally (Weitlauf and Greenwald, 1965, 1968; Menke and McLaren, 1970). Uptake of exogenous amino acids is higher in nondelayed mouse blastocysts than in delayed ones, a response apparently imposed by the uterus (Weitlauf, 1973). In rats, it is suggested that a direct inhibitory effect by a uterine substance is responsible for retarding the embryos during embryonic diapause (Psychoyos, 1969, 1973a,b; Gulyas and Daniel, 1969; Weitlauf, 1974). Progesterone and estrogen affect CO_2 production (Torbit and Weitlauf, 1974) and also the uptake and use of exogenous materials during delay in the mouse (Hensleigh and Weitlauf, 1974). In the mouse, delayed embryos steadily increase in weight during diapause (Hensleigh and Weitlauf, 1974). Total protein content, however, is unchanged in delayed mouse embryos (Weitlauf, 1973), and it is suggested that metabolic activation of these blastocysts requires the synthesis of new RNA (Daniel, 1970; Weitlauf, 1974). Certainly, during diapause, the mouse embryo itself enters a state of relative metabolic quiescence (Pike and Wales, 1975; Wales, 1975).

The level of RNA synthesis is low in delayed-implanting blastocysts, but when placed *in vitro* they "escape" from an inhibitory effect on the uterus (Weitlauf, 1974). Weitlauf (1976) showed that uterine flushings of ovariectomized mice given estrogen after progesterone, or progesterone alone, inhibit [^{3}H]uridine incorporation *in vitro*. The inhibitory activity is dialyzable and heat resistant. It is interesting, however, that uterine flushings from "implanting" mice were as effective as flushings from "delayed-implanting" mice in inhibiting RNA synthesis *in vitro*. Psychoyos (1973a,b) showed similar results with flushings or extracts of "delayed" uteri from rats and postulated that this inhibitory effect caused the delay since blastocysts transferred to extrauterine sites resumed development. However, it is difficult to explain why the flushings from both "delayed" and "implanting" uteri inhibited growth of mouse blastocysts *in vitro*. Weitlauf (1976) postulated that his results and those of Psychoyos (1973b) may be caused by an inhibitory factor that is ineffective when coupled to another molecule, and the coupling was disturbed in the collection procedure.

All these studies on blastocyst metabolism during delay have been carried out on rats and mice, and further studies should be made on blastocysts of the other groups of mammals that show the phenomenon. It seems reasonable, however, to assume that a marked depression of embryonic metabolism during embryonic diapause, probably mediated by the uterus, will be found in most species. The dependence of the embryo on the uterus for its activation and nourishment indicates the importance of synchronization between blastocyst and uterus. Tyndale-Biscoe, *et al.* (1974) view diapause as an adaptation that maintains the embryo in a viable state while awaiting the appropriate signal that will commit it to the irreversible process of differentiation. The results from rats and mice (Finn, 1974; Weitlauf, 1974, 1976; Aitken, 1977a, b; Psychoyos, 1973a, b; Surani, 1975) are difficult to interpret, since the rise in protein content of uterine flushings observed in activation is abolished by actinomycin D, and yet flushings (whether from "active" or "delayed" uteri) inhibit blastocysts *in vitro*. There is, as yet, no evidence with which to clearly distinguish between the two hypotheses of withdrawl of an inhibitor or synthesis of an embryotropic secretion in the resumption of blastocyst activity after diapause.

B. Diapause as a Strategem for Adaptation and Survival

The nature and controlling mechanisms of delay are difficult to interpret, especially since the range of species showing it is so wide. The studies of rodents describe above and the work on wild species (the skunk, roe deer, and tammar wallaby) demonstrate that the mechanisms are highly complex, and the evolutionary purpose of embryonic diapause is still obscure. It is, however, an extremely efficient mechanism for storing the blastocyst, and there is little if any loss during the period of diapause.

The main factors that control reproductive timing in vetebrates are nutrition, temperature, photoperiod, and the sexual cycle. However, among mammals showing embryonic diapause, the only factor common to all the types of delay is that it is a mechanism that ensures synchrony. Embryonic diapause synchronizes the development of the blastocyst with that of the uterus or with cessation of lactation, male with female breeding cycles, and animals with their environment. It is a stratagem employed during development that results in one or more of these factors being synchronized with the other. The underlying assumption is that there is an optimal period for reproduction. This optimal period may differ for differ species or even in different populations of the same species. However, it represents a *fixed* point in the reproductive cycle around which the other events can occur, and it is ultimately dependent on environmental conditions. Bigg (1973) discussed this concept:

> Selection within each population of harbour seal (*Phoca vitulina*) would be for adaptations which set the timing of the whole cycle so that weaning could occur at the appropriate date. The data suggest that one of these adaptations involves setting the time of onset of estrus for each population independently, thereby setting the time of conception, birth and subsequently the best time for weaning.

For example, to ensure the weaning of young at a particular optimal time, or alternatively to ensure that males and females of solitary species come together at appropriate times for mating, a number of strategies have been employed. One of these has been to extend the period of intrauterine care. This can be achieved in a number of ways, but in some species (those given in Tables I and II) it has been attained by a delay of development at the blastocyst stage: embryonic diapause. There are obviously other ways in which synchrony can be achieved, just as there are many different ways of fertilization.

If diapause confers an adaptive advantage by ensuring synchrony, then presumably in the evolution of diapause there would be a phase during which some members of the species would show delay and others would not. A population of that species might show a range of variation, changing over a period of time from a situation in which most of the members of the population would not have diapause to one in which most would have it. If this is so, then one would expect to find some individuals whose breeding cycles do not include a period of delay, whereas others show delay of different lengths. In support of this notion, we have observed that in a small captive population of tammars there are usually a few individuals who do not show the seasonal quiescence characteristic of the species. The blastocysts of these animals develop at the time when females in the rest of the population carry delayed embryos (M. B. Renfree and C. H. Tyndale-Biscoe, unpublished observations.) In the roe deer, one animal has recently been examined which mated at the normal time (August) and then went through a normal pregnancy without any delay (R. V. Short, personal communication). Observations like this have not been generally reported for other species but it seems likely that there are such individual variations in other populations as well. However, it has often been reported that breeding patterns change when an animal is brought into captivity (i.e., into a different environment), where food, for example, is no longer limiting. The quokka is one such species—when brought into captivity the viability of its dormant blastocyst is prolonged. In this regard it is interesting that the ferret does not show delay, although most of its close relatives do. Is it possible that the ferret has lost the delay because of a more stable environment resulting from long domestication? Published studies involve only domestic ferrets; a study of wild ferrets would be most interesting.

The importance of synchrony between uterus and blastocyst has been well established in mammals (e.g., Dickmann and Noyes, 1960; Noyes and Dickmann, 1960; Dickmann and De Feo, 1967; Tyndale-Biscoe, 1970) and is probably of great adaptive advantage in species in which facultative delay occurs during suckling. Many of these species have high fertility or are those in which the return to estrous activity is not suppressed by pregnancy (e.g., rodents, macropodid marsupials). Indeed, suppression of the estrous cycle could be viewed as an alternative to embryonic diapause, since the end result would be the same (Fig. 5). Tyndale-Biscoe (1968, 1973) believes that such embryo–uterine synchrony would have great selective value and might therefore occur in all marsupials but that only in some macropodids has the delay been extended to serve another, ecological function. This line of reasoning could be extended to all mammals, particularly if blastocysts could be delayed experimentally in nondelaying species. Sharman (1965a), however, concludes that the prime importance of diapause in macropodid marsupials is to prevent the occupation of the pouch by two young of different ages, an event that would otherwise occur because estrus is not inhibited and ovulation and fertilization take place postpartum.

Synchrony between animals and their environment has been dealt with in

Environmental   →   Endocrine   ←   Uterus-Embryo
Cues   Mediators   Responses

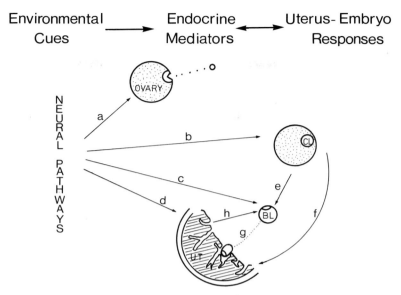

Fig. 5. Diagram summarizing alternate ways of preventing and reinitiating blastocyst development. The environmental cues can block reproductive activity at a number of levels: (a) preventing ovulation and estrus or, if ovulation has occurred, by (b) preventing further development of the corpus luteum, and/or (c) preventing further development of the blastocyst directly, and/or (d) preventing development of the luteal uterus. Reactivation of the blastocyst occurs when the corpus luteum secretions stimulate the growth of the blastocyst directly (e) or the uterus (f), which in turn stimulates the blastocyst (h) and causes implantation (g). Removal of any (b), (c), or (d) presumably allows stimulation to occur at (e) or (f) and causes reinitiation of development. Abbreviations: BL, blastocyst; CL, corpus luteum; UT, uterus.

considerable detail elsewhere in relation to diapause (e.g., Sadleir, 1969). Diapause is more prevalent in animals of high latitudes and in those that live in habitats where harsh conditions prevail (although not necessarily on a regular or seasonal basis). Diapause, then, frequently in combination with other physiological modifications such as occur during hibernation or torpor, ensures that the young are produced or reared when environmental conditions are most favorable for their survival. A wide range of diapausing species (e.g., bears, seals, bats, mustelids) probably fall into this category. The converse situation would involve species that are entirely, and perhaps obligatorily, opportunistic breeders. However, some of these species (e.g. the red kangaroo and the euro) also show delay, and in these cases it should perhaps be assumed that diapause was derived from a environmental situation that was once regular or seasonal but is no longer so. In the case of the red kangaroo, diapause has obvious selective advantages in synchronizing maternal "investment" in the young with good conditions. However, for seasonally breeding macropodids, the ecological advantage is not so obvious and in some the survival of diapause blastocysts is less prevalent (Tyndale-Biscoe, 1973; Russell, 1974).

Diapause may also be advantageous in allowing synchrony with the environment to be combined with the need to synchronize male and female breeding patterns as a means of maintaining species integrity, that is, as a species isolating mechanism. An example of this is found in the skunks. Females of the eastern population breed at 9–10 months of age in March/April, while females of the western group breed at 4–5 months old in September/October, although the embryos of both populations implant in April (Mead, 1968b). Since the males and females are fertile at the same time, the differences in timing presumably ensure little or no breeding overlap except in the sourthern latitudes, where cycles approximate each other (Mead, 1968b). However, the same cannot be said of eastern and western gray kangaroos for although one species (the eastern) generally shows diapause and the other does not, the breeding cycles overlap (Poole, 1973).

Diapause is a complex phenomenon, the prime importance of which seems to be to ensure synchrony at one or more of many levels and may be superimposed on other mechanisms to provide mammals with some additional evolutionary advantage. If diapause occurs, it must be accomplished in such a way that the reawakening of the dormant blastocyst is synchronized with the development of the uterus, as pointed out by Tyndale-Biscoe (1968, 1970, 1973). However, this is an all-or-none phenomenon in that failure of such synchrony leads to embryonic death: a lethal situation that acts as an obligate factor in selection. This does not constitute a reason for the occurrence of diapause in the first place; it merely defines conditions under which it can occur successfully. There is no physiological requirement of the mammalian embryo to delay its development other than the need for its synchronization with the external factors such as lactation, photoperiod, and nutrition that affect the maternal organism and, only indirectly, influence the blastocyst. Only about 65 species of the 100–1200 mammalian genera have been shown to use diapause. This fact, together with the physiological complexity and endocrine sophistication, speaks against diapause as the ancestral condition and suggests instead that it may be a later strategy in the evolution of the developmental processes of mammals.

ACKNOWLEDGMENT

I thank my colleagues, and Dr. R. V. Short in particular, for making numerous helpful comments during the preparation of this chapter. I am grateful to Dr. J. C. Daniel for allowing me to use his photographs of delayed blastocysts (Fig. 1), Mr. G. I. Wallace for the photograph of the bat blastocyst (Fig. 2A), and Dr. C. H. Tyndale-Biscoe for the photograph of the tammar wallaby blastocyst (Fig. 2B).

REFERENCES

Aitken, R. J. (1974a). *J. Reprod. Fertil.* **39**, 225–233.
Aitken, R. J. (1974b). *J. Reprod. Fertil.* **40**, 235–239.
Aitken, R. J. (1974c). *J. Reprod. Fertil.* **40**, 333–340.
Aitken, R. J. (1975). *J. Anat.* **119**, 369–384.
Aitken, R. J. (1976). *J. Reprod. Fertil.* **46**, 439–440.
Aitken, R. J. (1977a). *J. Reprod. Fertil.* **50**, 29–36.
Aitken, R. J. (1977b). *J. Reprod. Fertil.* **50**, 193–195.
Aitken, R. J. (1977c) *In "Development in Mammals* (M. H. Johnson, ed.), pp. North-Holland.
Aitken, R. J., Burton, J. H. W., and Steven, D. H. (1971). *J. Physiol. (London)* **217**, 13P–15P.
Aitken, R. J., Burton, J., Hawkins, J., Kerr-Wilson, R., Short, R. V., and Steven, D. H. (1973). *J. Reprod. Fertil.* **34**, 481–493.
Arthur, A. T., and Daniel, J. C. (1972). *Fertil. Steril.* **23**, 115–122.
Asdell, S. A. (1964). "Patterns of Mammalian Reproduction," 2nd ed. Constable, London. 670 pp.
Backhouse, K., and Hewer, H. R., (1956). *Nature (London)* **178**, 550.
Baevsky, U. B. (1963). *In "Delayed Implantation"* (A. C. Enders, ed.), pp. 141–154. Univer. of Chicago Press, Chicago Illinois.
Beier, H. M. (1970). *In "Ovo-implantation, Human-gonadotropins and Prolactin"* P. O. Hubinot *et al.*, eds. pp. 157–163. Karger, Basel.
Beier, H. M. (1974). *Adv. Biosci.* **13**, 199–219.
Berger, P. J. (1966). *Nature (London)* **211**, 435–436.
Berger, P. J. (1970). Ph.D. Thesis, Tulane University, New Orleans, Louisiana.
Berger, P. J., and Sharman, G. B. (1969), *J. Mammal.* **50** 630–632.
Bigg, M. A. (1973). *J. Reprod. Fertil., Suppl.* **19**, 131–142.
Bloch, S. (1958). *Experientia* **14**, 447.
Bodley, H. D. (1974). *Anat. Rec.* **178**, 313.
Bowley, E. A. (1939). *J. Mammal.* **20**, 499.
Bradshaw, G. V. R. (1962). *Science* **136**, 645–646.
Brambell, F. W. R. (1935). *Philos. Trans. R. Soc. London, Ser. B* **225**, 1–50.
Brambell, F. W. R. (1937). *Am. J. Obstet. Gynecol.* **33**, 942–953.
Brambell, F. W. R. and Hall, K. (1936). *Proc. Zool. Soc. London* p. 957.
Brambell, F. W. R., and Rowlands, I. W. (1936). *Philos. Trans. R. Soc. London, Ser. B* **226**, 71–97.
Buchanan, G. D. (1966), *Am. J. Anat.* **118**, 195.
Buchanan, G. D. (1969). *J. Reprod. Fertil.* **18**, 305–316.
Buchanan, G. D., Enders, A. C., and Talmage, R. V. (1956). *J. Endocrinol.* **14**, 121–129.
Calaby, J. H., and Poole, W. E. (1971). *Int. Zoo Yearb.* **11**, 5–12.
Canivenc, R. (1966). *Symp. Zool. Soc. London* **15**, 15–26.
Canivenc, R., and Bonnin Laffargue, M. (1963). *In "Delayed Implantation"* (A. C. Enders, ed.), p. 115. Univ. Of Chicago Press, Chicago, Illinois.
Canivenc, R., Cohere, G., and Brechenmacher, C. (1967). *C. R. Hebd. Seances Acad. Sci.* **264**, 1187–1189.
Canivenc, R., Bonnin-Laffargue, M., and Lajus-Boué, M (1971), *C. R. Hebd. Seances Acad. Sci.* **273**, 1855–1857.

Canivenc, R., Bonnin-Laffargue, M., Lamy, E., and Relexans, M. C. (1972). *C. R. Seances Soc. Biol. Ses. Fil.* **166**, 1645–1649.
Chang, M. C. (1968). *J. Exp. Zool.* **168**, 49–60.
Clark, M. J. (1966). *Aust. J. Zool.* **14**, 19–25.
Clark, M. J. (1967). *Aust. J. Zool.* **15**, 673–683.
Clark, M. J. (1968). *J. Reprod. Fertil.* **15**, 347–355.
Clark, M. J., and Poole, W. E. (1967). *Aust. J. Zool.* **15**, 441–459.
Cochrane, R. F. and Meyer, R. H. (1957) Proc. Soc. Exp. Biol. Med. **96**, 155–159.
Cochrane, R. L., and Shackelford, R. M. (1962), *J. Endocrinol.* **25**, 101–106.
Courrier, R. (1927). *Arch. Biol.* **37**, 173–334.
Craighead, J. J., Hornoker, M. G., and Craighead, F. C. (1969). *J. Reprod. Fertil. Suppl.* **6**, 447–476.
Daniel, J. C. (1967). *Comp. Biochem. Physiol.* **24**, 297–300.
Daniel, J. C. (1970). *BioScience* **20**, 411–415.
Daniel, J. C. (1971). *Dev. Biol.* **26**, 316–322.
Daniel, J. C. (1972). *Fertil. Steril.* **23**, 78–80.
Daniel, J. C., and Krishnan, R. S. (1969). *J. Exp. Zool.* **172**, 267–282.
Deanesly, R. (1943). *Nature (London)* **151**, 365.
Dickmann, Z., and De. Feo, V. J. (1967). *J. Reprod. Fertil.* **13**, 3–9.
Dickmann, Z., and Dey, S. K. (1973). *J. Reprod. Fertil.* **35**, 615–617.
Dickmann, Z., and Noyes, R. W. (1960). *J. Reprod. Fertil.* **1**, 197–212.
Dittrich, L., and Kronberger, H. (1962). *Z. Saeugertierkd.* **28**, 129.
Dukelow, W. R. (1966). *Nature (London)* **211**, 211.
Dwyer, P. D. (1963). *Aust. J. Zool.* **11**, 219–240.
Dwyer, P. D. (1968). *Aust. J. Zool.* **16**, 49–68.
Ealey, E. H. M. (1963). *In* "Delayed Implantation" (A. C. Enders ed.), Univ. of Chicago Press. Chicago, Illinois.
Eckstein, P., Shelesnyak, M. C., and Amoroso, E. C. (1959). *Mem. Soc. Endocrinol* **6**, 3–12.
Enders, A. C. (1961). *Anat. Rec.* **139**, 483–497.
Enders, A. C. (1962). *J. Anat.* **96**, 39–48.
Enders, A. C., ed. (1963). "Delayed Implantation. Univ. of Chicago Press, Chicago, Illinois.
Enders, A. C. (1966). *Symp. Zool. Soc. London* **15**, 295–310.
Enders, A. C. (1967). *In* "Cellular Biology of the Uterus" (R. M. Wynn, ed.), pp. 151–190. Appleton, New York.
Enders, A. C., and Buchanan, G. D. (1959). *Tex. Rep. Biol. Med.* **17**, 323–340.
Enders, A. C., and Given, R. L. (1977). *In* "Biology of the Uterus" R. M. Wynn, pp. 203–243. Plenum, New York.
Enders, R. K. (1952). *Proc. Am. Philos. Soc.* **96**, 691–755.
Enders, R. K., and Enders, A. C. (1963). *In* "Delayed Implantation" (A. C. Enders, ed.), pp. 129–140. Univ. of Chicago Press, Chicago, Illinois.
Finn, C. A. (1974). *J. Endocrinol.* **60**, 199–200.
Fisher, H. C. (1954). *Nature (London)* **173**, 879.
Fleming, T. H. (1971). *Science* **171**, 402–404.
Flynn, T. T. (1930). *Proc. Linn. Soc. N.S.W.* **55**, 506–531.
Foresman, K. R., and Mead, R. A. (1973). *J. Mammal.* **54**, 521–523.
Foresman, K. R., and Mead, R. A. (1974). *Biol. Reprod.* **11**, 475–480.
Foresman, K. R., Reeves, J. J., and Mead, R. A. (1974). *Biol. Reprod.* **11**, 102–107.
Frith, H. J., and Sharman, G. B. (1964). *CSIRO Wildl. Res.* **9**, 86–114.
Gulyas, B. J., and Daniel, J. C. (1969). *Biol. Reprod.* **1**, 11–20.
Hamilton, W. J. (1962). *J. Mammal.* **43**, 486–504.

Hamilton, W. J., and Eadie, W. R. (1964). *J. Mammal.* **45,** 242–251.

Hamlett, G. W. D. (1932). *Anat. Rec.* **53,** 283–303.

Hamlett, G. W. D. (1935). *Q. Rev. Biol.* **10,** 432–447.

Hansson, A. (1947) Acta Zoologica. **28** 1–136.

Harrison, R. J. (1960). *Mammalia* **24,** 372–385.

Harrison, R. J. (1963). *In* "Delayed Implantation" (A. C. Enders, ed.), pp. 99–114. Univ. of Chicago Press, Chicago, Illinois.

Harrison, R. J. (1969). *In* "The Biology of Marine Mammals" (H. T. Anderson, ed.), Chapter 8. Academic Press, New York.

Harrison, R. J., and Kooyman, G. L. (1968). *In* "Behaviour and Physiology of Pinnipeds" (R. J. Harrison *et al.*, eds.), Chapter 7. N.Y. Appleton-Century-Crofts.

Harrison, R. J., Matthews, L. H., and Roberts, J. M. (1952). *Trans. Zool. Soc. Lond.* **27,** 437–540.

Hartman, C. G. (1940). *J. Mammal.* **21,** 213.

Heap, R. B., Perry, J. S., Burton, R. D., Gadsby, J. E., Wyatt, C., and Jenkin, G. (1977). *In* "Reproduction and Evolution" J. H. Calaby and C. H. Tyndale-Biscoe eds., pp. 341–348. Aust. Acad. Sci., Canberra.

Hearn, J. P. (1972a). Ph.D. Thesis, Australian National University, Canberra, A.C.T.

Hearn, J. P. (1972b). *J. Reprod. Fertil.* **28,** 132.

Hearn, J. P. (1972c). *Aust. Mammal.* **1,** 40–42.

Hearn, J. P. (1973). *Nature (London)* **241,** 207–208.

Hearn, J. P. (1974). *J. Reprod. Fertil.* **39,** 235–241.

Hearn, J. P. (1975). *J. Endocrinol.* **64,** 403–416.

Hensleigh, H. C., and Weitlauf, H. M. (1974). *Biol. Reprod.* **10,** 315–320.

Hill, J. P. (1900). *Proc. Linn. Soc. N.S.W.* **25,** 519–532.

Hill, J. P. (1910). *Q. J. Microsc. Sci.* **56,** 1–134.

Huff, R. L., and Eik-Nes, K. B. (1966). *J. Reprod. Fertil.* **11,** 57–63.

Hughes, R. L. (1962). *Aust. J. Zool.* **10,** 193–224.

Hughes, R. L. (1969a). *J. Anat.* **104,** 407.

Hughes, R. L. (1969b). *J. Reprod. Fertil.* **19,** 387.

Kennaway, D. J., and Seamark, R. F. (1975). *J. Reprod. Fertil.* **46,** 503–504.

Kirkpatrick, T. H. (1965). *Queensl. J. Agric. Anim. Sci.* **22,** 319–328.

Labhsetwar, A. P., and Enders, A. C. (1969). *J. Reprod. Fertil.* **18,** 383–389.

Laws, R. M. (1956). *Falkland Isl. Depend. Surv., Sci. Rep.* **15,** 1–66.

Lemon, M. (1972). *J. Endocrinol.* **55,** 63–71.

Lincoln, G. A., and Guinness, F. E. (1972). *J. Reprod. Fertil.* **31,** 455–457.

McLaren, A. (1968). *J. Endocrinol.* **42,** 453–463.

McLaren, A. (1971). *J. Endocrinol.* **50,** 515–526.

McLaren, A. (1973). *In* "The Regulation of Mammalian Reproduction" S. J. Segal, *et al.*, eds.), p. 321. Thomas, Springfield, Illinois.

Mantalenakis, S., and Ketchel, M. (1966). *J. Reprod. Fertil.* **12,** 391–394.

Mayer, G. (1963). *In* "Delayed Implantation" (A. C. Enders, ed.), pp. 213–231. Univ. of Chicago Press, Chicago Illinois.

Maynes, G. M. (1973). *Aust. J. Zool.* **21,** 331–351.

Mead, R. A. (1968a). *J. Zool.* **156,** 119–136.

Mead, R. A. (1968b). *J. Mammal.* **49,** 373–390.

Mead, R. A. (1971). *Biol. Reprod.* **5,** 214–220.

Mead, R. A. (1972). *J. Reprod. Fertil.* **30,** 147–150.

Mead, R. A. (1973). *J. Mammal.* **54,** 521–523.

Mead, R. A. (1975). *Biol. Reprod.* **12,** 526–533.

Mead, R. A., and Eik-Nes, K. B. (1969a). *J. Reprod. Fertil., Suppl.* **6,** 397–403.
Mead, R. A., and Eik-Nes, K. B. (1969b). *J. Reprod. Fertil.* **18,** 351–353.
Meckley, P. E., and Ginther, O. J. (1971). *J. Anim. Sci.* **33,** 1160.
Meckley, P. E., and Ginther, O. J. (1972). *Am. J. Vet. Res.* **33,** 1247–1252.
Medway, Lord (1971). *J. Zool.* **165,** 261–273.
Menke, T. M., and McLaren, A. (1970). *J. Endocrinol.* **47,** 287–294.
Merchant, J. C. (1976). *Aust. Wildl. Res.* **3,** 93–103.
Mitchell, G. C. (1965) M. Sc. Thesis., University of Arizona.
Møller, O. M. (1973). *Endocrinology* **56,** 121–132.
Moore, G. P. M. (1975). *J. Reprod. Fertil.* **46,** 504.
Moors, P. J. (1975). *Aust. J. Zool.* **23** 355–362.
Murphy, B. D. (1972). *Anat. Rec.* **172,** 372.
Mutere, F. (1965). *Nature (London)* **207,** 780.
Mutere, F. A. (1967). *J. Zool.* **153,** 153–161.
Neal, E. G., and Harrison, R. J. (1958). *Trans. Zool. Soc. London* **29,** 67–131.
Newsome, A. E. (1965). *Aust. J. Zool.* **13,** 735–759.
Newsome, A. E. (1966). *CSIRO Wildl. Res.* **11,** 187–196.
Newsome, A. E. (1971). *Aust. Zool.* **16,** 32–50.
Nilsson, O. (1974). *J. Reprod. Fertil.* **39,** 87–194.
Norris, M. L., and Adams, C. E. (1971). *J. Reprod. Fertil.* **27,** 486–487.
Noyes, R. W., and Dickmann, Z. (1960). *J. Reprod. Fertil.* **1,** 186–196.
Nutting, E. F., and Meyer, R. K. (1963). *In* "Delayed Implantation" (A. C. Enders, ed.), pp. 233–252. Univ. of Chicago Press, Chicago, Illinois.
Pearson, A. K., and Enders, R. K. (1944). *J. Exp. Zool.* **95,** 21–35.
Pearson, A. K., and Enders, R. K. (1951). *Anat. Rec.* **111,** 695–712.
Perry, J. S., and Heap, R. B. (1973). *Acta Endocrinol. (Copenhagen), Suppl.* **177,** 178.
Perry, J. S., Heap, R. B., and Amoroso, E. C. (1973). *Nature (London)* **245,** 45–47.
Peyre, A., and Herlant, M. (1967). *C. R. Seances Soc. Biol. Ses. Fil.* **161,** 1779–1782.
Pike, I. L., and Wales, R. G. (1975). *J. Reprod. Fertil.* **46,** 531–532.
Poole, W. E. (1973). *Aust. J. Zool.* **21,** 183–212.
Poole, W. E. (1975). *Aust. J. Zool.* **23,** 333–354.
Poole, W. E., and Catling, P. C. (1974). *Aust. J. Zool.* **22,** 277–302.
Poole, W. E., and Pilton, P. E. (1964). *CSIRO Wildl. Res.* **9,** 218–234.
Psychoyos, A. (1969). *Adv. Biosci.***4,** 275–290.
Psychoyos, A. (1973a). *In* Handbook of Physiology *Sect. 7: Endocrinol.* **2,** Part 2, 187–215. (R. O. Greep, ed.) Am. Society of Physiologists.
Psychoyos, A. (1973b). *Adv. Res. Appl.* **31,** 201–256.
Racey, P. A. (1969) *J. Reprod. Fertil.* **19,** 465–474.
Racey, P. A. (1973) *J. Reprod. Fertil. Suppl.* **19,** 175–189.
Rand, R. W. (1954). *Proc. Zool. Soc. London* **124,** 717.
Renfree, M. B. (1970). *J. Reprod. Fertil.* **22,** 483–492.
Renfree, M. B. (1972a). Ph.D. Thesis, Australian National University.
Renfree, M. B. (1972b). *Nature (London)* **240,** 475–477.
Renfree, M. B. (1973a). *Dev. Biol.* **32,** 41–49.
Renfree, M. B. (1973b). *Dev. Biol.* **33,** 62–79.
Renfree, M. B., and Heap, R. B. (1977). *Theriogenology* **6,** 164.
Renfree, M. B., and Tyndale-Biscoe, C. H. (1973a). *Dev. Biol.* **32,** 28–40.
Renfree, M. B., and Tyndale-Biscoe, C. H. (1973b). *J. Reprod. Fertil.* **32,** 113–115.
Ride, W. D. L., and Tyndale-Biscoe, C. H. (1962). *West. Aust. Fish. Dep. Fauna, Bull.* **2,** 54–97.

Russell, E. M. (1974). *Mamm. Rev.* **4,** 1–59.
Sadleir, R. M. F. S. (1965). *Proc. Zool. Soc. London* **145,** 239–261.
Sadleir, R. M. F. S. (1969). "The Ecology of Reproduction in Wild and Domestic Mammals." Methuen, London.
Sadleir, R. M. F. S., and Shield, J. W. (1960). *Nature (London)* **185,** 335.
Sadleir, R. M. F. S., and Tyndale-Biscoe, C. H. (1977). *Biol. Reprod.* **16,** 605–608.
Samson, J. C. (1971). Ph.D. Thesis, University of Western Australia.
Schlafke, S. J., and Enders, A. C. (1963). *J. Anat.* **97,** 353–360.
Seamark, R. F., and Lutwak-Mann, C. (1972). *J. Reprod. Fertil.* **29,** 147–148.
Shapira, N., Kali, J., Amir, S., and Schindler, H. (1974). *J. Reprod. Fertil.* **36,** 295–300.
Sharman, G. B. (1954). *Nature (London)* **173,** 302–303.
Sharman, G. B. (1955a). *Aust. J. Zool.* **3,** 44–55.
Sharman, G. B. (1955b). *Aust. J. Zool.* **3,** 56–70.
Sharman, G. B. (1963). *In* "Delayed Implantation" (A. C. Enders, ed.), pp. 3–14. Univ. of Chicago Press, Chicago, Illinois.
Sharman, G. B. (1965a). *Viewpoints Biol.* **4,** pp. 1–28.
Sharman, G. B. (1965b). *Proc. Int. Congr. Endocrinol., 2nd, 1964,* Excerpta Med. Found. Int. Cong. Ser. No. **83,** pp. 669–674.
Sharman, G. B., and Berger, P. J. (1969). *Adv. Reprod. Physiol.* **4,** 211–240.
Sharman, G. B., and Calaby, J. H. (1964). *CSIRO Wildl. Res.* **9,** 58–85.
Sharman, G. B., Calaby, J. H., and Poole, W. E. (1966). *Symp. Zool. Soc. London* **15,** 205–232.
Sheldon, R. M. (1972). *J. Reprod. Fertil.* **31,** 347–352.
Sheldon, R. M. (1973). *J. Endocrinol.* **92,** 638–641.
Shield, J. W., and Woolley, P. (1963). *Proc. Zool. Soc. London* **141,** 783–790.
Short, R. V. (1967) *In* "Fetal Homeostasis" Vol. 2. (R. M. Wynn, ed.) New York Acad. Sci.
Short, R. V., and Hay, M. F. (1965). *J. Reprod. Fertil.* **9,** 372–374.
Short, R. V., and Hay, M. F. (1966). *Symp. Zool. Soc. London* **15,** 173–194.
Sinha, A. A., and Mead, R. A. (1975). *Cell Tissue Rep.* **164,** 179–192.
Sinha, A. A., Conaway, C. H., and Kenyon, K. W. (1966). *J. Wildl. Manage.* **30,** 121–130.
Smith, D. M., and Biggers, J. D. (1968). *J. Endocrinol.* **41,** 1–9.
Smith, M. J., and Sharman, G. B. (1969). *Aust. J. Biol. Sci.* **22,** 171–180.
Surani, M. A. H. (1975). *J. Reprod. Fertil.* **43,** 411–417.
Svihla, A. (1932). *Univ. Mich. Mus. Zool., Misc. Publ.* **24.**
Torbit, C. A., and Weitlauf, H. M. (1974). *J. Reprod. Fertil.* **39,** 379–382.
Tyndale-Biscoe, C. H. (1963a), *In* "Delayed Implantation" (A. C. Enders, ed.), pp. 15–32. Univ. of Chicago Press, Chicago, Illinois.
Tyndale-Biscoe, C. H. (1963b). *J. Reprod. Fertil.* **6,** 25–40.
Tyndale-Biscoe, C. H. (1963c). *J. Reprod. Fertil.* **6,** 41–48.
Tyndale-Biscoe, C. H. (1965). *Aust. J. Zool.* **13,** 255–267.
Tyndale-Biscoe, C. H. (1968). *Aust. J. Zool.* **16,** 577–602.
Tyndale-Biscoe, C. H. (1970). *J. Reprod. Fertil.* **23,** 25–32.
Tyndale-Biscoe, C. H. (1973). "Life of Marsupials." Arnold, London.
Tyndale-Biscoe, C. H., and Hawkins, J. (1977a). *In* "Reproduction and Evolution" (J. H. Calaby and C. H. Tyndale-Biscoe, eds.), pp. 245–252 Aust. Acad. Sci., Canberra.
Tyndale-Biscoe, C. H., and Hawkins, J. (1977b). *Theriogenology* (in press).
Tyndale-Biscoe, C. H., Hearn, J. P., and Renfree, M. B. (1974). *J. Endocrinol.* **63,** 589–614.
Wales, R. G. (1975). *Biol. Reprod.* **12,** 66–81.
Wallace, G. I. (1975). Honours Thesis, University of New England (Australia)
Wallace, G. I. (1978). *J. Zool.* (in press).

Weitlauf, H. M. (1973). *J. Exp. Zool.* **183,** 303–308.
Weitlauf, H. M. (1974). *J. Reprod. Fertil.* **39,** 213–224.
Weitlauf, H. M. (1976). *Biol. Reprod.* **14,** 556–571.
Weitlauf, H. M., and Greenwald, G. S. (1965). *J. Reprod. Fertil.* **10,** 203–208.
Weitlauf, H. M., and Greenwald, G. S. (1968). *J. Exp. Zool.* **169,** 463–470.
Whitten, W. K. (1955). *J. Endocrinol.* **13,** 1–6.
Whitten, W. K. (1958). *J. Endocrinol.* **16,** 435–440.
Wimsatt, W. A. (1963). *In* "Delayed Implantation" (A. C. Enders, ed.), pp. 49–76. Univ. of
    Chicago Press, Chicago, Illinois.
Wimsatt, W. A. (1975). *Biol. Reprod.* **12,** 1–40.
Wright, P. L. (1963). *In* "Delayed Implantation" (A. C. Enders, ed.), pp. 77–95. Univ. of
    Chicago Press, Chicago, Illinois.
Wright, P. L. (1966). *Symp. Zool. Soc. London* **15,** 27–45.
Wurtman, R. J., Axelrod, J., and Kelly, D. W. (1968). "The Pineal," pp. 58–107. Academic
    Press, New York.
Yadav, M. (1973). *Lab. Anim.* **7,** 89–92.
Yoshinaga, K., and Adams, C. E. (1966). *J. Reprod. Fertil.* **12,** 593–595.

2

Insect Dormancy

ARTHUR M. JUNGREIS

I. INTRODUCTION

The study of insect dormancies is a study of misstatements and misunderstandings. Just as the apocryhphal blind men of India, who upon touching

47

an elephant could not envision the beast in its entirety, so also scientists studying insect dormancies, since few employ the same criteria to define dormancy. To the endocrinologist, dormancy represents "obligatory diapause" in giant Lepidoptera. To the physiologist, dormancy represents a reduction in respiration with its associated decline in anabolic capacity. To the molecular biologist, dormancy represents a basal period that is studied in the context of events preceding the resumption of development. To the phenologist, dormancy is a state that can be described only in terms of behavioral ecology, but which also must be causally related to environmental cues. To the photoperiodist, dormancy represents reproductive diapause, which is used as a tool to "dissect" the properties of a "biological clock" without the necessity of messy internal inspection. All of these investigators fail to see the forest for the trees. The nature of the insect dormancy being studied by them is usually undefined, and, without defining the phenomenon being studied, it is impossible to compare the responses of various insect species with and to one another.

In this chapter, the characteristics of insect dormancies are examined more closely. Selected homeostatic mechanisms associated with the steady-state regulation of a variety of substrates and ions present in hemolymph are studied before, during, and after development of insect dormancies. The general thrust of this chapter is that basic mechanisms responsible for dormancy induction and maintenance cannot be revealed on the basis of external examination alone and that, in the absence of simultaneous measurements of changes in internal function, elucidation of the mechanisms responsible for insect dormancies will prove illusory.

II. ONTOGENY OF INSECT DORMANCIES

A. History of Insect Dormancies

The subject of diapause development and its control in insects has been reviewed periodically by investigators who were primarily concerned with the contribution of circadian rhythms and photoperiod (Andrewartha, 1952; Andrewartha and Birch, 1954; Beck, 1962, 1968b; Bonnemaison, 1945; Danilevsky, 1965; Danilevsky et al., 1970; de Wilde, 1962; Harker, 1961; Lees, 1955, 1968; Müller, 1965, 1970; Saunders, 1974b), endocrines (Braune, 1974; Chippendate, 1977; Church, 1955; de Wilde, 1970; Doane, 1973; Harvey, 1962; Maslennikova, 1974; Salt, 1961; Williams, 1969; Wyatt, 1972), and phenology (Cousin, 1932; de Wilde, 1962; Lees, 1968; Mansingh, 1971; Thiele, 1973; Tauber and Tauber, 1976b). However, the term "diapause" as currently used in the literature lacks precision and has been used synonymously to mean either (1) obligatory metabolic arrest, (2) a condition characterized by the failure to deposit eggs, (3) a reduced capacity to respond to environmental cues, or (4) alterations in behavioral response.

The term "diapause" was first introduced by Wheeler (1893) in the context of "arrested development between anatripsis and catatripsis during blastokinesis" in grasshoppers. The meaning of diapause was immediately obscure to all but a small number of etymologists, but this problem was solved by expanding the definition of diapause to include all cases of arrested development in insects (Henneguy, 1904). Recognition by entomologists of differences in the nature of the inductants of diapause led to the development of three parallel two-component taxonomies for diapause: (1) homodynamic (induced by changes in phenology) versus heterodynamic (independent of changes in phenology) (Roubaud, 1919), (2) obligatory (univoltine species—one generation per year—responding homo- or heterodynamically) versus facultative (multivoltine species responding homo- or heterodynamically) (Steinberg and Kamensky, 1936), and (3) quiescence (phenologically initiated interruptions in growth) versus diapause (spontaneous arrest of development) (Shelford, 1929). Unfortunately, these taxonomies are predicated upon the incorrect premise that the diapause response per se is homogenous rather than heterogeneous. The concept of diapause has been placed in proper perspective within the context of insect dormancies by Müller (1965, 1970), Mansingh (1971) and Thiele (1973). A discussion of dormancy follows.

B. Taxonomy of Dormancies

Dormancies can be broadly defined to incorporate the contributions from physiological, ecological, and phenological influences as "evolved physiological adaptations to overcome adverse environmental conditions of a particular climatic zone" (Mansingh, 1971). Dormancies can result in either growth retardation or arrest, conditions frequently characterized by reductions in DNA synthesis (Krishnakumaran *et al.*, 1967), RNA synthesis (Berry *et al.*, 1967), protein synthesis (Stevenson and Wyatt, 1962), or ATP-generating capacity (Shappirio, 1974; Shappirio and Harvey, 1965; Shappirio and Williams, 1957a,b).

Dormancies can be broadly classified as hibernations, aestivations, or athermopauses, each of which can be subdivided to include quiescence, oligopause, and diapause proper. Hibernations are adaptive responses that have evolved primarily in response to reduced temperatures during ontogeny, such that little or no growth occurs during temperature "stress." Thus, hibernations are physiological states characterized by growth retardation or arrested development and torpidity (i.e., an inability to respond).

Quiescence is defined as a period of growth retardation (Shelford, 1929) occurring in response to noncyclic deviations of one or more phenological factors which are normally of short duration. Insects can experience quiescence throughout ontogeny and in response to seasonal changes. However, in response to evolved natural histories, individual species exhibit only stage-specific and seasonal growth retardation. Quiescence occurs without prior acclimation (i.e., predormancy) and endows individual species with only minor but critical adaptive advantages over nonquiescent species. Quiescence ends immediately upon restoration of favorable environmental conditions. Insects experiencing quiescence during one stage in their life cycles may still experience other types of dormancy at other times during ontogeny.

The second type of hibernation is called "oligopause," a term coined by Müller (1965) and reintroduced at a later date with a different meaning by Mansingh (1971; Mansingh *et al.*, 1972; see also Thiele, 1973). The definition of Mansingh is employed throughout this chapter. "Oligopause" refers to types of hibernations (or aestivation or athermopause) that are better defined in terms of the inductant and the terminating phenological stimuli than is quiescence but less so than is diapause. Oligopause is a cyclic, long-term response to adversity that is initiated just before the seasonal adversity (Mansingh, 1971). It is normally preceded by a short period of preacclimation. Oligopause differs from quiescence in both severity and duration of the insect's response to adversity. Whereas quiescence can be experienced at any stage in the life cycle, oligopause is stage specific. In oliopause, following termination of adverse environmental conditions, growth is reinitiated after a brief lag period, in contrast to both quiescence, which lacks a lag period, and diapause, which has a prolonged lag period. Oligopausing insects are unable to withstand environmental extremes for prolonged periods, must feed occasionally (in contrast to continuously feeding quiescent insects), and normally lack the capacity to experience supercooling (Asahina, 1969). Examples of insects experiencing oligopause are the tiger moth, *Halisidota argentata* (Peck) (Silver, 1958); the dermestid (Kharpa) beetle, *Trogoderma granarium* Everts (Burges, 1959); the Indian meal moth, *Plodia interpunctella* Hübner (Tsuji, 1963); and the fleshfly, *Sarcophaga argyrostoma* (Fraenkel and Hsiao, 1968a,b).

Diapause is a genetically determined state (as is oligopause) of extreme complexity which either is always initiated in a species at the same stage in development (obligatory) or is initiated following exposure to various phenological cues (facultative). It is a long-term adaptation to adversity which lasts up to 2 years, (e.g., in the spruce beetle, *Dendroctonus obesus* (Mannerheim) (Dyer, 1969). Physiological adaptations preliminary to diapause are extremely complex (see Chen, 1971; Gilbert, 1967a,b; Jungreis

1973a, 1976a; Jungreis and Tojo, 1973; Jungreis and Wyatt, 1972; Jungreis *et al.*, 1973, 1974, 1975; Salt, 1961; Wyatt, 1967). Insects experiencing diapause are characterized by the absence of growth, the absence of feeding, and disrupted endocrine control. Diapause is followed by a relatively lengthy period of activation, in turn followed by resumption of growth and development. Despite the reduction in DNA synthesis (Krishnakumaran *et al.*, 1967), RNA synthesis (Berry *et al.*, 1967), protein synthesis (Stevenson and Wyatt, 1962), cytochromes (see Shappirio, 1974), and a seeming loss of capacity to metabolize sugars (Jungreis, 1976a; Wyatt, 1967), the diapause state is actively maintained, as is discussed in Section III.

"Aestivation" refers to the responses of insects that respond "negatively" to long-day photoperiod (primarily) and high temperatures (secondarily). Insects experience aestivation during the summer rather than the winter and have periods of active development synchronized with fall and winter rather than spring and summer (Mansingh, 1971). Just as with hibernation, there are three types of aestival responses, namely, quiescence, oligopause, and diapause. However, during aestivation phenological cues are inductive for dormancies only when they are opposite in direction to those that were inductive during hibernation. For example, the dragonfly, *Tetragoneura cynosura* (Say), ceases molting in response to long-day summer photoperiods but reinitiates development following exposure to shorter photoperiods (Lutz and Jenner, 1964). The adaptive physiology of insects experiencing "heat-hardiness" has not been studied.

The effects of temperature and photoperiod, the major inductants of hibernation and aestivation, have systematically been studied with regard to diapause induction in a large number of families from many insect orders: the elm spanworm egg parasitoid, *Ooencyrtus* (Anderson and Kaya, 1974); the emperor dragonfly, *Anax imperator* Leach (Corbet, 1956); flies of the families Sarcophagidae (Denlinger, 1972a,b; Gibbs, 1975; Saunders, 1971, 1974b; Vinogradova and Zinovjeva, 1971) and Muscidae (Stoffolano and Matthysse, 1967); other Diptera (Kappus and Venard, 1967; Saunders, 1966a); Odonata (Lutz, 1968, 1974; Proctor, 1973; Sawchyn and Church, 1973; Schaller, 1968); Lepidoptera (Jolly *et al.*, 1971; King and Benjamin, 1965; Roach and Adkisson, 1970; Sullivan and Wallace, 1967; Teetes *et al.*, 1969; Thurston, 1972); and representatives of other orders (Goryshin and Kozlova, 1967; Smith and Newsom, 1970; Sømme, 1967; Villacorta *et al.*, 1971). These insects were frequently studied in the laboratory under abnormal conditions of temperature, humidity, or photoperiod, thereby making generalizations from the laboratory to the field more difficult (see Tauber and Tauber, 1976b). However, a more serious criticism that can be leveled at the majority of these studies is their lack of proper controls. Apparently entomologists are unaware that no conclusions regarding the effects of, for example, photoperiod can be drawn following exposure of an insect species to, say, LD 17:7 at 26°C on diapause induction when contrasted with the response exhibited by the same species studied at LD 12:12 at 21°C.

The last category of insect dormancies, and the least understood, deals with the effects of phenological conditions other than photoperiod or temperature, such as light intensity (see Table I in Saunders, 1974b), humidity, nutrition, and latitude, which are nonetheless influenced by photoperiod and/or temperature. Whereas long-day photoperiods and high temperatures promote growth during hibernations, they have the opposite effect during aestivations. Clear-cut causal effects during athermopauses cannot be assigned because of the difficulty in ascribing to specific inductants specific responses. For example, at constant temperature and humidity but at variable latitudes, long-day species such as *Agrotis occulenta* (Danilevskii, 1965) or *Sarcophaga argyrostoma* (Denlinger, 1972a,b) develop more rapidly under long-day than short-day photoperiods, whereas a short-day species such as *Agrotis triangulum* (Danilevskii, 1965) has retarded development at long- versus short-day photoperiods.

The existence of gradations in response to phenological factors that result in quiescence, oligopause, and diapause cannot be doubted. However, documentation of these responses as has been done during aestivation and hibernation is more difficult, since it is impossible to separate the effects of photoperiod and temperature (when uncontrolled) from those of water availability, light intensity, latitude, genetics, or other factors.

The effects of water deficiency (hygric quiescence) in initiating dormancies in eggs of developing grasshoppers, *Chorthippus brunneus* (Moriarty, 1969), and fleshflies, *Sarcophaga bullata* (Wilkens, 1968), or in terminating diapause in the larval pink bollworm, *Pectinophora gossypiella* (Raina and Bell, 1974a,b), the corn borer, *Ostrinia nubilalis* (Beck, 1967), and the spinach leaf miner, *Pegomyia hyosciami* (Panz) (Zabirov, 1961), are well known. The mechanisms whereby water initiates these responses have not been studied.

Differences in latitude influence the rate of diapause induction in response to variations in day length in various insect species (Ankersmit and Adkisson, 1967; Bradshaw, 1969, 1973; Bondarenko and Hai-Yuan, 1958; Danileviski, 1965; Denlinger, 1972a,b; Depner and Harwood, 1966; Hong and Platt, 1975; Keeley, *et al.*, 1977; McMullen, 1967; Rabb, 1969; Wallace and Sullivan, 1966). Differences in diapause induction attributed to latitude must be adaptations to local phenology, since strains collected initially in divergent geographical areas exhibit identical responses to photoperiod following several generations of maintenance in the laboratory. Some responses are genetically determined and remain fixed despite several generations of inbreeding (Ankersmit, 1964; Barry and Adkisson, 1966; Denlinger, 1972a,b, 1974; House, 1967; Katsumata, 1968; Oldfield, 1970; Stoffolano, 1968, 1973; Waldbauer and Sternburg, 1973). One interesting example of an insect maintaining genetic versus phenological differences is the fly, *Poecilometopa punctipennis* (Denlinger, 1974). Latitude and heredity have caused strains of this species to rely on different phenological cues as inductants for diapause. In the tropical strain, temperature is the most important inductant, with photoperiod having no effect, whereas the temperature strain regulates diapause induction in response to photoperiod (primarily) and temperature (secondarily) (Denlinger, 1974). However, temperature can affect the photoperiodic response of both strains, since shorter photoperiods induce diapause at lower temperatures.

The effects of diet or heat stress on insect development cannot be neglected (Hogan, 1961, 1962, 1964, 1965; Lockshin *et al.*, 1975; Mansingh, 1974; Sawchyn, 1971; Stoffolano, 1974; Stoffolano *et al.*, 1974), but physiological adaptations in response to these factors have not been studied.

C. Induction of Dormant States

The previous sections on insect dormancies provided information on the characteristics of hibernations, aestivations, and athermopauses. The specific contributions of photoperiod and other phenological variables to the regulation of dormancy maintenance, dormancy termination, and post (oligo)diapause development were not included. This section deals with these topics.

Photoperiodic control of dormancy initiation was first reported by Sabrosky, Larson and Nabours (1933) and by Kogure (1933), and its role in dormancy termination was first described by Baker (1935). Unfortunately, little progress has been made in understanding the nature of photoperiodic responses by insects in the period after Bünning (1936; Bünning and Jörrens, 1960, 1962) first proposed that photoperiod influences internal circadian clocks. The role of the action spectrum of light in inducing dormancies has been studied (see citations in Saunders, 1974), but the physiological basis for these responses is unknown (see, however, Barker and Herman, 1976; Chippendale *et al.*, 1976; Chippendale and Yin, 1976). A spate of articles that rely primarily upon behavioral criteria (egg laying, yes! egg laying, no!) for defining reproductive "diapause" (oligopause and diapause) have been published (Adkisson *et al.*, 1963; Adkisson, 1964, 1966; Adkisson and Roach, 1971; Ankersmit, 1968; Barker *et al.*, 1963, 1964; Beards and Strong, 1966; Beck and Hanec, 1960; Beck *et al.*, 1963; Beck and Alexander, 1964; Beck, 1968a; Bell, 1967, 1972; Bell and Adkisson, 1964; Bell *et al.*, 1975; Belozerov, 1969; Benschoter, 1968; Bonnemaison, 1967, 1970; Boulay, 1967; Bradshaw, 1971,

1972a,b, 1973, 1974; Bradshaw and Holzapfel, 1975; Bradshaw and
Lounibos, 1972; Bünning and Jörrens, 1962; Butterfield, 1976; Cantelo, 1974;
Clark and Platt, 1969; Denlinger, 1971; Depner, 1962; de Wilde and de Boer,
1969; Earle and Newsom, 1964; Englemann and Shappirio, 1965; Fraenkel
and Hsiao, 1966, 1968a; Glass, 1970; Goryshin and Tyschenko, 1969; Gustin,
1974; Hamner and Takimoto, 1964; Harris *et al.*, 1969; Hayes *et al.*, 1970;
Hidaka and Hirai, 1970; Hidaka *et al.*, 1971; Hillman, 1973; Hodek, 1968,
1971; Hoy, 1975; Hoy and Flaherty, 1970; Knerer and Marchant, 1973;
Kono, 1970; Lees, 1965, 1966, 1967b, 1968, 1971, 1972, 1973; Lutz and Jenner, 1964; MacLeod, 1967; Mangum *et al.*, 1968; Mansingh and Smallman,
1966, 1967; McLeod and Beck, 1963; Minis, 1965; Müller, 1964, 1965;
Naton, 1966; Norris, 1965; Norris *et al.*, 1969; Pantyukhov, 1968; Peterson
and Hamner, 1968; Phillips and Newsom, 1966; Pittendrigh, 1966; Pittendrigh and Minis, 1964, 1971; Pittendrigh *et al.*, 1970; Propp *et al.*, 1969;
Rabb, 1966; Ring, 1967a,b, 1971; Rock *et al.*, 1971; Rosenthal and Koehler,
1968; Saunders, 1965, 1966b, 1969, 1970, 1974a,b, 1976; Saunders *et al.*,
1970; Schaller, 1965; Seeley, 1966; Sicker, 1964; Smith and Brust, 1971;
Sullivan and Wallace, 1965; Tadmor and Applebaum, 1971; Tauber and
Tauber, 1969, 1970a,b, 1976a,b; Tauber *et al.*, 1970a,b; Theron, 1943; Thiele,
1966, 1968a,b; Truman, 1972; Wellso and Adkisson, 1966; Williams, 1963,
1969; Williams and Adkisson, 1964; Williams *et al.*, 1965; Winfree, 1972;
Wingfield and Warren, 1972; Wright, 1970; Wright and Venard, 1971).
Although these articles deal primarily with reproductive "diapause," no men-
tion is made of the induction, synthesis, and release of vitellogenins from fat
body or of uptake by, growth of, or release of developing oocytes (Pan, 1969,
1971, 1976, 1977, Pan *et al.*, 1969; Pan and Wyatt, 1971; Kunkel and Pan,
1976; Wyatt and Pan, 1978). The criterion of egg laying (i.e., reproductive
diapause) is not sufficient to define the synergistic contributions of
photoperiod and other phenological factors as "reproductive diapause," since
egg deposition is only the last of many complex responses by the insect to
changes in reproductive capacity.

Instead of determining the mechanisms involved in the induction of
reproductive dormancies, many scientists employing well-constructed
experimental frameworks have been satisfied with splitting hairs in describing photoperiodic phenomena to nonphysiological photoperiods (48–72 hrs)
using the same fundamental approach that medieval monks used in determining the number of teeth in a horse's mouth, ergo, internal and external
coincidence models, zeitgebers, etc. (Figure 1) (Pittendrigh, 1960, 1964; Pittendrigh and Minis, 1964; Saunders, 1974b). Other investigators have
looked at hormonal and other homeostatic changes occuring during and
after photoperiodic induction of dormancy, but to date these studies have
yielded few data on the mechanisms of induction, maintenance, and termi-

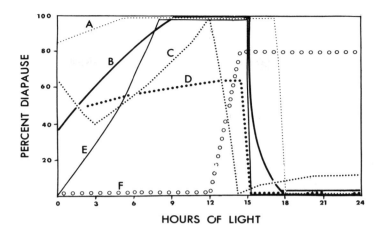

Fig. 1. Photoperiodic response curves for six insect species. Note the wealth of unique patterns that are observed. The notation "diapause" on the ordinate refers to a variety of dormancy responses including egg diapause, hibernal diapause, reproductive diapause (oligopause), and oligopause preliminary to hibernal diapause (A) *Megoura viciae*, (B) *Acronycta rumicis*, (C) *Pectinophora gossypiella*, (D) *Nasonia vitripennis*, (E) *Pieris brassicae*, and (F) *Bombyx mori*. (Modified after Saunders, 1974.)

nation of dormancies (Andrewartha, Miethke and Wells, 1974; Beck and Shane, 1969; Beck *et al.*, 1969; Bowers and Blickenstaff, 1966; Clay and Venard, 1971; Cymborowski and Dutkowski, 1969; Denlinger *et al.*, 1972; de Wilde and de Boer, 1961; Ferenz, 1975; Fraenkel and Hsiao, 1968b; Fukuda, 1951, 1952; Harker, 1956; Hasegawa *et al.*, 1972, 1974; Hasegawa and Yamashita, 1965, 1967; Hoffman, 1970; Ichimasa and Hasegawa, 1973; Ingram, 1975; Isobe *et al.*, 1973, 1975; Kayser-Wegmann, 1975; Kono, 1971, 1973a,b, 1975; Lambermont, Blum and Schrader, 1964; Lees, 1964; Loher, 1974; MacFarlane and Hogah, 1966; Madhavan, 1973; Mansingh and Steele, 1973; Marks *et al.*, 1973; McNeil and Rabb, 1973; Morohoshi and Oshiki, 1969a,b; Nair, 1974; Nishiitsutsuju-Uwo and Pittendrigh, 1968; Nishiitsutsuju-Uwo *et al.*, 1967; Normann, 1973; Ogawa and Hasegawa, 1975; Ohtaki and Takahashi, 1972; Park, 1973; Park and Seong, 1976; Park and Yoshitake, 1970a,b, 1971; Raina and Bell, 1974a,b; Roberts, 1965; Siew, 1965a,b,c, 1966; Siverly, 1972; Somme and Velle, 1968; Sonobe and Ohnishi, 1970; Takeda and Hasegawa, 1975; Truman, 1971; Valder *et al.*, 1969; Vinogradova, 1974; Whitmore *et al.*, 1973; Wilhelm *et al.*, 1961; Williams, 1969; Williams and Adkisson, 1964; Yamashita and Hasegawa, 1964, 1966; Yamashita *et al.*,1972; Yin and Chippendale, 1973a,b, 1974; Zaslavskii and Bogdanova, 1968).

The simple, clear definitions of insect dormancies proposed by Mansingh (1971, or possibly Müller, 1965) would lead one to believe that insect dormancies would now be classified properly, thereby facilitating explanations of dormancies in terms of quantitative changes in systemic function. Such is not the case. Examination of the articles listed above would lead one to the conclusion that failure to deposit eggs is synonymous with "reproductive diapause" and that distinctions between diapause and other types of dormancies are trivial. Unfortunately, the differences are not trivial, and little progress can be made if perturbations in systemic function are ascribed to a broad spectrum of insect dormancies under the nongeneric heading "reproductive diapause."

D. Maintenance of Dormant States

Hibernation and aestivation are classes of insect dormancies that are initiated primarily by alterations in photoperiod but that are frequently coupled to changes in the diurnal temperature rhythm. It is therefore not surprising that temperature and photoperiod (and infrequently moisture) become the major diapause-maintaining stimuli. Although the interaction between these factors is complex, Tauber and Tauber (1976b) have outlined five patterns of interactions with regard to photoperiod:

1. An all-or-none response in which all photoperiods less than the critical day length promote maintenance of dormancies equally, whereas all photoperiods above the critical day length promote renewed growth and development. Examples of insects experiencing this type of diapause maintenance are the culicid, *Wyeomyia smithii* (Smith and Brust, 1971), and the neuropteran, *Meleoma signoretti* (Tauber and Tauber, 1975a).

2. A situation wherein short-day photoperiods induce, but long-day photoperiods terminate, dormancy. An example of an insect exhibiting this type of response is *Nemobius yezoensis* (Masaki and Oyama, 1963).

3. A situation wherein short-day photoperiods induce and maintain dormancy, but increases in temperature terminate dormancy despite continued maintenance of the dormancy-inducing photoperiod. An insect exhibiting this type of response is the lacewing, *Chrysopa harrisii* (Tauber and Tauber, 1974).

4. A situation wherein short-day photoperiods induce dormancies, but insects lose the ability to respond, thereby terminating the dormant state. Examples of insects exhibiting this type of development are the lacewing, *Chrysopa carnea* (Tauber and Tauber, 1973); the sweet clover weevil, *Sitona cylindricollis* (Hans, 1961); the European corn borer, *Ostrinia nubilalis* (McLeod and Beck, 1963); the odonate *Tetragoneuria cynosura* (Lutz, 1974; Lutz and Jenner, 1964); and other insects (Tauber and Tauber, 1976b).

5. A response induced by short-day photoperiods but terminated by lengthening photoperiods. Examples are the lacewing, *Chrysopa downesi* (Tauber and Tauber, 1975b), and the dermestid beetle, *Anthrenus verbasci* (Blake, 1963).

Elevated temperatures can accelerate the rate at which dormant species resume development following termination of dormancy-maintaining conditions (Denlinger, 1972b; de Wilde, 1969; Fraenkel and Hsiao, 1968a). Temperature plays a vital role in maintenance of oligopause-type dormancies by preventing premature diapause development in species that experience late spring or early summer oligopauses, followed in turn in late summer and fall by hibernal diapause (Andrewartha, 1952; Danilevskii *et al.*, 1970). An example of an insect exhibiting this property is the European red mite, *Panonychus ulmi*, which experiences accelerated development at 0°–5°C during the transition between oligopause and diapause but fails to grow following "termination" of diapause at temperatures less than 9°C (Cranham, 1972; see also Sawchyn and Church, 1973).

E. Termination of Dormant States

Discussion of dormancy-terminating stimuli is complicated by a lack of knowledge regarding the etiology of dormant states: oligopause versus diapause. Additional complications arise when insects experience oligopause before diapause. Measurements of systemic output or capacity during dormancy must be quantified in ways other than total body respiration if the nature of the dormant state and its mode of initiation and termination are to be understood.

Termination of dormancy is complicated by an inability to find one and only one ecological variable that will "instantly" terminate diapause in a specific species. This is because diapause is terminated slowly and because individuals within a diapausing population initiate diapause termination only after differing periods of exposure to diapause-terminating sitmuli. In addition to temperature, diapause termination is accelerated by changes in photoperiod, moisture, and internal stimuli, whereas oligopause is terminated by food, photoperiod, moisture, and internal stimuli (Tauber and Tauber, 1976b). As previously mentioned, increasing the critical photoperiod above threshold initiates development. Infrequently, the terminating stimulus is an increasing change in photoperiod above threshold rather than simple fixed photoperiod durations above threshold (Tauber and Tauber, 1975a,b). Moisture or humidity is most influential in terminating dormancy in oligopausing species. Food can play a role only in oligopausing and quiescent types of dormancies. One interesting area of interaction is between an insect parasite and its insect host. During such interactions, either host or parasite initiates termination of dormancy, with consequent induction in the other following hormonal stimulation (Schneiderman and Horowitz, 1958; Schoonhoven, 1962).

III. PHYSIOLOGY OF HIBERNAL DIAPAUSE IN LEPIDOPTERA

In the sections that follow, the regulatory process and regulatory mechanisms in normal insects and in those experiencing various facets of dormancy are compared. The major conclusion to be drawn is that homeostatic mechanisms are as finely tuned during dormancy as during periods of normal growth and development.

A. Regulation of Cations

Regulatory mechanisms involved in maintenance of selected cations at steady state-levels before, during, and after insect dormancies have not been described. The reader is referred to the excellent review by Florkin and Jeuniaux (1974) for a discussion of stage-specific compositional differences present in all insect orders, including Lepidoptera other than *Hyalophora cecropia*. In this section, information regarding cation regulation in the silkmoth, *Hyalophora cecropia* (Lepidoptera; Saturniidae), a holometabolous species (i.e., undergoing complete metamorphosis), before, during, and after hibernal diapause is discussed.

1. SODIUM

A minor component in hemolymph of phytophagous insects, Na^+ is nonetheless necessary for maintenance of bioelectric potentials in electrically excitable tissues such as nerve and muscle (Pichon, 1974; Hoyle, 1974; Rivera, 1975; Jungreis, 1977; Jungreis and Vaughan, 1976, 1977; Vaughan and Jungreis, 1976, 1977). The concentration of Na^+ in hemolymph of foliage-reared *H. cecropia* (Quatrale, 1966; Jungreis *et al.*, 1973; Harvey *et al.*, 1975) is less than 2 mM despite dramatic changes in hemolymph volume during the larval–pupal and pupal–adult transformations (Fig. 2) (see Jungreis and Tojo, 1973; Levenbook *et al.*, 1971; Fyhn and Saether, 1970). In larvae reared on the synthetic diet of Riddiford (1968), the concentration of Na^+ in hemolymph reflects that of the diet (Fig. 2) (Jungreis *et al.*, 1973), ergo, a value considerably higher than that for wild black cherry foliage (*Prunus serotina*) (Quatrale, 1966). In preparation for the larval–pupal transformation, the mechanism of regulating hemolymph Na^+ shifts from "passive" uptake and "passive" excretion via

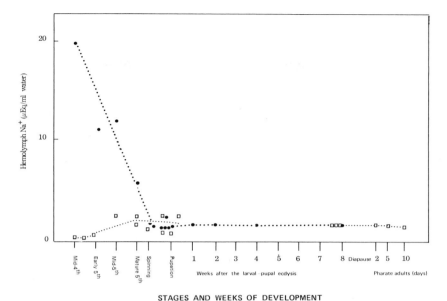

STAGES AND WEEKS OF DEVELOPMENT

Fig. 2. Changes in the concentration of hemolymph sodium during ontogeny of the silk-moth, *Hyalophora cecropia*, reared either on wild black cherry foliage, *Prunus serotina* (open squares), or on the synthetic diet of Riddiford (1968) (shaded circles). (Modified after Quatrale, 1966; Jungreis *et al.*, 1973; Jungreis, 1974; Harvey *et al.*, 1975; and A. M. Jungreis, unpublished.)

the Malpighian tubules to restricted uptake, as evidenced by a decline in hemolymph Na$^+$ without a concomitant decline in dietary Na$^+$ (Fig. 2). Evidence for passive movement of Na$^+$ across the midgut in the direction midgut lumen $\rightarrow$ hemolymph is the existence of an ion pump capable of actively transporting Na$^+$ in the direction hemolymph $\rightarrow$ gut lumen (Harvey and Zerahn, 1971) and the observation that the bulk of the Na$^+$ derived from the diet remains in the lumen of the midgut (Table I). The level of Na$^+$ in hemolymph of foliage-reared animals is so low that is has been virtually impossible to relate changes in this cation with developmental events preliminary to or during diapause.

2. POTASSIUM

Regulation of hemolymph K$^+$ is complicated by a dietary input in excess of 100 mM (Quatrale, 1966; Harvey *et al.*, 1975) (Table II), a restricted capacity to excrete K$^+$ via the Malpighian tubules (Irvine, 1969), and a complex relationship between K$^+$ and urate metabolism and storage (Jungreis and Tojo, 1973) (see also Section III,B). Despite these apparently insur-

TABLE I

Leaf Ion Concentrations in Gut Contents and Fecal Pellets of Feeding Fifth Larval Instar Cecropia Silkmoths, *Hyalophora cecropia*[a,b]

Compartment	Cation levels in leaves (%)			
	Na⁺	K⁺	Ca²⁺	Mg²⁺
Midgut contents	50.7 ± 6.0	127.3 ± 1.0	13.7 ± 2.9	14.4 ± 0.9
Hindgut contents	266.0 ± 73.9	77.7 ± 7.8	50.2 ± 3.2	64.6 ± 10.3
Fecal pellet	231.4 ± 94.2	130.2 ± 17.7	140.4 ± 34.7	127.7 ± 0.9

[a] Modified after Harvey *et al.* (1975).

[b] Sodium and potassium are regulated by exclusion, whereas calcium and magnesium are regulated by uptake into hemolymph followed by secretion in the Malpighian tubules.

mountable problems, K^+ is maintained at the same steady-state level throughout feeding and postfeeding stages in development (Table II; Fig. 3). Examination of the relationship between the concentration of K^+ in the wild black cherry foliage and the gut contents quickly leads to the conclusion that, *in vivo*, K^+ does not pass across the midgut to any appreciable extent (Table I). The apparent failure of K^+ to pass across the midgut results from a rather small passive influx across the midgut coupled with a large active transport capacity in the opposite direction: hemolymph $\rightarrow$ midgut lumen (Harvey and Nedergaard, 1964). The low passive rate of K^+ transport is due in part to the presence within the goblet cells of the midgut epithelium (Anderson and Harvey, 1966) of a viscous "plug," whose proposed role in

TABLE II

Steady-state Maintenance of Potassium and Magnesium in Hemolymph of Feeding Fifth Instar Larval Cecropia Silkmoths, *Hyalophora cecropia*[a]

Stage	Type of diet	K⁺ (μEq/ml)		Mg²⁺ (μEq/ml)	
		Hemolymph	Diet	Hemolymph	Diet
Early fifth	Foliage[b]	34	145	104	108
instar larva	Synthetic diet[c]	35	57	66	15
	Supplemented diet[d]	44	82	66	65
Mid-fifth	Foliage	39	165	111	104
instar larva	Synthetic diet	32	57	69	15
	Supplemented diet	35	82	76	65

[a] Modified after Quatrale (1966), Jungreis *et al.* (1973), Harvey *et al.* (1975), and A. M. Jungreis (unpublished).

[b] *Prunus serotina*.

[c] Diet of Riddiford (1968).

[d] Diet of Riddiford (1968) supplemented with 25 m*M* KCl and 25 m*M* MgCl₂.

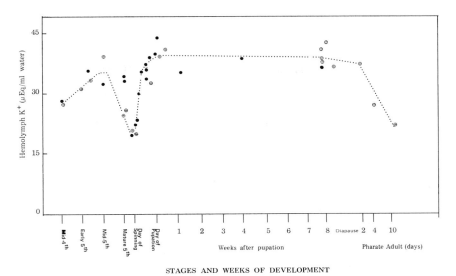

STAGES AND WEEKS OF DEVELOPMENT

Fig. 3. Changes in the concentration of hemolymph potassium during ontogeny of the silk-moth, *Hyalophora cecropia*, reared either on wild black cherry foliage, *Prunus serotina* (hatched circles), or on the synthetic diet of Riddiford (1968) (shaded circles). (Modified after Quatrale, 1966; Jungreis *et al.*, 1973; Jungreis and Tojo, 1973; Jungreis, 1974; Harvey *et al.*, 1975; A. M. Jungreis, unpublished.)

ion regulation across the midgut is to restrict movement of K$^+$ (Schultz and Jungreis, 1977a,b).

The decline in and restoration of the hemolymph steady-state K$^+$ concentration during the larval–pupal transformation (Fig. 3) is related to continued operation of the K$^+$ pump in midgut following evacuation of the midgut contents (extrapolated from data in Haskell *et al.*, 1968), dehydration resulting from loss of the gut contents and spinning of the silk cocoon (Jungreis and Tojo, 1973), formation by the integumentary epithelium of molting fluid, a potassium salt solution (Jungreis 1973b, 1974; Jungreis and Harvey, 1975; Jungreis, 1978) incorporation of K$^+$ into fat body in the form of potassium urate (Jungreis and Tojo, 1973; S. Tojo, unpublished data, cited in Wyatt, 1975). The decline in hemolymph K$^+$ upon initiation of pharate adult development probably results from cell proliferation, since the intracellular concentration of K$^+$ (90+ mM) far exceeds that present in hemolymph (see Quatrale, 1966; Harvey *et al.*, 1975). The close relationship during diapause between "excess" hemolymph K$^+$ and storage of urates in fat body would lead one to predict that hemolymph K$^+$ is regulated during diapause by an interaction between the degree of dehydration (i.e., duration of the pupal diapause and postdiapause states) and the rate of potassium urate formation in fat body with subsequent uptake by fat body tissue of K^+ from hemolymph.

3. CALCIUM

Calcium is maintained in hemolymph of *H. cecropia* at a steady-state concentration of 10 mEq/liter ($= 5\,\mathrm{m}M = 20$ mg%) (Fig. 4), a value double that of Ca^{2+} in mammalian blood (10 mg%). Little hemolymph Ca^{2+} protein is bound (A. M. Jungreis, unpublished), but some may be chelated to the organic acids normally present in hemolymph at high concentrations (see Levenbook and Hollis, 1961). The virtual absence of Na^+ in hemolymph coupled with the presence of Na^+-K^+-ATPases in nervous tissues (Vaughan and Jungreis, 1976, 1977; Jungreis and Vaughan, 1976, 1977) leads to a perplexing problem of regulation, since Na^+ presumably required to maintain the responsiveness of electrically excitable tissue is absent. This problem is solved in *H. cecropia* and presumably other phytophagous insects either by utilizing Ca^{2+} in lieu of Na^+ to facilitate optimization of neuronal integration (McCann, 1971) or by a neuronal sheath that maintains Na^+ at artificially high concentrations (Vaughan and Jungreis, 1977; Treherne, 1976). This dependence on Ca^{2+} as a substitute

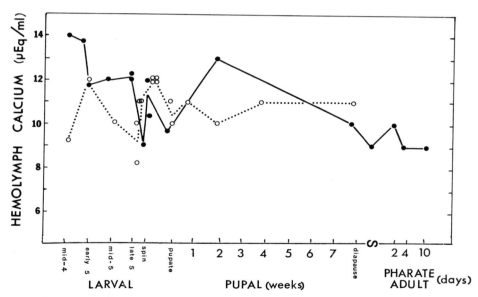

Fig. 4. Changes in the concentration of hemolymph calcium during ontogeny of the silk-moth, *Hyalophora cecropia*, reared either on wild black cherry foliage, *Prunus serotina* (shaded circles), or on the synthetic diet of Riddiford (1968) (open circles). (Modified after Quatrale, 1966; Jungreis et al., 1973; Jungreis, 1974; Harvey et al., 1975; A. M. Jungreis, unpublished.)

for Na^+ probably explains the more stringent regulation of hemolymph Ca^{2+} relative to that of other hemolymph cations.

The concentrations of Ca^{2+} in the diet is well above 300 mEq/liter, yet the steady-state level in hemolymph is only 10 mEq/liter (Fig. 4). Since Malpighian tubules in this species process only small volumes of fluid per day (Irvine, 1969), exclusion of Ca^{2+} from hemolymph rather than elimination following uptake should mirror the actual mode of regulation. Data comparing the concentrations of Ca^{2+} in the midgut contents with those in leaves (Table I) are at variance with this interpretation. The midgut epithelium of larval cecropia silkmoths is reported to retain the capacity to actively transport Ca^{2+} in the direction hemolymph → gut lumen (Wood and Harvey, 1976). Since the concentration of Ca^{2+} in midgut epithelium, integument, and fat body is quite low (Quatrale, 1966; A. M. Jungreis, unpublished data), it seems unlikely that Ca^{2+} passes from gut lumen to hemolymph before secretion by the Malpighian tubules (Table I). The fate of Ca^{2+} derived from leaves before its presence in the hindgut contents has yet to be determined. One possibility is that Ca^{2+} binds to the peritrophic membrane.

The fate of Ca^{2+} during and after "dehydration" associated with the larval–pupal transformation has been determined (Table III). When total quantities rather than concentrations of Ca^{2+} in the individual tissues were measured, the same respective quantities of Ca^{2+} were detected in larval and diapause pupal gut and integument tissues, whereas hemolymph experienced a major loss. The quantity of Ca^{2+} in pupal integument represents a substantial increase in concentration over that in larval integument, due to drastic differences in total tissue weight (Jungreis and Tojo, 1973; Quatrale, 1966). Removal of Ca^{2+} from hemolymph could be caused by retention as calcium urate with subsequent storage in Malpighian tubules. Instead, Ca^{2+} is sequestered in the contents of the midgut lumen. The presence of Ca^{2+} in the pupal gut lumen at a concentration 2.3 times that of hemolymph, coupled with the higher concentration of Ca^{2+} in pupal relative to larval gut tissues (Quatrale, 1966; Harvey et al., 1975; Jungreis and Tojo, 1973) (Table III), leads one to the conclusion that despite the loss in K^+ transporting capacity (Haskell et al., 1968), the pupal midgut epithelium which is cytologically different from the larval epithelium (Judy and Gilbert, 1970), retains the capacity to secrete Ca^{2+} during the diapause condition.

When Ca^{2+} is administered to feeding larvae at a concentration of 25 mM, the animals fail to respond within 6 hr after "recovery" from CO_2 anesthesia (A. M. Jungreis, unpublished), whereas diapause pupae administered 15 mM Ca^{2+} immediately recover from the anesthetic. Furthermore, pupae retain all the added Ca^{2+} in hemolymph even after 48

TABLE III

Fate of Calcium Initially Present in Feeding Fifth Instar Larvae of *Hyalophora cecropia* but Redistributed in Tissues of the Diapause Pupa[a,b]

Tissue	Late fifth instar larva (A)			Diapause pupa (B)			A − B
	Weight (g)	Concentration (μEq/ml)	Total quantity (μEq/animal)	Weight (g)	Concentration (μEq/ml)	Total quantity (μEq/animal)	
Hemolymph	5.6	12.0	67.2	3.3	9.6	31.7	−35.5
Gut tissue	0.7	7.1	5.0	0.24	9.1	2.2	−2.8
Integument	3.0	14.8	44.4	0.3	128.4	38.5	−5.9
Midgut lumenal contents	—	—	—	1.0	27.9	27.9	+27.9
Unassigned	—	—	—	—	—	—	−16.3

[a] Modified after Quatrale (1966), Harvey *et al.* (1975), and A. M. Jungreis (unpublished).

[b] Note that calcium initially present in hemolymph of feeding larvae is sequestered by the pupal midgut epithelium into the lumen of the midgut.

hr (A. M. Jungreis, unpublished data). (The criterion used to determine responsiveness in diapause pupae was tactile; i.e., squeeze the head–thorax and see if the abdomen wriggles.)

4. MAGNESIUM

Of the cations present in hemolymph of *H. cecropia*, regulation of the steady-state level of Mg^{2+} appears to be poorest. Feeding larvae retain as much as 110 mEq/liter (55 mM) Mg^{2+}, whereas the level in diapause pupae is as little as 27 mEq/liter (Harvey *et al.*, 1975; Jungreis *et al.*, 1973; Quatrale, 1966; Michejda and Thiers, 1963) (Fig. 5). Magnesium in hemolymph is neither protein bound (Carrington and Tenney, 1959; Jungreis, unpublished) nor can the concentration of free Mg^{2+} exceed 20 mEq/liter if intact muscle is to remain electrically excitable (Weevers, 1966). Problems caused by excess Mg^{2+} are obviated if the bulk of hemolymph Mg^{2+} is chelated to the organic acids and organophosphates in hemolymph (Levenbook and Hollis, 1961; Wyatt *et al.*, 1963; Jungreis , 1978).

The question of why hemolymph contains so much Mg^{2+} is difficult to answer. Magnesium is present in leaves of *Prunus serotina* at concentrations

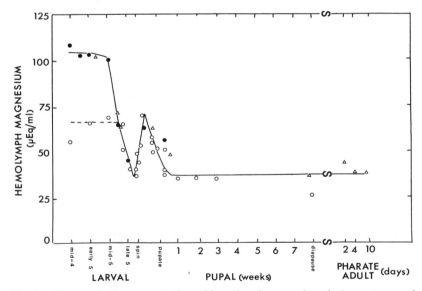

Fig. 5. Changes in the concentration of hemolymph magnesium during ontogeny of the silkmoth, *Hyalophora cecropia*, reared either on wild black cherry foliage, *Prunus serotina* (shaded circles and open triangles), or on the synthetic diet of Riddiford (1968) (open circles). Shaded circles are data of Quatrale (1966). (Modified after Quatrale, 1966; Jungreis *et al.*, Wyatt, 1973, 1974; Jungreis, 1973a, 1974; Harvey *et al.*, 1975; Jungreis, 1978.)

not in excess of 120 mEq/liter (Quatrale, 1966). The concentration of Mg^{2+} in hemolymph thus seems to reflect that which is present in the diet (Table II). However, examination of Mg^{2+} levels in larvae reared on the synthetic diet of Riddiford (1968), which contains Mg^{2+} at a level of only 15 mEq/liter, leads one to a different conclusion (Table II). At concentrations under 65 mEq/liter, Mg^{2+} derived from the gut contents and transported into hemolymph via active processes (Wood et al., 1975) is sequestered in hemolymph. One suspects that a high concentration of Mg^{2+} is needed to chelate the large quantity of anionic organic acids present, chloride and bicarbonate contributing little to the overall electroneutrality of hemolymph in this species (Jungreis et al., 1973; Johnston and Jungreis, 1977; Jungreis, 1974; Jungreis, 1978). At concentration above 65 mEq/liter, Mg^{2+} continues to be sequestered in hemolymph (Table I), but its rate of excretion by the Malpighian tubules is too slow to permit removal of much Mg^{2+} (Irvine, 1969). The presence of Mg^{2+} at levels higher than 65 mEq/liter can thus be explained if the steady-state concentration reflects the balance between passive and active fluxes across the midgut. I propose that massive Mg^{2+} fluxes across the midgut thus represent the major mechanism for regulating hemolymph Mg^{2+}.

Two questions can then be raised regarding regulation of Mg^{2+} during ontogeny: How is excess hemolymph Mg^{2+} eliminated when larval silkmoths initiate physiological adaptations associated with early dormancy? Is the mechanism whereby the steady-state level of Mg^{2+} is maintained the same in larvae and diapause pupae?

The fate of Mg^{2+} in H. cecropia during the larval–pupal transformation has been studied by Jungreis (1973a), who noted that excess Mg^{2+} is sequestered by the pupal midgut epithelium via a mechanism involving accelerated hydrolysis of hemolymph-derived α-glycerol phosphate (Jungreis et al., 1975; see also Turbeck, 1974) (Table IV). Regulation of Mg^{2+} sequestration in the midgut is discussed in Section III, C.

Differences in steady-state regulation between the larval and diapause pupal stages in development are readily detected. When the concentration of Mg^{2+} in larval hemolymph is artificially raised by some 50 mEq/liter, although larvae remain sluggish, about 35% (17 out of 50 mEq/liter) of the added Mg^{2+} is removed within 2 hr (Table V) (Jungreis et al., 1974; A. M. Jungreis, unpublished data). The rate of Mg^{2+} removal occurs at a sufficiently rapid rate that secretion by the Malpighian tubules is precluded (Irvine, 1969). Sequestration in fat body is also excluded (Table V). Only passive movement across the midgut surface, ca. 10 cm² surface area, could account for the observed rapid rate of removal. When diapause papae are administered sufficient Mg^{2+} to raise the hemolymph level by 30 mEq/liter, it is retained in its entirety even after 48 hr (Table V). Magnesium can not

TABLE IV

Fate of Magnesium Initially Present in Feeding Fifth Instar Larvae of *Hyalophor acecropia* but Redistributed in Tissues of the Diapause Pupa[a,b]

Tissue	Late fifth instar larva (A)			Diapause pupa (B)			A − B
	Weight (g)	Concentration (μEq/ml)	Total quantity (μEq/animal)	Weight (g)	Concentration (μEq/ml)	Total quantity (μEq/animal)	
Hemolymph	5.6	63.8	357.0	3.3	42.0	139.0	−218.0
Gut tissue	0.7	36.2	25.3	0.24	1062.0	255.0	229.6
Integument	3.0	9.5	28.5	0.3	43.2	13.3	−15.2
Fat body	0.9	35.6	32.0	2.0	20.0	40.0	8.0
Midgut lumenal contents	—	—	—	1.0	4.4	4.4	4.4
Unassigned	—	—	—	—	—	—	−8.8

[a] Modified after Quatrale (1966), Jungreis (1973a), Jungreis *et al.* (1974), Harvey (1975), and A. M. Jungreis (unpublished).
[b] Note that magnesium initially present in hemolymph is sequestered by the pupal midgut epithelium and stored as insoluble $Mg_3(PO_4)_2$ within the midgut epithelium (see Jungreis *et al.*, 1975).

TABLE V

Fate of Administered Mg^{2+} in Developing Cecropia Silkmoths,
Hyalophora cecropia[a,b]

	Concentration (μEq/ml)	
Stage	Hemolymph	Fat body
Feeding fifth instar larva		
Initial	113.6	36.7
Expected	163.6[c]	36.2[d]
Observed	146.3	40.4
Quantity eliminated	+17.3	−4.2
Diapause pupa		
Initial	27.1	19.1
Expected	57.1[c]	26.8[d]
Observed	57.5	19.1
Quantity eliminated	−0.4	+7.7

[a] From A. M. Jungreis (unpublished), Jungreis (1973a), and Jungreis *et al.* (1974).

[b] Fifth instar larvae were measured 2–3 hr after magnesium administration, whereas diapause pupae were measured 48 hr after administration.

[c] Assuming 80% body wet weight of water, 50% of body weight was hemolymph, and all administered Mg^{2+} remained in hemolymph.

[d] Assuming that administered Mg^{2+} was distributed throughout hemolymph and fat body.

be sequestered in the pupal midgut epithelium, as was observed during the pharate pupal stages in development, because uptake during those stages was predicted upon (1) the availability of inorganic phosphate to form insoluble $Mg_3(PO_4)_2$ (Jungreis *et al.*, 1975) and (2) elevated α-glycerol phosphatase/glycerol kinase enzyme ratios (Jungreis *et al.*, 1975). Such a pool of inorganic phosphate is unavailable during the pupal stage in development, while the enzyme ratios greatly favor synthesis of α-glycerol phosphate. Magnesium does not appear to diffuse across the gut epithelium, as evidenced by the greatly reduced level of Mg^{2+} in the pupal gut contents (Jungreis, 1974), perhaps due in part to the presence of magnesium phosphate deposits or spherules in this tissue (see Turbeck, 1974).

B. Regulation of Uric Acid Metabolism

The formation, storage, and excretion of uric acid is a subject about which most entomologists know "a lot" but about which there is little documentation in the literature. Some investigators have asked the question, Is uric acid the primary nitrogen excretory product in insects of the order Dictyoptera (cockroaches) as well as other orders? (Cochran, 1975; Mullins and Cochran, 1976). However, apart from the commercial silkworm, *Bombyx mori*, which is known to store uric acid during feeding stages as a complex with chromoproteins (Aruga *et al.*, 1953; Grassé and Lesperon, 1936; Isaka, 1952; Tsujita and Sakurai, 1965a,b; Yoshitake and Aruga, 1952), and butterflies of the genus *Papilio* and *Pieris*, which incorporate urate into wing pigments during pharate adult development (Mauchamp and LaFont, 1975; LaFont and Pennetier, 1975; Tojo and Yushima, 1972), other insects and especially the larger lepidopteran species that have been studied to date all utilize the synthesis and excretion of uric acid as the primary vehicle for the detoxification of "excess" nitrogen (Barrett and Friend, 1970; Bursell, 1967; Corrigan, 1970; Fyhn and Saether, 1970; Jungreis and Tojo, 1973; Levenbook *et al.*, 1971; Tojo, 1971; Tojo and Hirano, 1968). In this section, uric acid metabolism during diapause development is discussed.

Three questions are of primary concern: What is the origin of uric acid appearing in fat body at the end of the larval–pupal transformation and during diapause? Is the capacity for synthesizing and storing uric acid in fat body consistent with the appearance of uric acid in fat body? What is the function of fat body urate synthesis and storage?

Uric acid that accumulates in fat body of *H. cecropia* and *Mandura sexta* during the larval–pupal transformation originates in response to *de novo* synthesis in fat body (major) and other tissues (very minor) (Jungreis and Tojo, 1973; Williams-Boyce 1977; Williams-Boyce and Jungreis, 1977) (Table VI; Fig. 6). It is not derived from stored urate in the larval integument (see Jungreis and Tojo, 1973). Increases in fat body urate are observed during the larval–pupal transformation in *Bombyx mori*, *Manduca sexta*, *H. cecropia*, and *Antheraea pernyi* and appear to be a universal feature of diapause development in Lepidoptera.

The capacity for the uptake (Fig. 7) and synthesis (Table VII) of urate by fat body declines between the late feeding fifth instar larval and pupal stages in development. The ability of fat body to sequester urate that may be present in hemolymph (and originating in tissues other than fat body) declines during the larval–pupal transformation, with further reductions noted during early pupal and diapause pupal stages in development (Fig. 7). This decline can result from changes in fat cell membrane composition, since permeability to sugars and other compounds across intact fat cells is temperature insensitive during diapause but is temperature sensitive during feeding stages in development (see Jungreis and Wyatt, 1972).

The significance of urate storage in fat body is conjectural and in part reflects the question, Which came first, the chicken or the egg? Urate

TABLE VI

Distribution of Urate in Selected Tissues of Developing Silkmoths, *Hyalophora cecropia*[a,b]

Stage of development	Hemolymph	Integument	Midgut	Fat body
Larval				
Feeding fifth instar	<0.03	13.8 ± 1.9	2.8 ± 0.14	3.7 ± 0.7
Larval–pupal transformation				
Day of spinning	<0.03	8.2 ± 1.9	2.0 ± 0.3	11.0 ± 11.4
Day of apolysis	<0.02	5.1 ± 0.9	0.93 ± 0.13	13.1 ± 10
Two days after apolysis	<0.02	5.6 ± 0.7	0.88 ± 0.13	51.3 ± 26
Four days after apolysis	<0.02	7.4 ± 1.4	0.60 ± 0.09	137 ± 56
Pupal				
Day of ecdysis	<0.02	0.21 ± 0.03	0.56 ± 0.08	338 ± 232
Diapause	<0.02	—	—	680 ± 170

[a] Modified after Jungreis and Tojo (1973), Jungreis (1973a), and A. M. Jungreis (unpublished).

[b] Values in micromoles per animal.

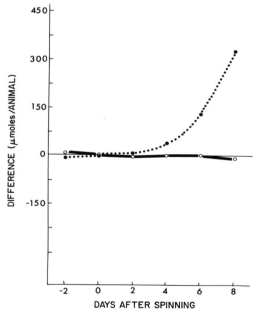

Fig. 6. Origin of uric acid appearing in fat body of the silkmoth, *Hyalophora cecropia*, during the larval–pupal transformation. The larval–pupal ecdysis occurs on day 8. Shaded circles, fat body; open circles, integument. (Insufficient uric acid was present in midgut epithelium or hemolymph to warrant inclusion in this figure.) (Modified after Jungreis and Tojo, 1973.)

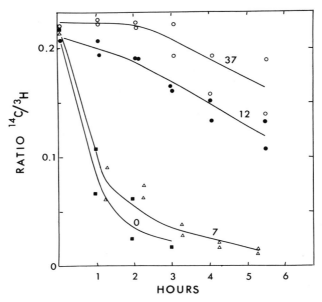

Fig. 7. Changes in the *in vivo* rate at which [¹⁴C]urate is taken into fat body of developing silkmoths, *Hyalophora cecropia*, between 0 and 37 days after initiation of spinning. Urate uptake was determined indirectly by comparing its concentration in hemolymph with that of [³H]inulin and determining changes in [¹⁴C]urate/[³H]inulin isotope ratios over a 5.5-hr period of incubation. (From Jungreis and Tojo, 1973.)

storage is intimately related to storage excretion of K^+ (Jungreis and Tojo, 1973) (Fig. 8). Potassium derived from the musculature of the integument is released into hemolymph at a time when the midgut epithelium has lost its capacity to actively secret potassium across the midgut (hemolymph → lumen) (Haskell *et al.*, 1968; Judy and Gilbert, 1970). During the

TABLE VII

Rate of [¹⁴C]Formate Incorporation *in Vitro* into Urate by Intact Fat Cells of *Manduca sexta*[a]

Stage of development	Micromoles [¹⁴C]formate incorporated into urate per gram fat cells per hour at 30°C
Feeding fifth instar	2.20 ± 0.45 ($N = 6$)
One day before larval– pupal ecdysis	0.50 ± 0.23 ($N = 9$)

[a] (From Williams-Boyce, 1977; Williams-Boyce and Jungreis, 1977).

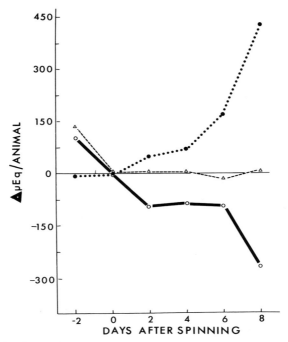

Fig. 8. Origin of potassium appearing in fat body of the silkmoth, *Hyalophora cecropia*, during the larval–pupal transformation. Dotted curve, fat body; dashed curve, hemolymph; solid curve, integument. (Modified after Jungreis and Tojo, 1973.) Δ = difference.

larval–pupal transformation, potassium is sequestered in fat body within urate granules (Tojo, Betchaku, and Wyatt, cited in Wyatt, 1975) in the form of potassium urate, whereas at later stages in development urate is stored without potassium (Jungreis and Tojo, 1973). Thus, elimination of excess K^+ and removal of waste nitrogen are effected by formation of potassium urate storage granules. Fat body continues to synthesize and store uric acid after all excess K^+ has been removed from hemolymph (Fig. 9). The increased storage of uric acid has two potential functions. (1) It provides additional capacity for the storage of K^+ during early pupal and diapause pupal stages in development, and (2) it ensures the availability of protein for selective synthesis during late pharate adult stages in development.

An analysis of urate granule composition finds some 24% to be protein, whereas protein granules are virtually devoid of uric acid (Table VIII). The ratio of urate granules to protein granules in diapause pupal fat body is 2:1 (S. Tojo, personal communication). Therefore, 33% of the total protein

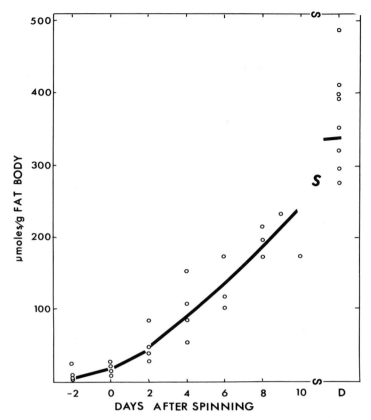

Fig. 9. Time course for uric acid accumulation in *Hyalophora cecropia* fat body before, during, and after the larval–pupal transformation. (Modified after Jungreis and Tojo, 1973.)

TABLE VIII

Compositional Analysis of Urate and Protein Granules in Diapause Pupal Fat Body[a]

Component	Protein granules	Urate granules
Protein (%)	98	24
Urate (%)	0.5	75

[a] Modified after Tojo and Wyatt (cited in Wyatt, 1975).

stored in the form of granules in fat body is inaccessible until the urate granules are "solubilized." Since protein granules are degraded slowly throughout diapause and early pharate adult development, protein in urate granules can be available both as a nutritive source and as a source of amino acids for, say, vitellogenin synthesis only during late pharate adult development (day 16 out of 21 days in *H. cecropia*), urate being solubilized and transferred to the meconial fluid at this time (see Levenbook *et al.*, 1971).

C. Regulation of Trehalose

Trehalose is a nonreducing sugar found in hemolymph of virtually all insects except honeybees (see citations in Wyatt, 1967), whose concentration varies within and among insect orders, families, genera, and species and during ontogeny of individual species (Wyatt, 1967; Jungreis and Wyatt, 1972; Jungreis *et al.*, 1973; Chang *et al.*, 1964; Duchateau-Bosson *et al.*, 1963). In the transition between larval and pupal stages in development, species exhibiting either "facultative" or "nonfacultative" hibernal pupal diapause exhibit a reduction in hemolymph trehalose. A frequently reproduced figure typifying these changes is that of the commercial silkworm, *Bombyx mori* (Duchateau-Bosson *et al.*, 1963), (Fig. 10), although

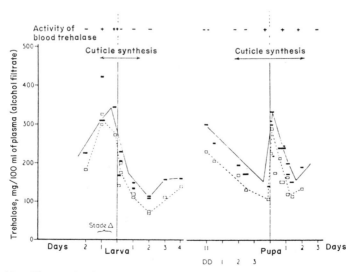

Fig. 10. Changes in the concentration of hemolymph trehalose and trehalase during larval–larval and larval–pupal transformations in the silkworm, *Bombyx mori*. Note the parallel increases in both trehalose and trehalase activity at the times of the larval–larval and larval–pupal ecdyses. (From Duchateau-Bosson *et al.*, 1963.)

the relationship in the eri silkworm is virtually identical (Chang *et al.*, 1964). In this species, the concentration of trehalose in hemolymph is presumed to change in response to alterations in hemolymph trehalase activity. Since trehalase is activated before each molt to facilitate "cuticle synthesis," hemolymph trehalose declines during periods of transition (e.g., larva → pupa). In this model hemolymph trehalose declines as a secondary response to an increase in trehalose activity. This model is untenable for a number of reasons. First, it implies that in *B. mori* (and by inference all species) homeostatic regulation of trehalose is so fragile that even long-term increased demands for trehalose [as opposed to short-term demands as in flight (see Sacktor, 1975)] result in drastic alterations in steady-state concentrations in hemolymph. Since levels of hemolymph trehalose are elevated during periods of enhanced hemolymph trehalose activity, and the converse, *B. mori* appears to exhibit considerable regulatory control over the concentration of trehalose in hemolymph. Second, since hemolymph is neither consuming nor producing trehalose, changes in hemolymph trehalose can only reflect changes in rates either of synthesis and release by fat body, and/or of uptake by other tissues. Last, the relationship between hemolymph trehalose and trehalase is tenuous at best. Trehalose is synthesized only in fat body. Hemolymph trehalase activity would then have little effect on the steady-state concentration of circulating trehalose in that glucose liberated upon hydrolysis of trehalose would either (1) diffuse back into fat cells, where as glucose it would be resynthesized into trehalose and released into hemolymph, or (2) enter tissues other than fat body, thereby depressing the rate at which trehalose was being removed from hemolymph. The net effect of hydrolyzing trehalose in hemolymph would be a slight depression in the steady-state level, and even this is predicated upon the assumption that no compensatory increases in fat body trehalose synthesis would occur.

An alternate model was initially proposed by Wyatt (1967, personal communication) to explain the dramatic decline in hemolymph trehalose observed in *H. cecropia* before (50 mM), during (5 mM), and after obligatory hibernal diapause (30 mM) (Fig. 11) (see also Jungreis and Wyatt, 1972; Jungreis *et al.*, 1973; Jungreis, 1976a,b). In this model the decline in hemolymph trehalose between larval and pupal stages in development results from (1) reduced metabolic requirements and capacity during pupal diapause (see Schneiderman and Williams, 1953; Avise and McDonald, 1976) and (2) a concomitant decline in trehalose-synthesizing capacity, as evidenced by an inability of diapause pupal fat body to incorporate glucose into trehalose either *in vitro* or *in vivo* (Table IX). Closer scrutiny of this model revealed its inadequacy to account for the developmental changes in hemolymph trehalose preliminary to diapause. When the rate of fat body

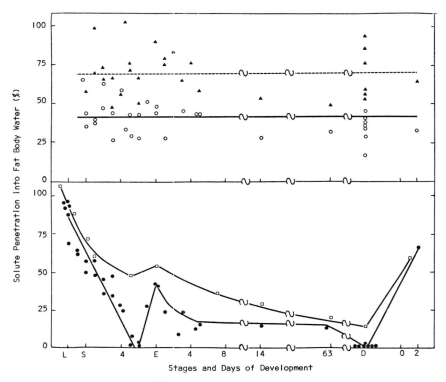

Fig. 11. Changes in *Hyalophora cecropia* hemolymph trehalose and in the penetrability of fat body to trehalose between feeding larval and pharate adult stages of development. Note that changes in trehalose penetrability occur independently of those of glucose and glycerol. Key: ▲, glycerol penetration; O, glucose penetration; □, hemolymph trehalose level; ●, trehalose penetration; L, larva; S, spinning; E, larval–pupal ecdysis, D, diapause pupal stage. Numbers after E are days of pupal development, after D are days of pharate adult development. (From Jungreis and Wyatt, 1972.)

trehalose synthesis from endogenous precursors was measured *in vitro*, larval and diapause pupal fat body retained equivalent capacities to synthesize trehalose (Fig. 12) (Jungreis, 1972). Since the rate of trehalose synthesis from exogenous or endogenous precursors was also the same in larval fat body (Jungreis, 1972), and between 1° and 37°C larval and pupal fat body synthesized trehalose from endogenous precursors at the same rate, it became obvious that Wyatt's alternate model for *H. cecropia* was inadequate. A reexamination of the control mechanisms involved in the regulation of trehalose synthesis before, during,and after diapause was clearly warranted.

TABLE IX

Rates of [^{14}C]Glucose Incorporation into Trehalose by Fat Body Measured *in Vivo* and *in Vitro* in Larval and Diapause pupal *Hyalophora cecropia*[a]

Stage of development	*In vivo* (cpm/μl hemolymph)	*In vitro* (μmoles trehalose synthesized/g fat body at 25°C)
Late feeding fifth instar larva		
30 min	257	6.7
60 min	294	13.3
Diapause pupa		
30 min	—	0.00
60 min	6	0.00

[a] Modified after Jungreis (1972), A. M. Jungreis (unpublished), and Wyatt (unpublished).

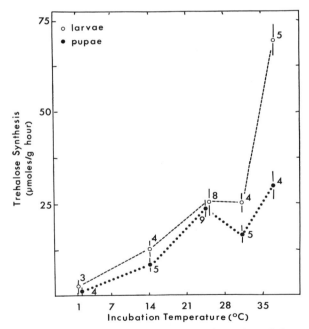

Fig. 12. Comparison between the *in vitro* rates of fat body trehalose synthesis from endogenous precursors in feeding fifth larval instar (open circles) and diapause pupal (shaded circles) *Hyalophora cecropia*. Fat body was incubated in Reddy, and Wyatt's medium (1967) modified after Jungreis and Wyatt (1972). (From A. M. Jungreis, unpublished).

Trehalose is synthesized according to pathways summarized in Fig. 13. Since the data of Table IX indicate that the capacity of fat body tissue to phosphorylate glucose during diapause is greatly reduced, hexokinase activity was measured between the feeding larval and pupal stages in development (Jungreis, 1976a). A 90% loss in the capacity to phosphorylate glucose was noted at pH 8.4 (Fig. 14). However, the loss in hexokinase activity was insufficient to account for the observed (virtual) absence both *in vitro* and *in vivo* of glucose conversion trehalose by diapause pupal fat cells (Table IX). The effects of glycolytic intermediates and precursors of trehalose (during its synthesis) on hexokinase activity were therefore examined (Jungreis, 1976b). Although glucose 6-phosphate and fructose 6-phosphate both depressed hexokinase activity, no single regulatory mechanism could be identified which would account for the absence of glucose uptake by diapause pupal fat body tissue.

The decline in hemolymph trehalose during pupal development and diapause could be attributed to an overall reduction in the availability of exogenous precursors. However, pharate adults, which exhibit elevated levels of trehalose relative to those in diapause pupae, are also without an exogenous source of monosaccharides. How, then, can the drastic shifts in

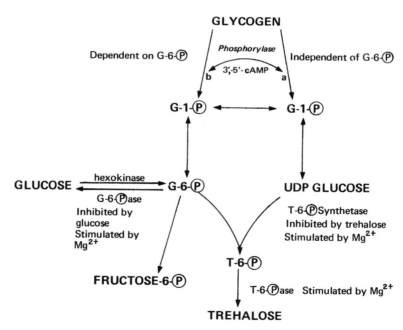

Fig. 13. A schematic diagram showing the flow of precursors in trehalose and their enzymatic regulation in insect fat body.

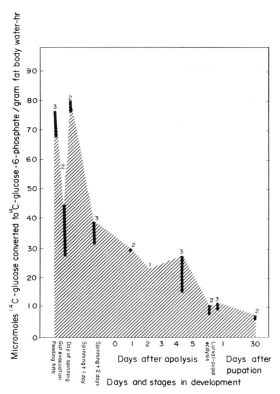

Fig. 14. Changes in *Hyalophora cecropia* fat body hexokinase activity between the feeding larval and pupal stages in development. (From Jungreis, 1976a.)

hemolymph trehalose observed in developing cecropia silkmoths be explained? Additional regulatory mechanisms have been elucidated and/or proposed and include alterations in the capacity of fat cells to release intracellular pools of trehalose (Jungreis and Wyatt, 1972), changes in the penetrability of fat cells to trehalose (Jungreis and Wyatt, 1972), alterations in the capacity to produce trehalose (Jungreis *et al.*, 1974), and changes in the availability of glycogen-derived endogenous precursors needed for trehalose synthesis (Wiens and Gilbert, 1967; Stevenson and Wyatt, 1964; Ziegler and Wyatt, 1975).

During diapause development, larval fat cells are unable to release intracellular trehalose at 1°C and have reduced capacities to release trehalose between 1° and 12°C relative to pupal fat cells (Fig. 15). The author therefore proposes that preliminary to diapause and as an adaptive response to diapause the composition of fat cell membranes shifts, presumably in the

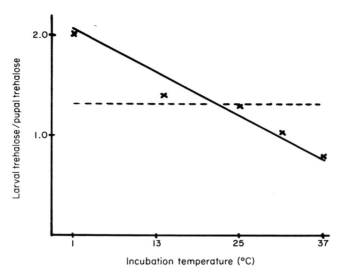

Fig. 15. Mean concentration ratios of trehalose remaining in larval/pupal *Hyalophora cecropia* fat body incubated *in vitro* at 1°–37°C for 1 hr in physiological saline. The dashed line is the mean ratio of retained trehalose between 1° and 37°C. (From A. M. Jungreis, unpublished.)

direction long-chain fatty acids → short-chain fatty acids. Accompanying this shift in the lipid composition of fat cell membranes is the acquisition of a capacity for macromolecules and as yet unidentified amino acids to interact on the surface of diapause pupal fat cells, such that egress of trehalose from these cells is restricted (Figs. 16 and 17). Without macromolecules (MW $\geq$ 17,000) and selected amino acids in the incubation medium, a rapidly exchangeable pool of intracellular trehalose is very quickly lost from both diapause pupal (Fig. 16) and larval fat cells (see Jungreis and Wyatt, 1972). However, in the presence of protein (Fig. 16), this rapidly exchangeable pool is retained by diapause pupal fat cells only. Furthermore, intact fat cells can readily restore the rapidly exchangeable intracellular trehalose pool following release if macromolecules are added to the incubation medium (Fig. 17). Associated with this macromolecular effect is a reduction in fat cell penetrability to trehalose (Fig. 11). The latter phenomenon is difficult to understand, since sugar movement across fat cells is by non-carrier-mediated passive diffusion (Jungreis and Wyatt, 1972), and dinitrophenol—an uncoupler of oxidative phosphorylation—is without effect on increasing or decreasing penetrability (Jungreis and Wyatt, 1972). Thus, the decline in hemolymph trehalose during diapause development in *H. cecropia* is the result of slow but continuous uptake of trehalose by tissue other than fat body, coupled with a drastic reduction in the rate at which trehalose that is needed to replace that which was consumed can be released from fat body.

Trehalose passively egresses from both larval and pupal fat cells. Since the rate of trehalose release during diapause is reduced, why instead of the observed decline does not the concentration of intracellular trehalose become elevated in pupal relative to larval fat cells? The most logical explanation is that allosteric inhibition by trehalose of the capacity to synthesize trehalose-6-phosphate (see Fig. 13) would prevent fat cell levels from becoming elevated. Although trehalose is reported to inhibit partially purified trehalose-6-phosphate synthetase (Murphy and Wyatt, 1965; McAlister and Fisher, 1971), trehalose failed to inhibit trehalose synthesis in intact fat cells even at a concentration of 100 mM (Table X). Murphy and Wyatt (1965) earlier noted that Mg^{2+} could both release the allosteric inhibition of the enzyme by trehalose as well as stimulate the activity of trehalose-6-phosphate synthetase. G. R. Wyatt (personal communication) first suggested that magnesium rather than trehalose could be directly

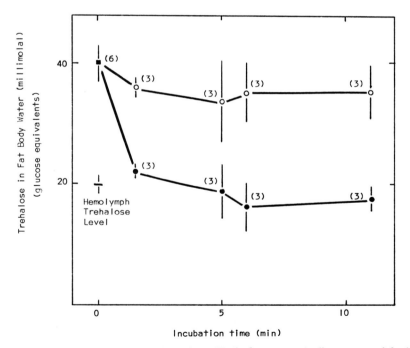

Fig. 16. Control of trehalose release from *Hyalophora cecropia* diapause pupal fat body incubated *in vitro*. Shaded square, unincubated fat body; open circles, fat body incubated in saline in the presence of 10% blood proteins; shaded circles, fat body incubated in saline in the absence of blood proteins. (From Jungreis and Wyatt, 1972.)

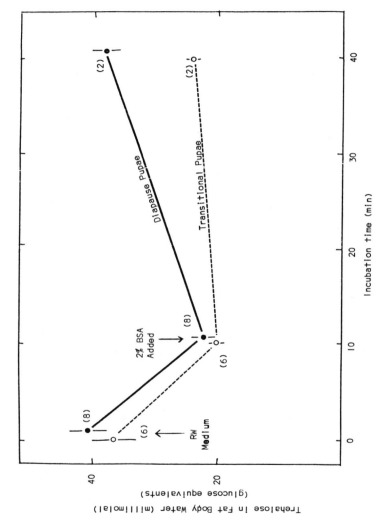

Fig. 17. Effect of added protein on restoration of *Hyalophora cecropia* fat body trehalose levels following release of trehalose by incubation in saline without added protein. Note that addition of protein to the incubation medium is without effect on postdiapause pupal fat body. Shaded circles, diapause pupal fat body; open circles, postdiapause, but prepharate adult fat body. RW, Reddy and Wyatt's medium; BSA, bovine serum albumen. (From Jungreis and Wyatt, 1972.)

TABLE X

Absence of an Effect of Exogenous Trehalose on [¹⁴C]Glucose Incorporation into Trehalose in *Hyalophora cecropia* Fat Body Incubated *in Vitro*[a]

Stage of development	Total [¹⁴C]glucose incorporated into trehalose (μmoles/g fat body water per hour at 25°C)		Inhibition (%)
	$-$Trehalose	$+100$ mM Trehalose	
Late feeding fifth instar	39.1	37.7	$+3.6$
Gut evacuation	41.9	39.3	$+6.2$
Initiation of spinning	14.5	15.6	-7.5

[a] Modified after Jungreis (1972) and Jungreis (unpublished).

responsible for regulating the rate at which fat cells synthesize trehalose. Changes in the concentration of Mg^{2+} in larval and diapause pupal fat body tissue development were therefore determined (Fig. 18) (Jungreis, 1973a; see also Jungreis *et al.*, 1974). The concentration of Mg^{2+} was noted to drop from 20 mM to 10 mM between the larval and pupal stages in development. This difference is of sufficient magnitude to account for the *reduced* pool size of trehalose in diapause pupal relative to larval fat cells. Although a reduction in fat cell Mg^{2+} slows down the rate at which trehalose is synthesized, when trehalose release from fat cells is restricted, the steady-

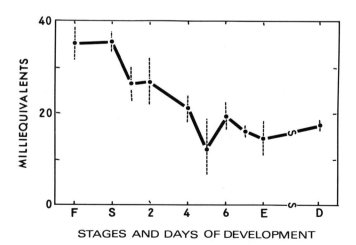

Fig. 18. Changes in the concentration of *Hyalophora cecropia* fat body magnesium during the larval–pupal transformation. Abbreviations: F, feeding fifth larval instar; S, day of spinning; E, day of the larval–pupal ecdysis; D, diapause. (Modified after Jungreis, 1973a.)

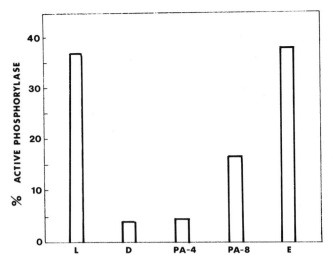

Fig. 19. Changes in the percentage of fat body phosphorylase present in the active form during ontogeny of the silkmoth, *Hyalophora cecropia*. Abbreviations: L, feeding fifth instar larva; D, diapause pupa; PA-4, day 4 of pharate adult development; PA-8, day 8 of pharate adult development; E, day of pupal–adult ecdysis. (Modified after Wiens and Gilbert, 1967; Stevenson and Wyatt, 1964.)

state level of trehalose within fat cells should be maintained or increased. An important factor effecting the concentration of intracellular trehalose is the pool size of intermediates needed for trehalose synthesis. The rate at which glycogen is enzymatically degraded is normally dependent on the proportion of the total phosphorylase activity present as active phosphorylase (Stevenson and Wyatt, 1964; Wiens and Gilbert, 1967; Ziegler and Wyatt, 1975). It is clear from Figs. 19 and 20 that the capacity to spontaneously hydrolyze glycogen is reduced during diapause but that in *H. cecropia* a compensatory regulatory mechanism has evolved which permits the rapid conversion of inactive to active phosphorylase in both diapause pupal (Zeigler and Wyatt, 1975) (Fig. 20) and larval (G. R. Wyatt, personal communication) fat body cells in response to elevations in temperature from those associated with hibernal diapause (1°–4°C) to those associated with postdiapause development (25°C).

D. Regulation of Glycerol

In invertebrates, the problems of regenerating NAD^+ and storing reducing equivalents of hydrogen during periods of high metabolic demand and/or reduced lactate dehydrogenase activity and/or reduced electron transport chain activity are frequently solved via the production and accumulation of metabolic intermediates of the tricarboxylic acid cycle, propionic acid, and the polyol glycerol (see Hochachka and Somero, 1973). In insects and other organisms capable of withstanding extended periods of extreme cold during hibernal diapause, glycerol can also be utilized as an antifreeze to retard the formation of ice crystals within cells (see Salt, 1961; Asahina, 1969). When the concentration of glycerol added to larval hemolymph of *H. cecropia* exceeds 0.15 *M*, a supercooling effect equivalent to the addition of 0.5 Osm of particles is accorded (Fig. 21). When the added concentration of glycerol exceeds 0.30 *M*, an osmotic equivalent in excess of 3.0 Osm is conferred. Thus, it appears that the concentration of glycerol in hemolymph need not exceed 0.30 *M* in order to confer the maximal supercooling capacity to the organism. This value of 0.30 *M* glycerol in hemolymph is consistent with that observed during diapause in

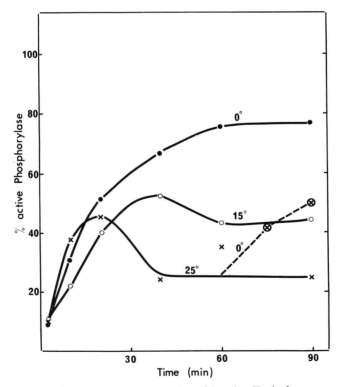

Fig. 20. Effects of temperature on conversion of inactive *Hyalophora cecropia* diapause pupal fat body phosphorylase to the active form. Key: ⊗, ●, fat body on ice; ○, fat body at 15°C; x, fat body at 25°C. (From Ziegler and Wyatt, 1975.)

86 Arthur M. Jungreis

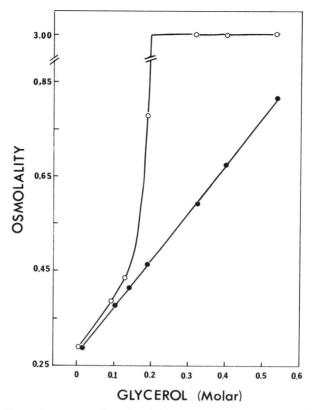

Fig. 21. Effects of exogenous glycerol on initiation of supercooling in larval hemolymph of the silkmoth, *Hyalophora cecropia*. At concentrations above 0.2 *M* glycerol, considerable supercooling is observed. (From A. M. Jungreis, unpublished.)

this species (see Wilhelm *et al.*, 1961; Wyatt and Meyer, 1959; Yamashita and Wyatt, 1976; Ziegler and Wyatt, 1975) (Fig. 22), as well as other Lepidoptera and insects from other orders (see Wyatt, 1967; Salt, 1961).

Several questions can be asked regarding regulation of glycerol synthesis, accumulation and metabolism before, during, and after diapause, as well as the control of pathways responsible for its appearance in hemolymph. Glycerol appearing in hemolymph during the larval–pupal transformation (LPT) is in good measure derived from hydrolysis in pharate pupal midgut epithelium of α-glycerol phosphate which originated in hemolymph (Jungreis *et al.*,1975). Hydrolysis of α-glycerol phosphate does not occur because α-glycerol phosphatase activity increases during the LPT (Fig. 23) but in response to an effective increase in hydrolytic activity brought about by more drastic declines in glycerol kinase activities in the same tissues (Fig. 24).

Hydrolysis of α-glycerol phosphate during the LPT occurs as a mechanism for removal of hemolymph Mg^{2+}. Hemolymph Mg^{2+} is present in excess quantities because of the dehydration characteristically experienced by Lepidoptera during the LPT (Jungreis and Tojo, 1973) (see Section III,A,4). Hydrolysis of α-gylcerol phosphate occurs selectively in gut relative to other tissues, because midgut is readily permeable to Mg^{2+} (see Wood et al., 1975, and Section III,A,4). α-glycerol phosphate is also the source of inorganic phosphate required for the removal of hemolymph

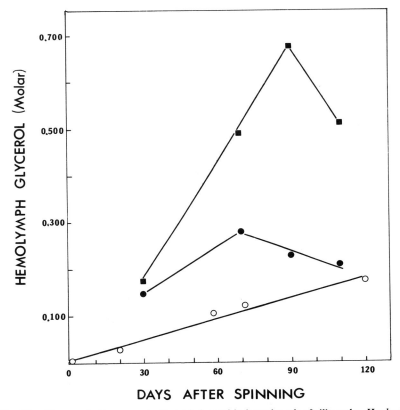

Fig. 22. Changes in the concentration of glycerol in hemolymph of silkmoths, *Hyalophora cecropia*, maintained at a variety of temperatures. Open circle, 25°C throughout (Wyatt and Meyer, 1959); shaded circles, 25°C throughout (Ziegler and Wyatt, 1975); shaded squares, 25°C for first 25 days, 4°C thereafter (Ziegler and Wyatt, 1975). (Modified after Wyatt and Meyer, 1959; Ziegler and Wyatt, 1975.)

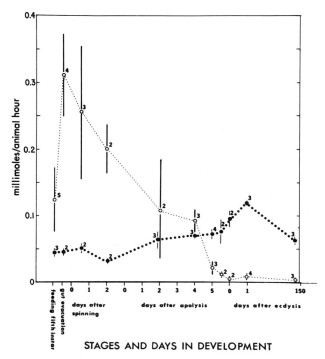

Fig. 23. Changes in the total activity of α-glycerol phosphatase in midgut and fat body tissues before, during, and after the larval–pupal transformation in *Hyalophora cecropia*. Shaded circles, fat body; open circles, midgut. (From Jungreis *et al.*, 1975.)

Mg^{2+} via the formation of insoluble $Mg_3(PO_4)_2$. Thus, one observes selective accumulation of inorganic phosphate and magnesium in midgut but not fat body when glycerol accumulation in hemolymph is first noted (Wyatt and Meyer, 1959; Jungreis, 1973a; Jungreis *et al.*, 1974, 1975) (Fig. 25).

An important corollary of diapause development is clearly illustrated by the accumulation of glycerol in hemolymph, namely, that the enzymatic capacity need not increase to facilitate an increase in the production and accumulation of a particular substrate (Table XI). Coupled with this first corollary is a second, which states that the appearance of high levels of selected substrates during diapause is frequently accompanied by an elevated enzymatic capacity for metabolism of the specific compounds. Accompanying this increase in enzyme capacity is a regulatory mechanism *in vivo* that acts to inhibit virtually all of this elevated enzymatic capacity. Accumulation of glycerol during diapause development illustrates these corollaries in an exemplary fashion since (1) the capacity to metabolize

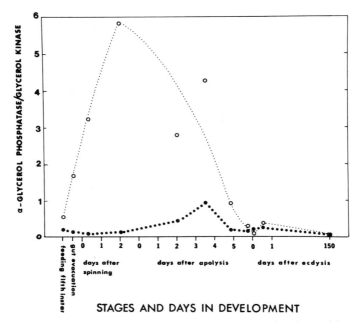

Fig. 24. Changes in midgut and fat body tissue α-glycerol phosphatase/glycerol kinase activity ratios before, during, and after the larval–pupal transformation in *Hyalophora cecropia*. Shaded circles, fat body; open circles, midgut. (From Jungreis *et al.*, 1975.)

glycerol via glycerol kinase (Fig. 26) is greatly elevated during a period when glycerol accumulation and retention in hemolymph are optimum (see also Yamashita and Wyatt, in preparation, cited in Wyatt, 1975) and (2) the enzymatic capacity to synthesize glycerol during diapause is greatly reduced relative to that of feeding stages when glycerol failed to accumulate (Table

TABLE XI

Relationship between Enzyme Capacity to Produce Glycerol and Glycerol Accumulation in Hemolymph of Prediapause and Diapause Pupae of *Hyalophora cecropia*[a]

Stage of development	*In vivo* rate of glycerol production (μmoles/mg N day)	Concentration of hemolymph glycerol (mM)
Pupal		
2 Days after pupation	0.54	15
60 Days after pupation	0.065	126
Ratio 2 days/60 days	8:3:1	1:8:4

[a] Modified after Wilhelm *et al.* (1961) and Wyatt and Meyer (1959).

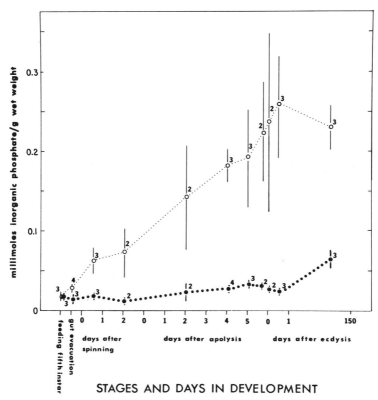

Fig. 25. Changes in the concentration of inorganic phosphate in midgut and fat body tissue before, during,and after the larval–pupal transformation in *Hyalophora cecropia*. Shaded circles, fat body; open circles, midgut. (From Jungreis *et al.*, 1975.)

XI). Glycerol synthesis may indeed be elevated during feeding stages, but glycerol is probably utilized in lipid synthesis, transport, and metabolism (see Gilbert, 1967b). These general principles characteristic of prediapause and diapause stages are also supported by measurements made during uric acid (Section III,B) and trehalose synthesis (Section III,C).

During diapause, glycerol plays a key role in regulating its own rate of synthesis as well as that of trehalose via control of glycogen phosphorylase (see Ziegler and Wyatt, 1975). When diapause pupal fat body from *H. cecropia* is incubated for 1 hr in saline lacking protein, the percentage of total phosphorylase present in the active form doubles (Ziegler and Wyatt, 1975) (Fig. 27). In response to injury, the precentage of phosphorylase in the active form also increases (Stevenson and Wyatt, 1964). The latter observation was also made by incubating fat body in the absence of glycerol. When glycerol is present at physiological concentrations of 0.35 M, no increase in activation is noted (interpolated and extrapolated from the data of Ziegler and Wyatt, 1975). Since fat body cells become leaky to trehalose following injury (Jungreis and Wyatt, 1972), the increase in hemolymph trehalose noted after injury results from the loss of feedback inhibition of trehalose on trehalose-6-phosphate synthetase. Since the level of trehalose in hemolymph of diapause pupae is lower than that in fat body (Jungreis and Wyatt, 1972), trehalose will egress from fat cells and an increase in hemolymph trehalose will be noted. However, the increase in trehalose will occur at the expense of stored glycogen rather than at the expense of glycerol, thereby maintaining the pupa's tolerance to cold. This

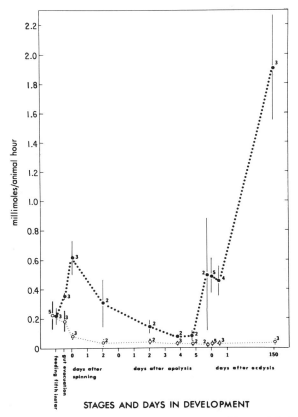

Fig. 26. Changes in total activity of glycerol kinase in midgut and fat body before, during, and after the larval–pupal transformation in *Hyalophora cecropia*. Shaded circles, fat body; open circles, midgut. (From Jungreis *et al.*, 1975.)

last conclusion is supported by data on glycerol metabolism following initiation of pharate adult development. Glycerol is primarily incorporated into glycogen rather than being quantitatively converted to trehalose (Fig. 28). In this regard, regulation of trehalose synthesis during postdiapause development differs in some fundamental aspects from that during feeding, since virtually 100% of exogenous glucose supplied to larval fat cells either *in vitro* or *in vivo* is converted to trehalose (Jungreis, 1972) (Section III,C), whereas glycerol is converted to glycogen via glucose without a dramatic increase in fat body trehalose production and release into hemolymph. The biochemical basis for this regulatory difference in glucose metabolism during larval and postdiapause stages in development is unknown and warrants further exploration.

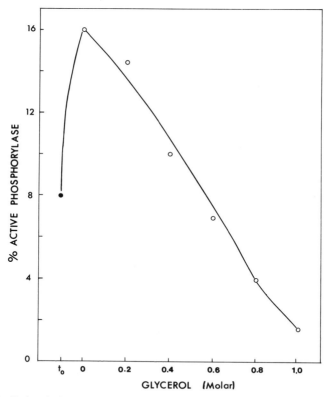

Fig. 27. Role of glycerol as an effector of glycogen phosphorylase activity in diapause pupal fat body of *Hyalophora cecropia*. Physiological quantities of glycerol are observed to prevent conversion of phosphorylase from the inactive to the active forms. Shaded circles, percent active phosphorylase. (Recalculated from data in Ziegler and Wyatt, 1975.)

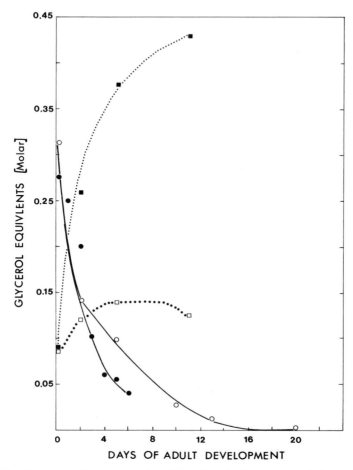

Fig. 28. Changes in the concentration of hemolymph glycerol and trehalose expressed as glycerol equivalents during pharate adult development in *Hyalorphora cecropia*. Shaded circles, glycerol (data from Yamashita and Wyatt, personal communication); open circles, glycerol (data from Wyatt and Meyer, 1959); open squares, observed trehalose concentrations expressed in glycerol equivalents (1 mM trehalose = 4 mM glycerol) (data from Jungreis *et al.*, 1973); shaded squares, predicted trehalose concentration (expressed in glycerol equivalents) if hemolymph glycerol were quantitatively converted to trehalose. (Modified after Wyatt and Meyer, 1959; Jungreis, Jatlow and Wyatt, 1973; O. Yamashita and G. P. Wyatt, personal communication; A. M. Jungreis, unpublished.)

In summary, the glycerol first appearing in hemolymph during the LPT is derived from previously synthesized α-glycerol phosphate which was present in hemolymph. Thereafter, the increase in glycerol accumulation is accompanied by a dramatic increase in the capacity to metabolize glycerol via glycerol kinase (Fig. 26) and a marked decline in the capacity to synthesize glycerol (Table XI). During diapause, glycerol functions as a supercooling agent and as a moderator for controlling active phosphorylase activity, especially during injury. Following diapause, glycerol is metabolized and converted via glucose to glycogen rather than trehalose, a response in marked contrast to the fate of glucose during the larval stages in development.

E. Hormonal Control

Endocrine control of insect development is an area that has been very active and exciting, but it has provided few generalizations for the vast majority of insects experiencing a broad spectrum of dormancies (see Wyatt, 1972; Morgan and Poole, 1977). The elegant model of Williams (1946, 1947, 1952, 1956) relating the activity of the brain to diapause development in giant silkmoths has stood the test of time and is now the accepted model for the control of hibernal diapause. However, one is unable to extend the generalizations and conclusions deriving from this model to other types of insect dormancies or even to hibernal diapause in other lepidopterans. For example, the role of the brain in the induction, maintenance, and terminatin of hibernal diapause in the tobacco hornworm, *Manduca sexta* (Judy, 1972; Wilson and Larsen, 1974; Lockshin *et al.*, 1975), appears to be quite different from the more internally consistent model of Williams (1946, 1947, 1952, 1956).

Several recent publications indicate that juvenile hormone may be responsible for the induction and maintenance of the hibernal diapause which occurs in fully grown larvae of the southwestern corn borer, *Diatraea grandiosella* Dyar, and the rice stem borer, *Chilo supressalis* Walker (Chippendale, 1977; Chippendale and Yin, 1973, 1975; Yin and Chippendale, 1973a,b, 1976; Yagi and Fukaya, 1974). In these species, under natural conditions fully grown larvae retain a moderate titer of juvenile hormone and may undergo stationary moults (larva → larva) especially during the early stages of diapause. If the titer of ecdysone is artificially elevated during this period of moderate juvenile hormone activity, insects immediately initiate and successfully complete the developmental sequence necessary for a stationary molt (see Blumenfield and Schneiderman, 1970; Chippendale and Yin, 1973; Yagi and Akaike, 1976, Chippendale, 1977). When the source of juvenile hormone is artificially (surgically) or naturally (age) depressed, these insects terminate diapause and reinitiate development. These same authors (Yin, Chippendale and Yagi) conclude that hibernal diapause is initiated in the late summer and is maintained throughout the fall and winter by the continued presence of juvenile hormone. Diapause terminates when the titer of juvenile hormone falls to the threshold for a pupal moult. However, post-diapause development is delayed until the temperature rises to the threshold for a pupal moult. Thus, only after the temperature and photoperiod reach specific levels during the following spring can the corn borers and rice stem borers complete development and emerge as adults (Chippendale, 1977).

In summary, the mechanisms for controlling hibernal diapause in *Diatraea grandiosella* and *Chilo suppressalis* may be typical of other insects experiencing hibernal diapause during apparently favorable conditions. The major problem with this model is the difficulty of determining whether various insect species are experiencing the same type of hibernal diapause as are *D. grandiosella* and *C. suppressalis*. For example, in the waxmoth, *Galleria mellonella* (Sehnal and Granger, 1975), surgical manipulations that result in a low titer of juvenile hormone at the end of the last larval instar will prolong this instar and prevent the initiation of pupal development. Criteria such as the presence of various stage-specific proteins (see, for example, Telfer and Williams, 1953) would facilitate analysis of dormancies in insects that are too small for surgical manipulation or that need to be rapidly surveyed to determine whether specific and characteristic patterns of dormancy are present.

F. Hemolymph Proteins during Ontogeny

One of the truly perplexing problems associated with the study of insect dormancies is the need to develop quantitative methods. The identification of stage-specific characteristics will facilitate recognition of the subtle changes that precede, occur simultaneously with, and follow the spectrum of insect dormancies. Although specific ultrasensitive assays for juvenile hormone (Fain and Riddiford, 1975) and ecdysone (Borst and O'Connor, 1972) have been developed, our understanding of the extremely complex interactions between these and other hormones (see Wyatt, 1972) that govern the various dormancies is extremely poor. Until the mechanisms that govern hormonal synthesis, release, and degradation during dormancies are better understood, reliance on other behavioral, physiological, and/or biochemical cues must continue.

One area of regulation under strict although often poorly understood hormonal control is that concerned with hemolymph proteins and their storage in fat body. The use of blood proteins as monitors for development has the advantage of permitting withdrawal of only small hemolymph volumes for quantitative studies, as well as permitting study of the specific sites of synthesis and uptake of blood proteins, thereby establishing a basis for quantitative measurement of dormancies. (Synthesis of hemolymph proteins is discussed in several recent reviews: Price, 1973; de Wilde and de Loof, 1973; Thomson, 1975; Wyatt and Pan, 1978.) In the discussion that follows, biochemical events associated with the synthesis, storage, release, circulation, uptake, and metabolism of blood proteins in saturniid silkmoths are described. Emphasis is on species that undergo hibernal diapause, although limited generalizations to oligopauses can be made.

One of the characteristic features of hibernal diapause in *H. cecropia* is a low level of blood protein synthesis (Fig. 29). From the pioneering work of Telfer and Williams (1960), one learned that the rate at which [^{14}C]glycine was incorporated into proteins appearing in hemolymph was statistically lower in animals exhibiting diapause levels of respiration than in those exhibiting postdiapause development. This observation has been confirmed in a wide variety of insect species (Price, 1973; Thomson, 1975). This work led to the elucidation of function of at least two such proteins, namely, fat body storage protein and vitellogenins.

1. STORAGE PROTEINS

Diapause and nondiapause species of holometabolous insects selectively increase the protein content of hemolymph beginning with the middle of the last larval instar. Accompanying this increase in hemolymph protein is the initiation of selective and characteristic storage of protein in fat body in histochemically identifiable granules. Sequestration of such proteins throughout the late larval, pharate pupal, and early pupal stages in development, their maintenance during diapause, and their subsequent although gradual disappearance from fat cells during pharate adult and adult development suggested that, in the absence of additional nutrient input, these storage proteins possibly were the basic "raw materials" for protein synthesis during adult development. The major components of these granules are immunochemically defined macromolecules (see Wyatt, 1975; Wyatt and Pan, 1978).

Storage proteins have been studied most exhaustively and fruitfully in the silkmoth, *H. cecropia* (R. Zicardi and W. H. Telfer, personal communication; Tojo, Betchaku and Wyatt, personal communication). In this species, as revealed by acrylamide gel electrophoresis, saline-soluble extracts of fat body contain two major proteins, which correspond immunochemically to blood antigens No. 2 and No. 6, as outlined in the definitive study of blood proteins by Telfer and Williams (1953). Curiously, fat body of diapausing females contains more of these two proteins than does male fat body,

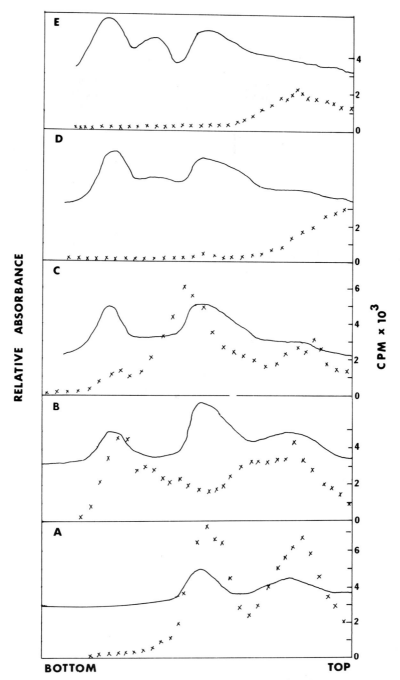

Fig. 29. Changes in the rate of [³H]leucine incorporation ( ) into hemolymph proteins (solid lines, absorbance) of female *Hyalophora cecropia* silkmoths between feeding and diapause pupal stages in development. (A) Feeding fifth instar larva; (B) day of the larval–pupal apolysis; (C) day of larval–pupal ecdysis; (D) diapause; (E) Postdiapause development (7 months of chilling at 4°C). (From Pan, 1978.)

whereas in hemolymph the converse is true. On a per unit volume basis, the quantity of protein extracted from fat body is less than that extractable from hemolymph. Both proteins are more or less simultaneously detected in larval hemolymph at 3–4 days before the onset of spinning. Their initial appearance is followed by a gradual increase in concentration, which reaches a maximum at about 1 day after cocoon construction or just before the larval–pupal apolysis. It is only at apolysis that accumulation of these proteins in fat body begins in earnest. As demonstrated with radioisotope techniques, protein sequestration in fat body is provided at the expense of hemolymph protein, whose concentration declines just as that in fat body is elevated. Why this unusual pattern of synthesis, release, and sequestration takes place is not obvious. At the time of spinning, 50% of the released label is found in these fat body proteins (Fig. 30). Thus, the synthesis of fat body proteins and the synthesis of vitellogenins (Section III, F, 2) are two major biochemical events that occur before the onset of diapause in *H. cecropia*.

Protein granules increase in both size and number throughout spinning and pharate pupal development, reaching a maximum shortly after the

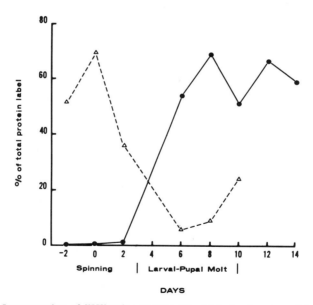

Fig. 30. Incorporation of [³H]leucine *in vitro* into fat body storage proteins (open triangles) and vitellogenins (shaded circles) released by fat body following incubation in physiological saline and measured during the larval–pupal transformation in female *Hyalophora cecropia*. Data are expressed as percentage of total label incorporated into protein. (From M. L. Pan, 1969 personal communication.)

larval–pupal ecdysis. Females were found to have a greater abundance of granules than were males, with between 30 and 50% contrasted with 10–30%, respectively, of the histological cross-sectional area of fat cells of females relative to males occupied by protein granules. A portion of the total storage protein is also cosequestered with uric acid in urate granules (Section III, B).

Examination of protein granules under the electron microscope (Tojo, Betchaku, and Wyatt, cited in Wyatt, 1975) revealed that these granules consist of a dense inner region surrounded by a zone of lower density. Isolation of the granules revealed a composition of 98+% protein, 0.5% uric acid, and a small quantity of nucleic acid. When solubilized, these protein granules yielded two major proteins, which were immunochemically and electrophoretically identical to the two soluble fat body proteins. It thus appears that fat body proteins sequestered from hemolymph, possibly together with newly synthesized proteins, are utilized in fat body during formation of the protein granules. The process whereby protein granules are formed in the cytosol is unclear in *H. cecropia* (for *Calpodes ethlius*, see Locke and Collins, 1965; Collins, 1974). The physical and chemical characteristics of these two fat body proteins are described by Wyatt and Pan (1978).

During diapause, the number of protein granules in fat body remains stable. However, upon termination of diapause and initiation of pharate adult development, the gradual loss of these granules becomes evident. During pharate adult development, differences present in the quantity of protein granules sequestered in diapause pupal male and female fat body persist during granule breakdown, females retaining greater quantities than males. However, this difference could reflect the increased number of ways that these proteins are utilized in females relative to males, with selective female-specific utilization occurring fairly late in pharate adult development relative to that in males. Newly emerged adult male fat body is virtually devoid of granules, whereas many persist in female fat body.

The concentration of soluble fat body proteins in hemolymph was determined following termination of diapause. The concentration of both proteins decreased rapidly during early pharate adult development, with the quantity remaining at the time of adult development almost undetected even by immunochemical methods. On the other hand, their concentration in the fat body extract initially rose, declining slowly thereafter. Both proteins are readily detected in adult fat body. During the pharate adult stage, no net synthesis of fat body protein is observed. Specific or selective utilization of these two fat body storage proteins is unknown, although they can be presumed to supply amino acids needed for the construction of new adult structures.

2. VITELLOGENINS

Sex-limited female hemolymph proteins synthesized by fat body and
selectively sequestered by oocytes to form the major yolk component are
called vitellogenins. Discovered by Telfer (1954) in *H. cecropia* and named
vitellogenin by Pan (Pan *et al.*, 1969), these proteins have been shown to
exist in all insects studied to date (see reviews in Telfer, 1965; Englemann,
1970; Wyatt, 1972, 1975; Wyatt and Pan, 1978; Doane, 1973). In *H.
cecropia*, vitellogenin appears in hemolymph shortly after cocoon spinning
is initiated, with a maximal concentration reached at the time of the
larval–pupal ecdysis (Fig. 30). The synthesis of the two fat body storage
proteins is also plotted in Fig. 30. Thus, it appears that synthesis of storage
proteins is turned off at the same time that vitellogenin synthesis is being
initiated. The concentration of vitellogenin in hemolymph remains constant
throughout diapause and declines during pharate adult development in
response to sequestration by the oocytes (data not shown). Incorporation of
[³H]leucine into vitellogenin indicates that in *H. cecropia* there are two
periods of synthesis, the first occurring during the late pharate pupal stages
in development and a second occurring during yolk deposition late in
pharate adult development (Pan, 1971) (Fig. 31). As is true for all other
blood proteins, scant synthesis of vitellogenin occurs during diapause (Fig.
29), nor does injury initiate vitellogenin synthesis (Telfer, 1965).

In summary, synthesis of both fat body storage proteins and vitellogenin
is readily measured during pre- and postdiapause development but not during diapause itself. Since the synthesis of these proteins can readily be moni-

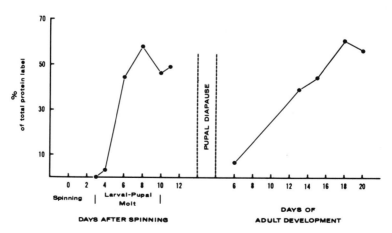

Fig. 31. *In vivo* incorporation of [³H]leucine into vitellogenins during ontogeny and
expressed as a percentage of the total label incorporated into blood proteins. (From Pan, 1971.)

tored, measurement of selective isotope incorporation into them accords an excellent means of determining in individuals of a population their exact point in dormancy development. Furthermore, it is an excellent way to study phenological influence on insect dormancy. The study of insects such as *Antheraea polyphemus*, which are bivoltine, in contrast to *H. cecropia*, which is normally univoltine, accords an opportunity to determine which developmental changes are associated with the pupal stage and which are selective adaptations to the diapause state. The control of vitellogenin induction, synthesis, uptake, and metabolism in lepidopterans other than *H. cecropia* and in other insect orders is extensively covered in reviews by Wyatt (1972, 1975), de Wilde and de Loof (1973), Doane (1973), and Englemann (1970) and in articles by Kelly and Davenport (1976), Kunkel and Pan (1976), Pan (1971), Pan and Wyatt (1971), and Wyatt and Pan (1978).

IV. SUMMARY

The most difficult problem encountered in the study of insect dormancy is a general lack of precision. Authors of diverse background, interests, and orientations consistently use ambiguous terminology when describing the presumed "physiological state" (e.g., reproductive diapause) and then compound the problem through an overreliance on qualitative behavioral criteria (egg laying, yes! egg laying, no!) to determine whether postdormancy development has been initiated. (see Fig. 1). In this chapter, I have shown how a broad spectrum of readily quantifiable variables can be used to monitor pre- and postdormancy development as well as the specific dormancy itself. One unexpected observation that appeared with surprising regularity was that an apparent increase in storage or synthetic activity before initiation of a dormant state was frequently accompanied by a decline in total enzyme activity. It thus appears that alterations in regulatory mechanisms are effected through subtle changes in the balance between antagonistic enzyme pathways rather than via enhancement of an absolute enzyme capacity.

Throughout this chapter, quantitative information obtained from *H. cecropia* was presented because of the wealth of unique features possessed by this species during its development. These include holometabolic development, univoltine life history, true hibernal diapause with known endocrine mechanisms for its induction and termination, as well as the availability of a wealth of data from physiological studies on this species. No attempt has been made nor should one attempt to apply generalizations drawn from *H. cecropia* to other species, even those that have similar life histories, since no insect species is sufficiently "typical" to warrant generalization to the class Insecta. However, if in the future quantitative versus qualitative measurements can be made in a broad spectrum of species from various orders, it may be possible to draw conclusions and make generalizations relating various facets of dormancy development in insects and to relate them to dormancy development in botanical flora and in noninsect animal fauna.

ACKNOWLEDGMENTS

I wish to thank Carole A. Price for typing and retyping this manuscript; her patience and indulgence are gratefully acknowledged. I also wish to thank Drs. M. L. Pan, A. Mansingh and G. M. Chippendale for critically reading the manuscript and for offering a wealth of excellent suggestions, which the author frequently chose to disregard at the expense of quality but in the interest of time and space limitations. This work was supported in part by grants from the National Science Foundation (PCM75-23456), Biomedical Sciences Support Grant (RR-07088), and a University of Tennessee Faculty Research Grant.

REFERENCES

Adkisson, P. L. (1964). *Am. Nat.* **98,** 357–374.
Adkisson, P. L. (1966). *Science* **154,** 234–241.
Adkisson, P. L., and Roach, S. H. (1971) *In* "Biochronometry" (M. Menaker, ed.), pp. 272–280. Nat. Acad. Sci., Washington, D.C.
Adkisson, P. L., Bell, R. A., and Wellso, S. G. (1963). *J. Insect Physiol.* **9,** 299–310.
Anderson, E., and Harvey, W. R. (1966). *J. Cell Biol.* **31,** 107–134.
Anderson, J. F., and Kaya, H. K. (1974). *Ann. Entomol. Soc. Am.* **67,** 845–849.
Andrewartha, H. G. (1952). *Biol. Rev. Cambridge Philos. Soc.* **27,** 50–107.
Andrewartha, H. G., and Birch, L. C. (1954). "The Distribution and Abundance of Animals." Univ. of Chicago Press, Chicago, Illinois.
Andrewartha, H. G., Miethke, P. M., and Wells, A. (1974). *J. Insect Physiol.* **20,** 679–701.
Ankersmit, G. W. (1964). *Meded. Landbouwhogesch. Wageningen* **64,** 1–60.
Ankersmit, G. W. (1968). *Entomol. Exp. Appl.* **11,** 231–240.
Ankersmit, G. W., and Adkisson, P. L. (1967). *J. Insect Physiol.* **13,** 553–564.
Aruga, H., Yoshitake, N., and Ishihara, R. (1953). *J. Seric. Sci. Jpn.* **22,** 11–18.
Asahina, E. (1969). *Adv. Insect Physiol.* **6,** 1–50.
Avise, J. C., and McDonald, J. F. (1976). *Comp. Biochem. Physiol. B* **53,** 393–397.
Barker, J. F., and Herman, W. S. (1976). *J. Insect Physiol.* **22,** 1565–1568.
Barker, R. J., Mayer, A., and Cohen, C. F. (1963). *Ann. Entomol. Soc. Am.* **56,** 292–294.
Barker, R. J., Cohen, C. F., and Mayer, A. (1964). *Science* **145,** 1195–1197.
Barrett, F. M., and Friend, W. G. (1970). *J. Insect Physiol.* **16,** 121–129.
Barry, B. D., and Adkisson, P. L. (1966). *Ann. Entomol. Soc. Am.* **59,** 122–125.
Beards, G. W., and Strong, F. E. (1966). *Hilgardia* **37,** 345–362.
Beck, S. D. (1962). *Biol. Bull. (Woods Hole, Mass.)* **122,** 1–12.
Beck, S. D. (1967). *J. Insect Physiol.* **13,** 739–750.

Beck, S. D. (1968a). *In* "Evolution and Environment" (E. T. Drake, ed.) pp. 279–296. Yale Univ. Press, New Haven, Connecticut.

Beck, S. D. (1968b). "Insect Photo-periodism," p. 135. Academic Press, New York.

Beck, S. D., and Alexander, N. (1964). *Biol. Bull.* (*Woods Hole, Mass.*) **126**, 175–184.

Beck, S. D., and Hanec, W. (1960). *J. Insect Physiol.* **4**, 304–318.

Beck, S. D., and Shane, J. L. (1969). *J. Insect Physiol.* **15**, 721–730.

Beck, S. D., Cloutier, E. J., and McLeod, D. G. R. (1963). *In* Insect Physiology. *Proc. Biology Colloquim*, 23rd ed. V. J. Brookes, 43–64. Corvallis, Oreg.: Oregon State Univ. Press.

Beck, S. D., Shane, J. L., and Garland, J. A. (1969). *J. Insect Physiol.* **15**, 945–951.

Bell, R. A. (1967). Ph.D. Dissertation, Texas A & M University, College Station.

Bell, R. A. (1972). *Proc. North Cent. Branch Entomol. Soc. Am.* **27**, 176–177.

Bell, R. A., and Adkisson, P. L. (1964). *Science* **144**, 1149–1151.

Bell, R. A., Rasul, C. G., and Joachim, F. G. (1975). *J. Insect Physiol.* **21**, 1471–1480.

Belozerov, V. (1969). *Moskva* **5**, 521–526.

Benschoter, C. A. (1968). *Ann. Entomol. Soc. Am.* **61**, 1272–1274.

Berry, S. J., Krishnakumaran, A., Oberlander, H., and Schneiderman, H. A. (1967). *J. Insect. Physiol.* **13**, 1511–1537.

Blake, G. M. (1963). *Nature* (*London*), **198**, 462–463.

Blumenfeld, M., and Schneiderman, H. A. (1970). *Biol. Bull.* (*Woods Hole, Mass.*) **138**, 466–475.

Bondarenko, N., and Hai-Yuan, K. (1968). *Dokl. Akad. Nauk SSSR* **119**, 295–298.

Bonnemaison, L. (1945). *Ann. Epiphyt.* **11**, 19–56.

Bonnemaison, L. (1967). *C. R. Hebd. Seances Acad. Sci.* **264**, 2661–2663.

Bonnemaison, L. (1970). *C. R. Hebd. Seances Acad. Sci.* **270**, 2829–2831.

Borst, D. W., and O'Connor, J. D. (1972). *Science* **178**, 418–419.

Boulay, G. (1967). *Ann. Soc. Entomol. Fr.* [N.S.] **3**, 845–858.

Bowers, W. S., and Blickenstaff, C. C. (1966). *Science* **154**, 1673–1674.

Bradshaw, W. E. (1969). *Biol. Bull.* (*Woods Hole, Mass.*) **136**, 2–8.

Bradshaw, W. E. (1971). *Bull. Ecol. Soc. Am.* **52**, 35.

Bradshaw, W. E. (1972a). *Ann. Entomol. Soc. Am.* **65**, 755–756.

Bradshaw, W. E. (1972b). *Science* **172**, 1361–1362

Bradshaw, W. E. (1973a). *Ecology* **54**, 1247–1259.

Bradshaw, W. E. (1973b) *Can. J. Zool.* **51**, 355–357.

Bradshaw, W. E. (1974). *Biol. Bull.* (*Woods Hole, Mass.*) **146**, 11–19.

Bradshaw, W. E., and Holzapfel, C. M. (1975). *Can. J. Zool.* **53**, 889–893.

Bradshaw, W. E., and Lonibos, L. P. (1972). *Can. J. Zool.* **50**, 713–719.

Braune, H. J. (1974). *In* "Effects of Temperature on Ectothermic Organisms: Ecological Implications and Mechanisms of Compensation." W. Wieser, pp. 233–238. Springer-Verlag, Berlin and New York.

Bünning, E. (1936). *Ber. Dtsch. Bot. Ges.* **54**, 590–608.

Bünning, E., and Jörrens, G. (1960). *Z. Naturforsch, Teil B* **15**, 205–213.

Bünning, E., and Jörrens, G. (1962) *Z. Naturforsch, Teil B* **17**, 57–61.

Burges, H. D. (1959). *Nature* (*London*) **186**, 1761–1762.

Bursell, E. (1967). *Adv. Insect Physiol.* **4**, 36–67.

Butterfield, J. (1976). *J. Insect Physiol.* **22**, 1443–1446.

Cantelo, W. W. (1974). *Ann. Entomol. Soc. Am.* **67**, 828–830.

Carrington, C. B., and Tenney, S. M. (1959). *J. Insect Physiol.* **3**, 402–413.

Chang, C. K., Liu, F., and Feng, H. (1964). *Acta Entomol. Sin.* **13**, 494–502.

Chen, P. S. (1971). "Biochemical Aspects of Insect Development." Karger, Basel.

Chippendale, G. M. (1977). *Annu. Rev. Entomol.* **22**, 121–138.

Chippendale, G. M., and Yin, C.-M. (1973). *Nature (London)* **246**, 511–513.
Chippendale, G. M., and Yin, C.-M. (1975). *Biol. Bull. (Woods Hole, Mass.)* **149**, 151–164.
Chippendale, G. M., and Yin, C.-M. (1976). *J. Insect Physiol.* **22**, 989–996.
Chippendale, G. M., Reddy, A. S., and Catt, C. L. (1976). *J. Insect Physiol.* **22**, 823–828.
Church, N. S. (1955). *Can. J. Zool.* **33**, 339–369.
Clark, S. H., and Platt, A. P. (1969). *J. Insect Physiol.* **15**, 1951–1957.
Clay, M. E., and Venard, C. E. (1971). *Ann. Entomol. Soc. Am.* **64**, 968–970.
Cochran, D. G. (1975). *In* "Insect Biochemistry and Function" D. J. Candy and B. A. Kilby, eds.), pp. 171–281. Wiky, New York.
Collins, J. V. (1974). *Can. J. Zool.* **52**, 639–642.
Corbet, P. S. (1956). *J. Exp. Biol.* **33**, 1–14.
Corrigan, J. J. (1970). *In* "Comparative Biochemistry of Nitrogen Metabolism" (J. W. Campbell, ed.), Vol. I, pp. 387–488. Academic Press, New York.
Cousin, G. (1932). *Bull. Biol. Fr. Belg. Suppl.* **15**, 1–341.
Cranham, J. E. (1972). *Ann. Appl. Biol.* **70**, 119–137.
Cymborowski, B., and Dutkowski, A. (1969). *J. Insect Physiol.* **15**, 1187–1197.
Danilevskii, A. S. (1965). "Photoperiodism and Seasonal Development of Insects." Oliver & Boyd, Edinburgh.
Danilevskii, A. S., Goryshin, N. I., and Tyshchenko, V. P. (1970). *Annu. Rev. Entomol.* **15**, 201–244.
Denlinger, D. L. (1971). *J. Insect Physiol.* **17**, 1815–1822.
Denlinger, D. L. (1972a). *Ann. Entomol. Soc. Am.* **65**, 410–414.
Denlinger, D. L. (1972b). *Biol. Bull. (Woods Hole, Mass.)* **142**, 11–24.
Denlinger, D. L. (1974). *Nature (London)* **252**, 223–224.
Denlinger, D. L., Willis, J. G., and Fraenkel, G. (1972). *J. Insect Physiol.* **18**, 871–882.
Depner, K. R. (1962). *Int. J. Biometeorol.* **5**, 68–71.
Depner, K. R., and Harwood, R. F. (1966). *Ann. Entomol. Soc. Am.* **59**, 7–11.
de Wilde, J. (1962). *Annu. Rev. Entomol.* **7**, 1–26.
de Wilde, J. (1969). *In* Dormancy and Survival. *Symp. Soc. Exp. Biol.*, 23rd, ed. H. W. Woolhouse, 263–284.
de Wilde, J. (1970). *Mem. Soc. Endocrinol.* **18**, 487–514.
de Wilde, J., and de Boer, J. A. (1961). *J. Insect Physiol.* **6**, 152–161.
de Wilde, J., and de Boer, J. A. (1969). *J. Insect Physiol.* **15**, 661–675.
de Wilde, J., and de Loof, A. (1973). *In* "The Physiology of Insecta," (M. Rockstein, ed.), 2nd ed., Vol. 1, pp. 11–157, Academic Press, New York.
Doane, W. W. (1973). *In* "Developmental Systems: Insects" (S. J. Counce and C. H. Waddington, eds.), vol.2, pp. 291–497. Academic Press, New York.
Duchâteau-Bosson, G., Jeuniaux, C., and Florkin, M. (1963). *Arch. Int. Physiol. Biochim.* **71**, 566–576.
Dyer, E. D. (1969). *J. Entomol. Soc. B.C.* **66**, 41–45.
Earle, N. W., and Newsom, L. D. (1964). *J. Insect Physiol.* **10**, 131–139.
Englemann, F. (1970). "The Physiology of Insect Reproduction." Pergamon, Oxford.
Englemann, W., and Shappirio, D. G. (1965). *Nature (London)* **207**, 548–549.
Fain, M. J., and Riddiford, L. M. (1975). *Biol. Bull. (Woods Hole, Mass.)* **149**, 506–521.
Ferenz, H.-J. (1975). *J. Insect Physiol.* **21**, 331–341.
Florkin, M., and Jeuniaux, C. (1974). *In* "The Physiology of Insecta" (M. Rockstein, ed.), 2nd ed., Vol. 2, pp. 255–307. Academic Press, New York.
Fraenkel, G., and Hsiao, C. (1966). *Am. Zool.* **6**, 576–577.
Fraenkel, G., and Hsiao, C. (1968a). *J. Insect Physiol.* **14**, 689–705.

Fraenkel, G., and Hsiao, C. (1968b). *J. Insect Physiol.* **14**, 707–718.
Fukuda, S. (1951); *Proc. Jpn. Acad.* **27**, 672–677.
Fukuda, S. (1952). *Annot. Zool. Jpn.* **25**, 149–155.
Fyhn, H. J., and Saether, T. (1970). *J. Insect Physiol.* **16**, 263–269.
Gibbs, D. (1975). *J. Insect Physiol.* **21**. 1179–1186.
Gilbert, L. I. (1967a). *Compr. Biochem.* **28**, 199–252.
Gilbert, L. I. (1967b). *Advn. Insect Physiol.* **4**, 69–211.
Glass, E. H. (1970). *Ann. Entomol. Soc. Am.* **63**, 74–76.
Goryshin, N. I., and Kozkova, R. N. (1967). *Zh. Obshch. Biol.* **28**, 278–290.
Goryshin, N. I., and Tyshchenko, V. P. (1969). *Dokl. Akad. Nauk SSSR* **29**, 481–498. In Russian.
Grassé, P.-P. and Lespéron, L. (1936). *C. R. Seances Soc. Biol. Ses Fil.* **122**, 1013–1015.
Gustin, R. D. (1974). *Ann. Entomol. Soc. Am.* **67**, 607–609.
Hamner, K. C., and Takimoto, A. (1964). *Am. Nat.* **98**, 295–322.
Hans, H. (1961). *Entomol. Exp. Appl.* **4**, 41–46.
Harker, J. E. (1956). *J. Exp. Biol.* **33**, 224–234.
Harker, J. E. (1961). *Annu. Rev. Entomol.* **6**, 131–146.
Harris, F. A., Lloyd, E. P., Lane, H. C., and Burt, E. C. (1969). *J. Econ. Entomol.* **62**, 854–857.
Harvey, W. R. (1962). *Annu. Rev. Entomol.* **7**, 57–80.
Harvey, W. R., and Nedergaard, S. (1964). *Proc. Natl. Acad. Sci. U.S.A.* **51**, 757–765.
Harvey, W. R., and Zerahn, K. (1971). *J. Exp. Biol.* **54**, 269–274.
Harvey, W. R., Wood, J. L., Quatrale, R. P., and Jungreis, A. M. (1975). *J. Exp. Biol.* **63**, 321–330.
Hasegawa, K., and Yamashita, O. (1965). *J. Exp. Biol.* **43**, 271–277.
Hasegawa, K., and Yamashita, O. (1967). *J. Seric. Soc. Jpn.* **36**, 297–301.
Hasegawa, K., Isobe, M., and Goto, T. (1972). *Naturwiss enschatten* **59**, 364–365.
Hasegawa, K., Isobe, M., Kubota, I., and Goto, T. (1974). *Zool. Jahrb., Abt. Allg. Zool. Physiol. Tiere* **78**, 327–332.
Haskell, J. A., Harvey, W. R., and Clark, R. M. (1968). *J. Exp. Biol.* **48**, 25–37.
Hayes, D. K., Sullivan, W. N., Cawley, B. M., Oliver, M. Z., and Schechter, M. S. (1970). *Life Sci.* Part II, **9**, 601–606.
Henneguy, L. F. (1904). "Les Insectes, Morphologie, reproduction, embryologenie. Ph.D. Dissertation, Paris, University of Paris.
Hidaka, T., and Hirai, Y. (1970). *Proc. Jpn. Acad.* **46**, 541–545.
Hidaka, T., Ishizuka, Y., and Sakagami, Y. (1971). *J. Insect Physiol.* **17**, 197–203.
Hillman, W. S. (1973). *Nature (London)* **242**, 128–129.
Hochachka, P. W., and Somero, G. N. (1973). "Strategies of Biochemical Adaptation," pp. 18–76. Saunders, Philadelphia, Pennsylvania.
Hodek, I. (1968). *Acta Entomol. Bohemoslov.* **65**, 422–435.
Hodek, I. (1971). *J. Insect Physiol.* **17**, 205–216.
Hoffmann, H. J. (1970). *J. Insect Physiol.* **16**, 629–642.
Hogan, T. W. (1961). *Aust. J. Biol. Sci.* **14**, 419–426.
Hogan, T. W. (1962). Aust. J. Biol. Sci. **15**, 538–542.
Hogan, T. W. (1964). *Aust. J. Biol. Sci.* **17**, 752–757.
Hogan, T. W. (1965). *Aust. J. Biol. Sci.* **18**, 81–87.
Hong, J. W., and Platt, A. P. (1975). *J. Insect Physiol.* **21**, 1159–1165.
House, H. L. (1967). *Can. J. Zool.* **45**, 149–153.
Hoy, M. A. (1975). *J. Insect Physiol.* **21**, 745–751.

Hoy, M. A., and Flaherty, D. L. (1970). *Ann. Entomol. Soc. Am.* **63,** 960–963.
Hoyle, G. (1974). *In* "The Physiology of Insecta" (M. Rockstein, ed.), 2nd ed., Vol. 4, pp. 176–236. Academic Press, New York.
Ichimasa, Y., and Hasegawa, K. (1973). *J. Seric. Soc. Jpn.* **42,** 380–392.
Ingram, B. R. (1975). *J. Insect Physiol.* **21,** 1909–1916.
Irvine, H. B. (1969). *Am. J. Physiol.* **217,** 1520–1527.
Isaka, S. (1952). *Zool. Mag.* **61,** 217–218.
Isobe, M., Hasegawa, K., and Goto, T. (1973). *J. Insect Physiol.* **19,** 1221–1239.
Isobe, M., Hasegawa, K., and Goto, T. (1975). *J. Insect Physiol.* **21,** 1917–1920.
Johnston, J. W., and Jungreis, A. M. (1977). *Am. Zool.* **17,** 861.
Jolly, M. S., Sinha, S. S., and Razdan, J. L. (1971). *J. Insect Physiol.* **17,** 753–760.
Judy, K. J. (1972). *Life Sci.* **11,** Part II, 605–611.
Judy, K. J., and Gilbert, L. I. (1970). *J. Morphol.* **131,** 277–300.
Jungreis, A. M. (1972). *Am. Zool.* **12,** 227 (abstr.).
Jungreis, A. M. (1973a). *Am. J. Physiol.* **224,** 27–30.
Jungreis, A. M. (1973b). *Am. Zool.* **13,** 270A.
Jungreis, A. M. (1974). *J. Comp. Physiol.* **88,** 113–127.
Jungreis, A. M. (1976a). *Comp. Biochem. Physiol. B* **53,** 201–204.
Jungreis, A. M. (1976b). *Comp. Biochem. Physiol. B* **53,** 405–413.
Jungreis, A. M. (1977). *In* "Water Relations in Membrane Transport in Plants and Animals" (A. M. Jungreis, *et al.*, ed.), pp. 89–96. Academic Press, New York.
Jungreis, A. M. (1978). *J. Insect Physiol.* **24,** 1–6.
Jungreis, A. M. and Harvey, W. R. (1975). *J. Exp. Biol.* **62,** 357–366.
Jungreis, A. M., and Tojo, S. (1973). *Am. J. Physiol.* **224,** 21–26.
Jungreis, A. M., and Vaughan, G. L. (1976). *Physiologist* **19,** 246.
Jungreis, A. M., and Vaughan, G. L. (1977). *J. Insect Physiol.* **23,** 503–509.
Jungreis, A. M., and Wyatt, G. R. (1972). *Biol. Bull. (Woods Hole, Mass.)* **143,** 367–391.
Jungreis, A. M., Jatlow, P., and Wyatt, G. R. (1973). *J. Insect Physiol.* **19,** 225–233.
Jungreis, A. M., Jatlow, P., and Wyatt, G. R. (1974). *J. Exp. Zool.* **187,** 41–46.
Jungreis, A. M., Dailey, J. C., and Hereth, M. L. (1975). *Am. J. Physiol.* **229,** 1448–1454.
Kappus, K. D., and Venard, C. E. (1967). *J. Insect Physiol.* **13,** 1007–1019.
Katsumata, F. (1968). *J. Seric. Soc. Jpn.* **37,** 453–461.
Kayser-Wegmann, I. (1975). *J. Insect Physiol.* **21,** 1065–1072.
Keeley, L. L., Moody, D. S., Lynn, D., Joiner, R. L., and Vinson, S. B. (1977). *J. Insect Physiol.* **23,** 231–234.
Kelly, T. J., and Davenport, R. (1976). *J. Insect Physiol.* **22,** 1381–1393.
King, L. L., and Benjamin, D. M. (1966). *Proc. North Cent. Branch Entomol. Soc. Am.* **20,** 139–140.
Knerer, G., and Marchant, R. (1973). *Can. J. Zool.* **51,** 105–108.
Kogure, M. (1933). *J. Dep. Agric., Kyushu Univ.* **4,** 1–5.
Kono, Y. (1970). *App. Entomol. Zool.* **5,** 213–223.
Kono, Y. (1971). *Jpn. J. Appl. Entomol. Zool.* **15,** 228–239.
Kono, Y. (1973a). *J. Insect Physiol.* **19,** 255–272.
Kono, Y. (1973b). *Appl. Entomol. Zool.* **8,** 50–52.
Kono, Y. (1975). *J. Insect Physiol.* **21,** 249–264.
Krishnakumaran, A., Berry, S. J., Oberlander, H., and Schneiderman, H. A. (1967). *J. Insect Physiol.* **13,** 1–57.
Kunkel, J. G., and Pan, M. L. (1976). *J. Insect Physiol.* **22,** 809–818.
LaFont, R., and Penneteir, J.-L. (1975). *J. Insect Physiol.* **21,** 1323–1336.

Lamberemont, E. N., Blum, M. S., and Schrader, R. M. (1964). *Ann. Entomol. Soc. Am.* **57,** 526–532.

Lees, A. D. (1955). "The Physiology of Diapause in Arthropods." Cambridge Univ. Press, London and New York.

Lees, A. D. (1964). *J. Exp. Biol.* **41,** 119–133.

Lees, A. D. (1965). *In* "Circadian Clocks" (J. Aschoff, ed.), pp. 351–356. North-Holland Publ., Amsterdam.

Lees, A. D. (1966). *Nature (London)* **210,** 986–989.

Lees, A. D., (1967). *J. Insect Physiol.* **13,** 1781–1785.

Lees, A. D. (1968). *Photophysiology* **4,** 47–137.

Lees, A. D. (1971). *In* "Biochronometry" (M. Menaker, ed.), pp. 372–380. *Nat. Acad. Sci.,* Washington, D.C.

Lees, A. D. (1972). *Circadian Rhythmicity, Proc. Int. Symp., 1971,* pp. 87–110.

Lees, A. D. (1973). *J. Insect Physiol.* **19,** 2279–2316.

Levenbook, L. R. and Hollis, V. W. (1961). *J. Insect Physiol.* **6,** 52–63.

Levenbook, L. R., Hutchins, F. N., and Bauer, A. C. (1971). *J. Insect Physiol.* **17,** 1321–1331.

Locke, M., and Collins, J. V. (1965). *J. Cell Biol.* **26,** 857–884.

Lockshin, R. A., Rosett, M., and Srokose, K. (1975). *J. Insect Physiol.* **21,** 1799–1802.

Loher, W. (1974). *J. Insect Physiol.* **20,** 1155–1172.

Lutz, P. E. (1968). *Ecology* **49,** 637–644.

Lutz, P. E. (1974). *Ecology* **55,** 370–377.

Lutz, P. E., and Jenner, C. E. (1964). *Biol. Bull. (Woods Hole, Mass.)* **127,** 304–316.

McAlister, R. O., and Fisher, F. M., Jr. (1971). *J. Parasitol.* **58,** 51–62.

McCann, F. V. (1971). *Comp. Biochem. Physiol. A.* **40,** 353–357.

MacFarlane, J. R., and Hogan, T. W. (1966). *J. Insect Physiol.* **12,** 1265–1278.

MacLeod, E. G. (1967). *J. Insect Physiol.* **13,** 1343–1349.

McLeod, D. G. R., and Beck, S. D. (1963). *Biol. Bull. (Woods Hole, Mass.)* **124,** 84–96.

McMullen, R. D. (1967). *Can. Entomol.* **99,** 42–49.

McNeil, J. N., and Rabb, R. L. (1973). *J. Insect Physiol.* **19,** 2107–2118.

Madhavan, K. (1973). *J. Insect Physiol.* **19,** 441–454.

Mangum, C. L., Earle, N. W., and Newsom, L. D. (1968). *Ann. Entomol. Soc. Am.* **61,** 1125–1128.

Mansingh, A. (1971). *Can. Entomol.* **103,** 983–1009.

Mansingh, A. (1974). *Can. J. Zool.* **52,** 629–637.

Mansingh, A., and Smallman, B. N. (1966). *Can. Entomol.* **98,** 613–616.

Mansingh, A., and Smallman, B. N. (1967). *J. Insect Physiol.* **13,** 1147–1162.

Mansingh, A., and Steele, R. W. (1973). *Can. J. Zool.* **51,** 611–618.

Mansingh, A., Steele, R. W., and Helson, B. V. (1972). *Can. J. Zool.* **50,** 31–34.

Marks, E. P., Holman, G. M., and Borg, T. K. (1973). *J. Insect Physiol.* **19,** 471–478.

Masaki, S., and Oyama, N. (1963). *Konty û* **31,** 16–26.

Maslennikova, V. A. (1974). *In* "Humoral Control of Growth and Differentiation" (J. LoBue and A. S. Gordon, eds.), Vol. 2, pp. 3–33. Academic Press, New York.

Mauchamp, B., and LaFont, R. (1975). *Comp. Biochem. Physiol. B* **51,** 445–449.

Michejda, J. W., and Thiers, R. E. (1963). *Proc. Int. Cong. Zool., 16th, 1963* Vol. 2, p. 39A.

Minis, D. H. (1965). *In* "Circadian Clocks" (J. Aschoff, ed.), pp. 333–343. North-Holland Publ., Amsterdam.

Morgan, E. D., and Poole, C. F. (1977). *Comp. Biochem. Physiol. B* **57,** 99–110.

Moriarty, F. (1969). *J. Insect Physiol.* **15,** 2069–2074.

Morohoshi, S., and Oshiki, T. (1969a). *J. Insect Physiol.* **15,** 167–175.

Morohoshi, S., and Oshiki, T. (1969b). *Proc. Jpn. Acad.* **45**, 308–313.

Müller, H. J. (1964). *Zool. Jahrb., Abt. Allg. Zool. Physiol. Tiere* **70**, 411–426.

Müller, H. J. (1965). *Zool. Anz., Suppl.* **29**, 192–222.

Müller, H. J. (1970). *Nova Acta Leopoldina* **35**, 1–27.

Mullins, D. E., and Cochran, D. G. (1976). *Comp. Biochem. Physiol. A* **53**, 393–399.

Murphy, T. A., and Wyatt, G. R. (1965). *J. Biol. Chem.* **240**, 1500–1508.

Nair, K. S. S. (1974). *J. Insect Physiol.* **20**, 231–244.

Naton, E. (1966). *Anz. Schaedlingskd.* **39**, 85–89.

Nishiitsutsuju-uwo, J., and Pittendrigh, C. S. (1968). *Z. Vergl. Physiol.* **58**, 14–46.

Nishiitsutsuju-uwo, J., Petropoulos, S. F., and Pittendrigh, C. S. (1967). *Biol. Bull.* (*Woods Hole, Mass.*) **133**, 679–696.

Normann, T. C. (1973). *J. Insect Physiol.* **19**, 303–318.

Norris, M. J. (1965). *J. Insect Physiol.* **11**, 1105–1119.

Norris, K. H., Howell, F., Hayes, D. K., Adler, V. E., Sullivan, W. N., and Schechter, M. S. (1969). *Proc. Natl. Acad. Sci. U.S.A.* **63**, 1120–1127.

Ogawa, H., and Hasegawa, K. (1975). *Insect Biochem.* **5**, 119–134.

Ohtaki, T., and Takahashi, M. (1972). *Jpn. J. Med. Sci. Biol.* **25**, 369–376.

Oldfield, G. N. (1970). *Ann. Entomol. Soc. Am.* **63**, 180–183.

Pan, M. L. (1969). The synthesis of vitellogenic blood protein in the cecropia moth. Univ. of Penn. Ph.D. Dissertation.

Pan, M. L. (1971). *J. Insect Physiol.* **17**, 677–689.

Pan, M. L. (1976). *Am. Zool.* **16**, 112.

Pan, M. L. (1977). *Biol. Bull.* (*Woods Hole, Mass.*) **153** (in press). 336–345.

Pan, M. L. (1978). In preparation.

Pan, M. L., and Wyatt, G. R. (1971). *Science,* **174**, 503–504.

Pan, M. L., Bell, W. J., and Telfer, W. R. (1969). *Science,* **165**, 393–394.

Pantyukhov, G. A. (1968). *Entomol. Rev.* (*Engl. Transl.*) **47**, 26–29.

Park, K. E. (1973). *J. Insect Physiol.* **19**, 293–302.

Park, K. E., and Seong, S. I. (1976). *J. Insect Physiol.* **22**, 201–206.

Park, K. E., and Yoshitake, N. (1970a). *J. Insect Physiol.* **16**, 1655–1663.

Park, K. E., and Yoshitake, N. (1970b). *J. Insect Physiol.* **16**, 2223–2239.

Park, K. E., and Yoshitake, N. (1971). *J. Insect Physiol.* **17**, 1305–1313.

Peterson, D. M., and Hamner, W. M. (1968). *J. Insect Physiol.* **14**, 519–528.

Phillips, J. R., and Newsom, L. D. (1966). *Ann. Entomol. Soc. Am.* **59**, 154–159.

Pichon, Y. (1974). *In* "The Physiology of Insecta" (M. Rockstein, ed.), 2nd ed., Vol. 4, pp. 101–174. Academic Press, New York.

Pittendrigh, C. S. (1966). *Z. Pflanzenphysiol.* **54**, 275–307.

Pittendrigh, C. S., and Minis, D. H. (1964). *Am. Nat.* **98**, 261–294.

Pittendrigh, C. S., and Minis, D. H. (1971). *In* "Biochronometry" (M. Menaker, ed.), pp. 212–247. *Natl. Acad. Sci.*, Washington, D.C.

Pittendrigh, C. S., Eichorn, J. H., Minis, D. H., and Bruce, D. G. (1970). *Proc. Natl. Acad. Sci. U.S.A.* **66**, 758–764.

Price, G. M. (1973). *Biol. Rev. Cambridge Philos. Soc.* **48**, 333–375.

Procter, D. L. C. (1973). *Can. J. Zool.* **51**, 1165–1170.

Propp, G. D., Tauber, M. J., and Tauber, C. A. (1969) *J. Insect Physiol.* **15**, 1749–1757.

Quatrale, R. P. (1966). Ph.D. Thesis, University of Massachusetts, Amherst.

Rabb, R. L. (1966). *Ann. Entomol. Soc. Am.* **59**, 160–165.

Rabb, R. L. (1969). *Ann. Entomol. Soc. Am.* **62**, 1252–1256.

Raina, A. K., and Bell, R. A. (1974a). *J. Insect Physiol.* **20**, 2171–2180.

Raina, A. K., and Bell, R. A. (1974b). *Environ. Entomol.* **3**, 316–318.

Riddiford, L. M. (1968). *Science* **160**, 1461–1462.
Ring, R. A. (1967a). *J. Exp. Biol.* **46**, 117–122.
Ring, R. A. (1967b). *J. Exp. Biol.* **46**, 123–136.
Ring, R. A. (1971). *Can. J. Zool.* **49**, 137–142.
Rivera, M. E. (1975). *Comp. Biochem. Physiol. B* **52**, 227–234.
Roach, S. H., and Adkisson, P. L. (1970). *J. Insect Physiol.* **16**, 1591–1597.
Roberts, S. K. (1965). *Science* **149**, 958–959.
Rock, G. C., Yeargan, D. R., and Rabb, R. L. (1971). *J. Insect Physiol.* **17**, 1651–1659.
Rosenthal, S. S., and Koehler, C. S. (1968). *Ann. Entomol. Soc. Am.* **61**, 531–534.
Roubaud, E. (1919). *Bull. Biol. Fr. Belg.* **56**, 455–544.
Sabrosky, C. W., Larson, I. and Nabours, R. K. (1933). *Trans. Kansas Acad. Sci.* **36**, 298–322.
Sacktor, B. (1975). *In* "Insect Biochemistry and Function" (D. J. Candy, and B. A. Kilby, eds.), pp. 1–88. Wiley, New York.
Salt, R. W. (1961). *Annu. Rev. Entomol.* **6**, 55–76.
Saunders, D. S. (1965). *J. Exp. Biol.* **42**, 495–508.
Saunders, D. S. (1966a). *J. Insect Physiol.* **12**, 569–581.
Saunders, D. S. (1966b). *J. Insect Physiol.* **12**, 899–908.
Saunders, D. S. (1969). *Symp. Soc. Exp. Biol.* **23**, 301–329.
Saunders, D. S. (1970). *Science* **168**, 601–603.
Saunders, D. S. (1971). *J. Insect Physiol.* **17**, 801–812.
Saunders, D. S. (1974a). *J. Insect Physiol.* **20**, 77–88.
Saunders, D. S. (1974b). *In* "The Physiology of Insecta" (M. Rockstein, ed.), 2nd ed., Vol. 2, pp. 461–533. Academic Press, New York.
Saunders, D. S. (1976). *Sci. Am.* **234**, 114–121.
Saunders, D. S. Sutton, D., and Jarvis, R. A. (1970). *J. Insect Physiol.* **16**, 405–416.
Sawchyn, W. W. (1971). Ph.D. Thesis, University of Saskatchewan, Saskatoon.
Sawchyn, W. W., and Church, N. S. (1973). *Can. J. Zool.* **51**, 1257–1265.
Schaller, F. (1965). *C. R. Seances Soc. Biol. Ses Fil.* **159**, 846–849.
Schaller, F. (1968). *J. Insect Physiol.* **14**, 1477–1483.
Schneiderman, H. A. and Horowitz, J. (1958). *J. Exp. Biol.* **35**, 520–551.
Schneiderman, H. A., and Williams, C. M. (1953). *Biol. Bull. (Woods Hole, Mass.)* **105**, 320–334.
Schoonhoven, L. M. (1962). *Arch. Neer. Zool.*, Ser. IV (B), 111–174.
Schultz, T. W., and Jungreis, A. M. (1977a). *J. Insect Physiol.* **23**, 29–32.
Schultz, T. W., and Jungreis, A. M. (1977b). *Tissue & Cell* **9**, 255–272.
Seeley, C. (1966). *J. Lepid. Soc.* **20**, 47–48.
Sehnal, F., and Granger, N. A. (1975). *Biol. Bull. (Woods Hole, Mass.)* **148**, 106–116.
Shappirio, D. G. (1974). *J. Insect Physiol.* **20**, 291–300.
Shappirio, D. G., and Harvey, W. R. (1965). *J. Insect Physiol.* **11**, 305–327.
Shappirio, D. G., and Williams, C. M. (1957a). *Proc. R. Soc. London, Ser. B* **147**, 218–232.
Shappirio, D. G., and Williams, C. M. (1957b). *Proc. R. Soc. London, Ser. B* **147**, 233–246.
Shelford, V. E. (1929). "Laboratory and Field Ecology." Williams and Wilkins, Baltimore.
Sicker, W. (1964). *Z. Morphol. Oekol. Tiere* **54**, 107–140.
Siew, Y. C. (1965a). *Nature (London)* **205**, 523–524.
Siew, Y. C. (1965b). *J. Insect Physiol.* **11**, 1–10.
Siew, Y. C. (1965c). *J. Insect Physiol.* **11**, 463–479.
Siew, Y. c. (1966). *Trans. Entomol. Soc. London* **118**, 359–374.
Silver, G. T. (1958). *Can. Entomol.* **90**, 65–80.
Siverly, R. E. (1972). *Environ. Entomol.* **1**, 526.

Smith, J. C., and Newsom, L. D. (1970). *Ann. Entomol. Soc. Am.* **63**, 460–462.

Smith, S. M., and Brust, R. A. (1971). *Can. J. Zool.* **49**, 1065–1073.

Sømme, L. (1967). *J. Insect Physiol.* **13**, 805–814.

Sømme, L., and Velle, W. (1968). *J. Insect Physiol.* **14**, 135–143.

Sonobe, H., and Ohnishi, E. (1970). *Dev., Growth & Differ.* **12**, 41–52.

Steinberg, D. M., and Kamensky, S. A. (1936). *Bull. Biol.* **70**, 165–169.

Stevenson, E., and Wyatt, G. R. (1962). *Arch. Biochem. Biophys.* **99**, 65–71.

Stevenson, E., and Wyatt, G. R. (1964). *Arch. Biochem. Biophys.* **108**, 420–429.

Stoffolano, J. G., Jr. (1968). *J. Insect Physiol.* **14**, 1205–1214.

Stoffolano, J. G., Jr. (1973). *J. Gerontol.* **28**, 35–39.

Stoffolano, J. G., Jr. (1974). *Can. J. Zool.* **52**, 981–988.

Stoffolano, J. G., Jr., and Matthysse, J. G. (1967). *Ann. Entomol. Soc. Am.* **60**, 1242–1246.

Stoffolano, J. G., Jr., Greenberg, S., and Calabrese, E. (1974). *Ann. Entomol. Soc. Am.* **68**, 518–519.

Sullivan, C. R., and Wallace, D. R. (1965). *Can. J. Zool.* **43**, 233–245.

Sullivan, C. R., and Wallace, D. R. (1967). *Can. Entomol.* **99**, 834–850.

Tadmor, U., and Applebaum, S. W. (1971). *J. Insect. Physiol.* **17**, 1211–1225.

Takeda, S., and Hasegawa, K. (1975). *J. Insect Physiol.* **21**, 1995–2003.

Tauber, M. J., and Tauber, C. A. (1969). *Can. Entomol.* **101**, 364–370.

Tauber, M. J., and Tauber, C. A. (1970a). *Science* **167**, 170.

Tauber, M. J., and Tauber, C. A. (1970b). *J. Insect Physiol.* **16**, 2075–2080.

Tauber, M. J., and Tauber, C. A. (1973). *J. Insect Physiol.* **19**, 1455–1463.

Tauber, M. J., and Tauber, C. A. (1974). *Can. Entomol.* **106**, 969–978.

Tauber, M. J., and Tauber, C. A. (1975a). *Chronobiologia* **1**, 126–131.

Tauber, M. J., and Tauber, C. A. (1975b). *Chronobiologia* **1**, 152–160.

Tauber, M. J., and Tauber, C. A. (1976a). *J. Insect Physiol.* **22**, 331–335.

Tauber, M. J., and Tauber, C. A. (1976b). *Annu. Rev. Entomol.* **21**, 81–107.

Tauber, M. J., Tauber, C. A., and Denys, C. J. (1970a). *Can. Entomol.* **102**, 474–478.

Tauber, M. J., Tauber, C. A., and Denys, C. J. (1970b). *J. Insect Physiol.* **16**, 949–955.

Teetes, G. L., Adkisson, P. L., and Randolph, N. M. (1969). *J. Insect Physiol.* **15**, 755–761.

Telfer, W. H. (1965). *Annu. Rev. Entomol.* **10**, 161–184.

Telfer, W. H. (1954). *J. Gen. Physiol.* **37**, 539–558.

Telfer, W. H., and Williams, C. M. (1953). *J. Gen. Physiol.* **36**, 389–413.

Telfer, W. H., and Williams, C. M. (1960). *J. Insect Physiol.* **5**, 61–72.

Theron, P. P. A. (1943). *J. Entomol. Soc. South. Afr.* **6**, 114–125.

Thiele, H. U. (1966). *Z. Angew. Entomol.* **58**, 143–149.

Thiele, H. U. (1968a). *Naturwiss. Rundsch.* **21**, 57–65.

Thiele, H. U. (1968b). *Zool. Anz., Suppl.* **31**, 358–364.

Thiele, H. U. (1973). *Can. Entomol.* **105**, 925–928.

Thomson, J. A. (1975). *Adv. Insect Physiol.* **11**, 321–398.

Thurston, R. (1972). *Environ. Entomol.* **1**, 638–640.

Tojo, S. (1971). *Insect Biochem.* **1**, 249–263.

Tojo, S., and Hirano, C. (1968). *J. Insect Physiol.* **14**, 1121–1133.

Tojo, S., and Yushima, T. (1972). *J. Insect Physiol.* **18**, 403–422.

Treherne, J. E. (1976). *Perspect. Exp. Biol., Proc. Anniv. Meet. Soc. Exp. Biol., 50th, 1974* Vol. 1, pp. 323–330.

Truman, J. W. (1971). *In* "Biochronometry" (M. Menaker, ed.), pp. 483–504. Natl. Acad. Sci., Washington D.C.

Truman, J. W. (1972). *Z. Vergl. Physiol.* **76**, 32–40.

Tsuji, H. (1963). Ph.D. Thesis, pp. 1–76. Kyushu University, Kynshu, India.

Tsujita, M., and Sakurai, S. (1965a), *Proc. Jpn. Acad.* **41,** 225–229.
Tsujita, M., and Sakurai, S. (1965b). *Proc. Jpn. Acad.* **41,** 230–235.
Turbeck, B. O. (1974). *Tissue & Cell* **6,** 627–640.
Valder, S. M., Hopkins, T. L., and Valder, S. A. (1969). *J. Insect Physiol.* **15,** 1199–1214.
Vaughan, G. L., and Jungreis, A. M. (1976). *Am. Zool.* **16,** 127A.
Vaughan, G. L., and Jungreis, A. M. (1977). *J. Insect Physiol.* **23,** 585–589.
Villacorta, A., Bell, R. A., and Callenbach, J. A. (1971). *Ann. Entomol. Soc. Am.* **65,** 419–422.
Vinogradova, E. B. (1974). *J. Insect Physiol.* **20,** 2487–2496.
Vinogradova, E. B., and Zinovjeva, K. B. (1971). *Proc. Int. Congr. Entomol. 13th, 1968* pp. 456–457.
Waldbauer, G. P., and Sternberg, J. G. (1973). *Biol. Bull. (Woods Hole, Mass.)* **145,** 627–641.
Wallace, D. R., and Sullivan, C. R. (1966). *Can. J. Zool.* **44,** 147.
Weevers, K. deG. (1966). *J. Exp. Biol.* **44,** 163–175.
Wellso, S. G., and Adkisson, P. L. (1966). *J. Insect Physiol.* **12,** 1455–1465.
Wheeler, M. W. (1893). *J. Morph.* **8,** 1.
Whitmore, D., Jr., Applebaum, S. W., and Gilbert, L. I. (1973). *J. Insect Physiol.* **19,** 349–354.
Wiens, A. W., and Gilbert, L. I. (1967). *Comp. Biochem. Physiol.* **21,** 145–159.
Wilhelm, R. C., Schneiderman, H. A., and Daniel, L. J. (1961). *J. Insect Physiol.* **7,** 273–288.
Wilkens, J. L. (1968). *J. Insect Physiol.* **14,** 927–943.
Williams, C. M. (1946). *Biol. Bull. (Woods Hole, Mass.)* **90,** 234–243.
Williams, C. M. (1947). *Biol. Bull. (Woods Hole, Mass.)* **93,** 89–98.
Williams, C. M. (1952). *Biol. Bull. (Woods Hole, Mass.)* **103,** 120–138.
Williams, C. M. (1956). *Biol. Bull (Woods Hole, Mass.)* **110,** 201–218.
Williams, C. M. (1963). *Science* **140,** 386.
Williams, C. M. (1969). *Symp. Soc. Exp. Biol.* **23,** 285–300.
Williams, C. M., and Adkisson, P. L. (1964). *Biol. Bull. (Woods Hole, Mass.)* **127,** 511–525.
Williams, C. M., Adkisson, P. L., and Walcott, C. (1965). *Biol. Bull. (Woods Hole, Mass.)* **128,** 497–507.
Williams-Boyce, P. K. (1977). "Regulation of fat body urate accumulation in the development of the tobacco hornworm, *Mandura sexta*, M. S. Thesis, Knoxville, Univ. of Tenn.
Williams-Boyce, P. K. and Jungreis, A. M. (1977). *Amer. Zool.* **17,** 862.
Wilson, G. R., and Larsen, J. R. (1974). *J. Insect Physiol.* **20,** 2459–2473.
Winfree, A. T. (1972). *J. Comp. Physiol.* **77,** 418–434.
Wingfield, M., and Warren, L. O. (1972). *Kans. Entomol. Soc.* **45,** 1–6.
Wood, J. L., and Harvey, W. R. (1976). *J. Exp. Biol.* **65,** 347–360.
Wood, J. L., Jungreis, A. M., and Harvey, W. R. (1975). *J. Exp. Biol.* **63,** 313–320.
Wright, J. E. (1970). *Ann. Entomol. Soc. Am.* **63,** 1273–1275.
Wright, J. E., and Venard, C. E. (1971). *Ann. Entomol. Soc. Am.* **63,** 11–13.
Wyatt, G. R. (1967). *Adv. Insect Physiol.* **4,** 287–360.
Wyatt, G. R. (1972). *In* "Biochemical Action of Hormones" (G. Litwack, ed.), Vol. 2, pp. 385–490. Academic Press, New York.
Wyatt, G. R. (1975). *Verh. Dtsch. Zool. Ges.* pp. 209–226.
Wyatt, G. R. and Meyer, W. L. (1959). *J. Gen. Physiol.* **42,** 1005–1011.
Wyatt, G. R., and Pan, M. L. (1978). *Annu. Rev. Biochem.* **47,** 779–817.
Wyatt, G. R., and Kropf, R. B., and Carey, F. G. (1963). *J. Insect Physiol.* **9,** 137–152.
Yagi, S., and Akaike, N. (1976). *J. Insect Physiol.* **22,** 389–392.
Yagi, S., and Fukaya, S. (1974). *Appl. Entomol. Zool.* **9,** 247–255.
Yamashita, O., and Hasegawa, K. (1964). *J. Seric. Soc. Jpn.* **33,** 115–123.
Yamashita, O., and Hasegawa, K. (1966). *J. Insect Physiol.* **12,** 957–962.

Yamashita, O., Hasegawa, K., and Seki, M. (1972). *Gen. Comp. Endocrinol.* **18**, 515–523.
Yin, C. -M., and Chippendale, G. M. (1973a). *Ann. Entomol. Soc. Am.* **66**, 943–947.
Yin, C. -M., and Chippendale, G. M. (1973b). *J. Insect Physiol.* **19**, 2403–2420.
Yin, C. -M., and Chippendale, G. M. (1974). *J. Insect Physiol.* **20**, 1833–1847.
Yin, C. -M., and Chippendale, G. M. (1976). *J. Exp. Biol.* **64**, 303–310.
Yoshitake, N., and Aruga, H. (1952). *J. Seric. Soc. Jpn.* **21**, 7–14.
Zabirov, S. M. (1961). *Entomol. Rev. (Engl. Transl.)* **40**, 148–151.
Zaslavskii, V. A., and Bogdanova, J. P. (1968). *Rev. Appl. Entomol.* **56**, 221.
Ziegler, R., and Wyatt, G. R. (1975). *Nature (London)* **254**, 622–623.

Oats, peas, beans and barley grow
Oats, peas, beans and barley grow
Will you or I or anyone know
How oats, peas, beans and barley grow?
Anonymous

3

Control Mechanisms for Plant Embryogeny

VIRGINIA WALBOT

DORMANCY AND DEVELOPMENTAL ARREST

I. INTRODUCTION

Despite the general trend of scientists embracing an ever more narrow focus in their work, my graduate and postgraduate years were recently spent in two laboratories where the large and often intractable questions of developmental biology have been the binding theme of diverse experimental approaches to the problem of the regulation of plant embryo development. Developmental biology asks the same question, in various forms, that is a stumping riddle to youngsters: Which came first, the chicken or the egg? To a developmental biologist, no plant, animal, or microbe can be perceived as existing solely at one point in time; there has been a preexisting state from which the organism progressed to its present form. Scientists engaged in studies of molecular biology view this progression as a series of discrete, steplike changes in the expression of particular genes. Changes in morphology, anatomy, and physiological behavior remain, however, the evidence that an organisms has indeed arrived at a new developmental stage.

Embryo and seed development is the feature of plant development most easily explained to other developmental biologists since a new individual progresses through an orderly series of morphological stages. Embryo development and embryo developmental arrest are particularly satisfying concepts to developmental biologists in comparison to the reiterative elaboration of organs and growth of vegetative plants.

Embryo and seed development in all species has several common processes. New cells are produced from a zygote, and these cells give rise to an organ- and tissue-containing plantlet. Typically the developed embryo contains all of the tissues of the vegetative plant, and through the various meristems retains the capability of producing additional tissues and organs during vegetative growth. Before expression of this capacity for growth and continued elaboration of the plant body, the plantlet–embryo typically enters a period of metabolic quiescence.

The purpose of developmental arrest is usually considered to be protection for the embryo during dissemination of seeds and during periods environmentally unfavorable to the successful germination and establishment of a seedling. The means by which embryo and seeds maintain this quiescent or dormant state is the subject of numerous investigations on mechanisms to promote germination or impose dormancy (reviewed by Amen, 1968; Barton, 1965a, b; Mayer and Shain, 1974; Koller et al., 1962; Wareing and Saunders, 1971; Berlyn, 1972; Villiers, 1972).

Developmental arrest in angiosperm embryos is prefaced by changes in physiology and anatomy, which are the concern of this chapter. Entry into developmental arrest requires regulation of three tandem developmental processes:

1. Completion of the developmental program, leading to a finished embryo–plantlet capable of germination and growth

2. Completion of the developmental program that prepares the completed embryo for survival during desiccation, anaerobiosis, or other adverse conditions imposed on the embryo by the surrounding seed tissue or by the environment

3. Completion of the developmental program that primes the embryo to perceive and respond to appropriate environmental signals, thus ensuring germination under favorable conditions

These three developmental programs adequately describe most angiosperm embryos; some exceptions are discussed in Section II,D. The primary aim of this chapter is to draw attention to the interplay between these developmental programs and the factors that regulate developmental arrest. As a prologue to the description of experimental results, the next section outlines a series of thematic questions important in the discussion of the existing data.

A. Questions Discussed in This Chapter

1. MATERNAL VERSUS EMBRYONIC CONTROL OF DEVELOPMENT

Within the brief outline of embryo development just presented are several interesting questions. First, what is the relationship between the developing embryo and the maternal tissue? Is maternal tissue required for the normal development of the embryo and for the imposition of dormancy? Can the biochemical and anatomical control of embryo development be duplicated *in vitro*?

That the vegetative plant plays a major role in providing for the normal development of the embryo should not be surprising. The embryo is dependent on the vegetative plant for photosynthetic products, water, and minerals and for the perception of photoperiod and other environmental cues. The successful development in culture of exceedingly young embryos is rare due to the apparent complexity of required nutrients. Excised fern embryos grown in culture usually germinate precociously.

Raghavan and Torrey (1963) postulated that the embryo becomes increasingly independent of the maternal plant and surrounding nutritive tissue as it develops and that it becomes more successful in autotrophic growth or in nutrient-supplemented culture. However, excised embryos or regenerating cells undergoing embryoid formation may conclude structural and biochemical differentiation requisite for germination before the expression of germinability. If immature embryos of *Phaseolus* are placed in a water and salts medium comparable to that encountered during normal postdormancy germination, there is a lower percentage of germination than if the medium is supplemented with sugars (Walbot *et al.*, 1972a); cotyledon-stage embryos germinate equally well without sugar supplementation. Immature embryos and embryoids may have an insufficient supply of energy reserves to support germination.

The regeneration *in vitro* of single cells to plantlets proves that a single zygote-like cell can divide to form a mass of cells capable of histodifferentiation and normal organ development. The morphology of induced embryogenesis is in many respects similar to that of normal embryo ontogeny. Therefore, the maternal tissues are not required for the completion of developmental program 1, the specification of an embryo. Induced embryos or excised embryos grown in culture typically germinate precociously after histodifferentiation. The cotyledons remain rudimentary as the embryo–plantlet is provided with nutrients directly by the medium and is therefore independent of the reserve materials usually contained in the cotyledons. Such embryos are incapable of withstanding stresses, such as severe dehydration, which the normal embryo can after the completion of developmental program 2, embryo maturation.

The maternal tissue within which an embryo develops and the nutrients and hormones supplied by the vegetative plant must promote the continued growth of the embryo as such and prevent germination. The role of maternally supplied abscisic acid (ABA), nutrients, and water will be discussed subsequently in relation to the regulation of precocious germination.

2. EMBRYO DEVELOPMENT AND DEVELOPMENTAL ARREST

Other questions concern the interrelationship among the three developmental programs. Are there embryos that do not enter a period of developmental arrest? Is the order of changes in embryo physiology preceding developmental arrest similar in different species? Does a single stimulus control all three developmental programs, resulting in their coordinate regulation? The enormous variability of seed types, dispersal mechanisms, requirements for germination, etc., may obscure the existence of basic principles common to all embryo types. Such principles might include a fixed temporal order in which major changes in embryo morphology and physiology occur or the exogenous inhibition of embryo growth resulting in quiescence. Much of the variability in the extent of embryo development, as in the contrast between monocotyledonous and dicotyledonous species, could be due to the length of embryo development before the imposition of dormancy. Similarly, the balance between imposition and exit from dormancy may be dependent on the length of embryogeny. This balance may determine the extent to which the embryo is constrained from germinating, i.e., may determine how much time or what treatments are required for germination.

The description in the next section of the experimental data on the regulation of embryo development addresses these primary questions. These data are selected from only a few examples and do not comprise an extensively comparative description of the events preceding developmental arrest. This selectivity is an effort in part to avoid duplication of existing comparative descriptions (Maheshwari, 1950; Bhatnagar and Johri, 1972) and in part to provide detailed support for several hypotheses. These hypotheses are briefly discussed here to provide an introduction to the descriptive section.

First, the existence of vivipary (*in vivo* precocious germination), *in vitro* precocious germination, and embryos such as mangrove which fail to show any quiescent metabolic period suggests that developmental arrest is an imposed condition and not an inevitable result of embryo development. Typically plant embryo development is conceived of as a linear progression from zygote formation through germination. A cascade regulation is often invoked to explain the transition from one stage to the next, and a certain amount of predestination is implied in the "inevitable" progression to the next developmental stage. Dormancy, in particular, is seen as the inevitable consequence of the preceding development and maturation of the embryo:

Histodifferentiation → Growth → Maturation → Dormancy → Germination

However, a hypothesis of this chapter is that dormancy or developmental arrest is really a bypass of the normal developmental program, which would proceed directly from histodifferentiation to germination if certain physical and environmental restraints were not placed on the embryo. The scheme used to designate this way of looking at embryo development has only a closed loop leading in and out of dormancy:

Embryos cannot move backward from germination to dormancy, but they can move directly from embryogeny to germination by bypassing dormancy altogether.

If we accept as inevitable the progression directly from embryo to seedling then only an antigermination agent can act as a negative control to induce temporary quiescence. Abscisic acid seems a likely compound to prevent germination in many seed systems, and consequently considerable attention is given to the possible modes and sites of action of this hormone during embryo and early postembryonic development. If ABA or a similar agent can impose dormancy then the timing of ABA action and ABA destruction will determine the length of embryogeny and the depth of the quiescent period. The existing data are discussed in light of the hypothesis that ABA may be a major regulatory molecule in embryo maturation and arrest.

B. Dormancy versus Developmental Arrest

The term "dormancy" is not really appropriate to describe many angiosperm embryos. Typically dormancy is considered a state in which some external stimulus is required for the embryo to germinate; by this definition many embryos are not truly dormant. However, many angiosperm embryos enter a period of developmental arrest in which growth is not possible until rehydration or gas exchange is possible. Such embryos germinate as soon as the environmental conditions are favorable and do not require special treatments designed to overcome seed dormancy and promote germination.

One hypothesis of this chapter is that the time of destruction of the agents imposing developmental arrest is highly variable. Such agents may never appear, as in the case of *in vitro* embryo development in which there is no quiescence. If quiescence precedes destruction of the agents imposing arrest, that is if developmental program 2 is finished before program 3, external factors such as temperature or leaching may be required to remove the restriction to germination. Alternatively, the conditions that elicited embryo quiescence may be destroyed before developmental arrest so that the embryo can germinate immediately. In this case, developmental program 3 promoting germination is activated before seed maturation is complete, but germination does not occur due to physical limitations such as lack of water.

Genetically controlled vivipary in citrus, corn, and other species may involve systems in which a precession of the timing of the destruction of agents imposing quiescence has occurred. The difference between dormant, quiescent, and viviparous embryos would therefore be solely a difference in the timing of developmental changes rather than an actual fundamental difference in the developmental processes taking place. The experimental evidence on the timing of developmental events will be reviewed with this hypothesis in mind.

II. DESCRIPTIONS OF PLANT EMBRYO DEVELOPMENT

A. Phaseolus

1. DEVELOPMENTAL TIMETABLES

A simple approach to the description of a developing system is to measure the concentration or level of a particular compound during development. This approach has been used extensively by physiologists to catalog the onset of synthesis and rise and fall of particular metabolites; data on seed physiology were recently reviewed by Dure (1975). This type of measurement, however, typically reveals little about the underlying mechanisms by which development is regulated. From such data, however, it is possible to construct a developmental timetable in which the order of changes can be more clearly seen. From the order in which metabolic processes change it is possible to deduce a scenario of cause and effect relationships, a useful first step in postulating how major shifts in developmental program might be regulated.

Figure 1 shows the developmental timetables for two interfertile bean species. These tables were constructed from measurements of gross morphological characters such as color, weight, and length. However, beans sorted by such characters can be matched within 1 day of their developmental age. Consequently, large amounts of tissue are available for biochemical measurement of stage-specific characters.

Using the developmental timetable we can divide development into stages in which major events occur. Typically embryo development is subdivided into the stages shown in the tabulation below:

Stage	Characteristics
1. Cleavage and histo-differentiation	Cell division with little growth; differentiation of all major tissues
2. Growth	Rapid cell expansion and division
3. Maturation	Little or no cell division or expansion; synthesis and storage of reserved materials
4. Dormacy	Developmental arrest
5. Germination	Renewed cell expansion and division; embryo growth

This division of embryo ontogeny into specific stages brings into sharper focus the following fundamental questions:

1. Why does development continue after histodifferentiation? Since all of the tissues found in the seedling are present (root and shoot meristems, root, shoot, true leaves, vascular tissues, etc.), why does the embryo continue to enlarge within the seed when histodifferentiation is complete?

2. What promotes the accumulation of stored reserve materials?

3. What triggers the maturing embryo to become quiescent?

4. What signal does the quiescent embryo receive that terminates dormancy?

A general difficulty with each of these questions is the problem of distinguishing cause and effect during critical transition periods. Rough division of development into a few stages does not allow us to conclude, for

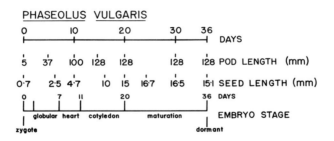

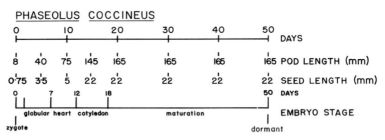

Fig. 1. Postfertilization developmental timetable for *Phaseolus* species. (From Walbot *et al.*, 1972a.)

example, whether changes in hormonal level occur before or after changes in the metabolism of the embryo. It is clear that samples must be collected as often as feasible in order to set an unambiguous temporal order to the developmental process.

Distinguishing between cause and effects is a major stumbling block to developmental biology, particularly in plant biology, in which the molecular basis of action of plant hormones is poorly understood. My own prejudice is that changes in water potential and membrane permeability, whether hormone mediated or not, will ultimately be shown to control embryo development. Many of the effects that can be easily measured, such as changes in weight, water content, macromolecular synthesis, etc., are secondary. Biophysical constraints on embryo growth have received less attention than biochemical description. Such factors as gas and water vapor partial pressure and physical restraint may be quite important in determining growth rate and pattern, but there is very little information available on these topics.

The expression of data collected at different developmental stages also has inherent problems. For example, water content can be expressed as a percentage of fresh weight, fresh weight, water potential, water content per organ, or water content per cell. Comparison of Figs. 2 and 5 shows that these ways of expressing the data yield substantially different pictures of the hydration state of developing *Phaseolus* cotyledons. Since the accumulation of dry matter is low during early development and substantial during maturation, the percent fresh weight falls whereas water content per organ, water content per cell, and fresh weight increase. The content of specific metabolites can also be expressed in several ways: molarity, cellular content, or per organ. In the case of phosphate ion these different means of expressing the data yield different graphic representations of phosphate availability (Fig. 3).

As Walker (1974) discussed, data calculation on a per cell basis is probably the most significant way of expressing developmental data when the data are used to understand regulation. The primary reason for preferring data on a per cell basis, either in concentration or amounts, is that

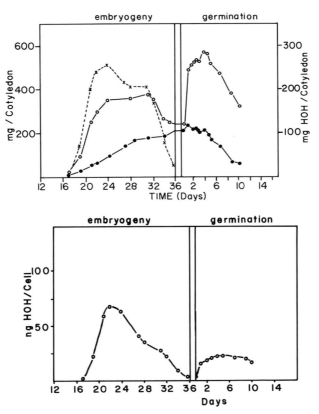

Fig. 2. Water content of *Phaseolus vulgaris* cotyledons. (A) Fresh weight (O), milligram per cotyledon; dry weight (●), milligram per cotyledon; and water content (×), milligram H_2O per cotyledon. (B) Cellular water (O), nanogram H_2O per cell. (From Walker, 1974.)

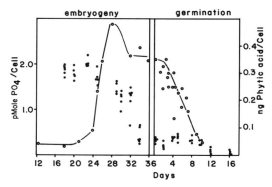

Fig. 3A. The cellular content of phytic acid (O) and inorganic phosphate (●) during embryogeny and germination. (From Walker, 1974.)

the changes in metabolism of the organ are a reflection of changes in cellular metabolism. In maturing embryos there is little cell division, so that changes in metabolite concentration reflect changes in the preexisting cell population. The per cell and per organ measurements are interchangeable if cell number is known at each stage. Percent water content as a basis for data reporting is the least accurate reflection of intracellular events. It is clear that despite the loss of percent fresh weight after day 20 shown in Fig. 5c, the amount of water per cell actually increases until day 22 (Fig. 2B). Phosphate ion concentration actually decreases due to dilution of the

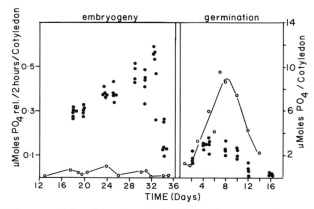

Fig. 3B. Phytase activity (O) and orthophosphate (●) content of the cotyledon during ontogeny. The phytase activities observed during embryogeny are nonsignificant because they are less than the sensitivity of the test. The lower limit of the test is 0.05 μmole PO_4 released per cotyledon under these conditions of assay. (From Walker, 1974.)

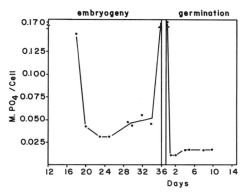

Fig. 3C. The molar concentration of intracellular phosphate in the cotyledon cell during embryogeny and germination. (From Walker, 1974.)

cellular contents, although the cells continue to accumulate phosphate. These observations are discussed in more detail in Section II,A,3.

2. MACROMOLECULAR SYNTHESIS

One approach to the cause and effect problem is to examine a particular developmental stage in sufficient detail so as to construct a timetable in which the order of biochemical events can be defined. In *Phaseolus vulgaris* the transition from the cotyledon stage to maturation occurs over a period

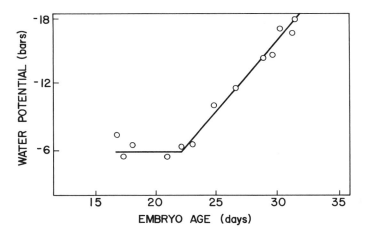

Fig. 4. Water potential of *Phaseolus vulgaris* axes during embryogeny. The water potential of axes after day 30 rapidly approaches that of the fully dehydrated seed, a value that has been estimated at greater than −1000 bars (Manohar, 1966). (Data from I. Sussex, personal communication.)

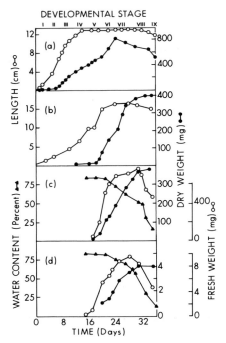

Fig. 5. Aspects of postfertilization development length and dry weight of *Phaseolus vul-garis* pods (a) and seeds (b) and water content and fresh and dry weight of cotyledons (c) and axes (d). (Adapted from Walbot *et al.*, 1972a.)

of several days centered on day 20. During this transition, pod length and seed length are fixed and maximal (Fig. 1 and 5), cell division in both the axis and cotyledonary cells is rare, and the embryo becomes capable of immediate germination when removed from its seed coat and exposed to water (Walbot *et al.*, 1972a). Starch and storage protein, although previously present at low levels, are now rapidly accumulated in the cotyledon (Öpik, 1968), and there is a linear increase in dry weight from day 16 onward until desiccation (Walker, 1974). Similarly the synthesis of phytic acid, which is codeposited in the protein bodies with storage protein (Ory and Henningsen, 1969; Tronier *et al.*, 1971; Hofstein, 1973), is detectable before day 20 at low levels; after day 20 there is a rapid increase in phytic acid synthesis (Fig. 6). Nucleotide, orthophosphate, and ATP pool sizes decrease during the transition from cotyledonary to maturation stage (Fig. 7 and 8). The rate of RNA synthesis reaches a maximum on day 18, just before the maturation, and falls rapidly after day 20 to nearly undectectable

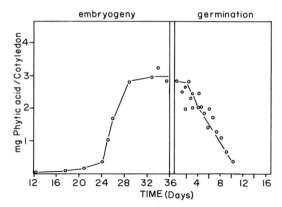

Fig. 6. Phytic acid content of *Phaseolus vulgaris* cotyledons during ontogeny. (From Walker, 1974.)

levels at day 28 (Fig. 1). Can these various parameters be related to one another?

The first parameter to stabilize is pod length, which is fixed from day 13 onward; the cavities within which the embryos develop continue to expand for several additional days, however. Not until day 18 do the developing seeds fill the pod cavities (Walbot *et al.*, 1972a). The maximal size of the

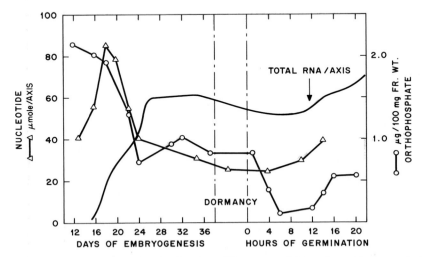

Fig. 7. Nucleotide (Δ) and orthophosphate (O) content of *Phaseolus vulgaris* axes during development. The pattern of total RNA per axis is also shown. (From Walbot, 1971.)

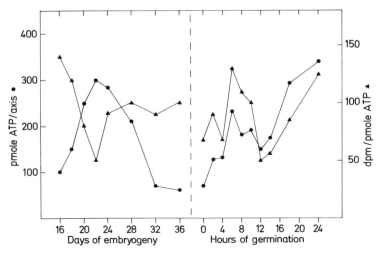

Fig. 8. The content of ATP per *Phaseolus vulgaris* axis (●) and the specific activity of the ATP pool after a 1-hr labeling period with [2-,8-³H]adenosine (▲). (From Walbot, 1972.)

seeds is reached some days later, however, due to a shape change during maturation. The change of shape occurs after the pod has softened and is beginning to degenerate, and the seeds are not really constrained within the seed cavities as they are at days 18–20.

There is also a considerable lateral expansion of the cotyledons during this period. It has not been determined whether cell division or expansion determines the changes in cotyledonary shape during maturation. The expansion of the embryo and seed before day 16 is due primarily to rapid cell proliferation in the cotyledon. Up to day 16 cotyledon cell number doubles approximately every 24 hrs based on the assumption that cotyledons containing approximately 10^5 cells each are differentiated on day 11 (Fig. 1) and that the cotyledonary cell number reaches 8×10^5 on day 15. At day 16 there is a dramatic change in the rate of cell division within the cotyledon as shown in Fig. 9. The next cell doubling requires 8 days (days 16–24), and a final cell doubling occurs between days 24 and 32. Cotyledonary dry weight doubles approximately every 4 days during maturation (Fig. 5) so that a substantial portion of the apparent weight increase is due to an increase in dry weight accumulation per cell.

Concomitant with the change in the rate of cell division at day 16 there is a dramatic increase in the amount of water per cotyledonary cell (Fig. 3C). The considerable hydration over days 16–22 serves to dilute many intracellular metabolites, i.e., nucleotides and phosphate. Water content per cell is approximately threefold higher on day 22 than during germination, when the cotyledons are turgid for approximately 1 week. This comparison suggests that considerable hydrostatic pressure should be exerted by the cotyledons against surrounding pod and seed coat tissue from day 16 to day 22. Calculation from the data on axis water content and cell number also suggests that hydration occurs. The rapid hydration over days 16–20 does not lead to a decrease in axis water potential, however, a parameter that remains constant at −6 bars (Fig. 4). Hydration must therefore be offset by similar increases in the content of osmotically active molecules such as precursors for starch, protein, phytic acid, and nucleic acid synthesis.

Likely osmotic molecules in developing *Phaseolus* include sugars, amino acids, inorganic ions, nucleotides, phytic acid, etc. As discussed in more detail later, both phosphate ion and nucleotide decrease in molarity rapidly during the hydration phase of days 16–20, although the amounts of these materials increase on a per organ or per cell basis. In other bean species, sugars and amino acids increase more rapidly on a fresh weight basis than do storage proteins or starch (Briarty *et al.*, 1969). Flinn and Pate (1968) have calculated that the pods, seed coat, and endosperm of maturing peas can contribute no more than 20% of the dry weight or 40% of the nitrogen found in the mature seeds. The rapid influx of small molecules is therefore from the vegetative plant, not from the degenerating fruit tissue. *In vivo* it is

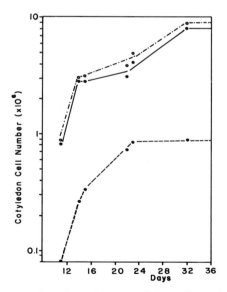

Fig. 9. The change in *Phaseolus vulgaris* cotyledon cell number during embryogeny. Epidermal cell number (---), internal cell number (——), and total cell number (—·—·). (From Walker, 1974.)

probable that the rapid uptake of these precursor molecules from the vascular tissue during days 16–20 is accompanied by and balanced by water uptake to maintain water potential at -6 bars, rather than an influx of water balanced by solute uptake. These solutes are rapidly absorbed since they are the precursor materials for the synthesis of the various stored reserves deposited in the cotyledons and axis tissues.

During later maturation stages water potential does increase despite the rapid polymerization of monomeric molecules, which should serve to maintain water potential by two distinct mechanisms: Water is a by-product of the polymerization reactions, and each polymerization removes an osmotically active small molecule. However, some solutes are not extensively polymerized in *Phaseolus* since nearly 10% of the nitrogen content of the seed is found in various amino acids and derivatives. These compounds, including pipecolic acid (6.8 mg/g dry weight) and γ-glutamyl-*s*-methylcysteine (3 mg/g dry weight), are found in extremely large amounts up to 14 mg/g dry weight (Zacharius, 1970). Phytic acid levels reach 3 mg per cotyledon and are probably matched by nearly an equal weight of calcium and magnesium complexed to the acid. In sum these molecules and the remaining unpolymerized precursors constitute approximately 10% of seed dry weight, resulting in the attainment of extremely negative water potential when seed dehydration begins.

Since water potential does fall during maturation the loss of water must exceed the rate of polymerization of solutes. The mechanisms by which this water loss, a real reduction in water content per cell and per organ, is achieved are not known. The rate of evaporative water loss through the seed and fruit tissues is not known. If evaporative loss is substantial throughout development, two events occur during maturation that provide for rapid, irreversible water loss from the seed. First, the pod dries earlier than the seed tissue, giving less protection against evaporative water loss from the seed and fruit air spaces. Second, the degeneration of the funicular vascular connection of the seed during maturation ultimately restricts the entry of water into the seeds; concurrently the formation of an abscision zone between the fruit and stem restricts water entry into the entire fruit.

The changes in water metabolism that occur during maturation are correlated with the completion of developmental programs 1 and 2. Early in maturation the embryo attains the capability for germination (program 1) and subsequently acquires the ability to withstand stress (program 2). In axes of *Phaseolus lunatus* at moderate water content (70%) the embryos have the ability to germinate and grow but cannot withstand desiccation (Klein and Pollock, 1968). However, this 70% water content is probably achieved during the accumulation of dry matter and represents a stage before actual water loss on a per cell basis. During the period of actual water loss from the seed system, *P. lunatus* axes acquire the ability to withstand moderate desiccation; water content must fall below 50–55% (Klein and Pollock, 1968).

The rate of DNA synthesis during *Phaseolus* development has not been measured at all stages. However, in *Vicia* (Millerd and Whitfeld, 1973) and *Gossypium* (Walbot and Dure, 1976) cotyledonary cell number is stabilized about midway through development, whereas DNA content per cell continues to increase; in each case this increase is due to endoreduplication of the genome (reviewed by Dure, 1975). The cotyledon cells of *Gossypium* also increase in size in the absence of cell division. In *Phaseolus*, axis cell number is stabilized by days 18–20, when the seed fills the pod cavity; the axis contains approximately 5×10^5 cells (Clutter *et al.*, 1974, and unpublished). Cotyledon cell number is shown in Fig. 9 for both the epidermal cells and the internal cells in which stored reserve materials are deposited. Smith (1974) has determined by microdensitometry that the epidermal cells of the bean cotyledon remain 2C whereas the internal storage parenchyma cells often reach a 16C ploidy level; whether this increase in DNA content is due to endoreduplication of the entire genome is not known. The continued cell division and DNA accumulation per cotyledon cell in the absence of cell division contrasts sharply with the pattern of RNA synthesis and accumulation. Although both nucleic acids presumably have the same precursors available, RNA synthesis ceases abruptly in the axis tissue (Walbot 1972) and cotyledonary tissue (Walbot, 1973a) after day 24.

Immature *Phaseolus* embryos synthesize RNA at a moderate rate expressed as a fraction of RNA synthesized (μg RNA/A_{260}/2 hr), but both the axis and cotyledons accumulate little total RNA during early development since there are few cells (Fig. 7 and 11). RNA accumulation is coordinate with increases in cell number; in *P. coccineus* and *P. vulgaris* the amount of RNA per cell does not change while the seed is elongating (Sussex *et al.*, 1973). The rate of RNA synthesis in the axis does increase to a maximum at day 18 (Fig. 10), and there is substantial RNA accumulation from day 18 to day 24 after cell number has stabilized. By day 28, however, there is no detectable RNA synthesis, and the rate must be at least 100-fold less than the rate on day 18.

The changes in RNA synthesis and accumulation are preceded by several days by changes in precursor pool sizes. Total nucleotide content of the embryonic axis reaches a maximum of 80 μM/axis on days 16–18 and subsequently falls to about 40 μM/axis by day 24 (Walbot, 1971). This change in total nucleotide content occurs in the absence of substantial cell division in the axis and thus represents a change in the amount of nucleotide per cell. The concentration of nucleotide (molarity) falls more than total

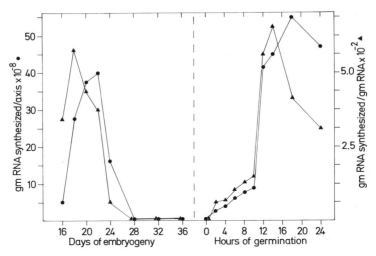

Fig. 10. The rate (▲) and amount (●) of RNA synthesized over a 2-hr period by excised axes of *Phaseolus vulgaris*. (From Walbot, 1972.)

content due to the continued and nearly linear increase in axis fresh weight from day 16 through day 28 (Fig. 7). In previous publications (Walbot, 1971) orthophosphate content is expressed on a fresh weight basis rather than in absolute amounts. This mode of expressing the data masks the fact that in absolute terms the amount of phosphate per axis, and per cell, increases during the period of 16–20 days although the concentration (molarity) of phosphate decreases.

The data previously presented on cotyledonary phosphate content (Fig. 3) clearly illustrate the pattern of phosphate metabolism. Despite increasing

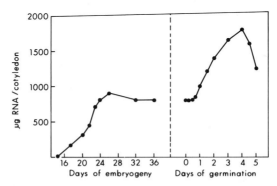

Fig. 11. Accumulation of RNA in cotyledons of *Phaseolus vulgaris*. (From Walbot, 1973b.)

phosphate content there is a rapid, fivefold dilution of phosphate per cell between days 16 and 18 (Fig. 3B). Since intracellular phosphate concentration is far more significant for metabolic regulation than absolute amounts of phosphate, it is clear that phosphate concentration may be rate limiting for some metabolic processes despite tissue accumulation of phosphate during this period (Fig. 3B) and a stable content of phosphate per cell (Fig. 3A).

Phytic acid, hexaphosphoinositol, is detectable in *Phaseolus* cotyledons as early as day 12 of embryogeny. However, rapid accumulation of phytic acid is not evident until after day 24; between day 24 and 30 nearly 90% of the final content of 3 mg phytic acid per cotyledon is deposited (Fig. 6). Similarly storage protein is detectable early in development (Kloz *et al.*, 1966), but the greatest accumulation is from day 24 to day 32 (Öpik, 1968; Walbot, 1973b). Cytological evidence indicates that starch synthesis is also most rapid during this period (Öpik, 1968). These three storage reserves are therefore synthesized coordinately and to a very late stage of development; desiccation is already evident during the final days of starch, phytic acid, and storage protein deposition.

Although there is no detectable RNA synthesis and a plateau in RNA content from day 28 onward, there is continued synthesis of protein until at least day 32 in the cotyledons and axis as seen by the continued increase in tissue dry weight (Fig. 5). This protein synthesis is presumably dependent on long-lived RNA's including the mRNA for storage protein. Both the cotyledons and axis of *P. vulgaris* contain protein bodies and a substantial amount of starch (Öpik, 1968); these materials contribute significantly to the continued increase in dry weight. During this period from day 28 to day 32 protein accumulation in the cotyledons in insensitive to inhibition by actinomycin D (V. Walbot, unpublished data) as expected for cells in which there is little, if any, RNA synthesis.

The decline in RNA synthesis may be explained by a shortage of precursors and the preferential use of existing precursors in other biosynthetic pathways. Axis ATP content undergoes only a twofold decline between days 22 and 28. Residual ATP may be used preferentially to support unabated protein and starch synthesis and continued DNA synthesis rather than RNA accumulation during maturation.

3. HORMONE LEVELS DURING EMBRYO DEVELOPMENT

The concept of regulation of embryo development based on a changing ratio of growth-promoting to growth-suppressing hormones has formed the central hypothesis of a number of studies in which hormone levels during embryo development have been correlated with changes in growth rate. These studies have focused nearly exclusively on the consequences of hormone concentration on growth rather than on morphogenesis. High levels of extractable auxins, cytokinins, and/or gibberellins are often correlated with periods of rapid embryo, seed, and fruit growth. Abscisic acid and unknown antigrowth factors often increase in concentration during seed maturation; high ABA levels are also correlated with abortion and abscission of young fruit.

Carr and Skene (1961) correlated a decline in extractable gibberellin activity measured by bioassay with a decrease in the rate of seed fresh weight increase. They reported that there are two periods of high gibberellin content. A twofold decline in gibberellin midway through development is coincident with an apparent transient slower rate of fresh weight increase. The final decline in gibberellin activity coincides with the beginning of seed maturation.

Eeuwens and Schwabe (1975) measured the hormonal content of the pod and embryo tissue in *Pisum sativum*. The pods contain only low levels of extractable hormone compared to the developing seed, and hormone levels in the seed may control pod development. In experiments in which embryos were killed *in situ* the pod tissue grew considerably less than normal. Exogenously supplied auxin/gibberellin mixtures or an aqueous extract of immature pea seeds could substitute for the killed embryo and restore growth of the pod to that of a seeded control.

The general pattern of hormone level in the pea seed is shown in Table I. Early in development the levels of extractable growth promoters are high, and seed growth is rapid; following histodifferentiation seed growth rate slows, and there is progressively less extractable growth promoter. At maturity the abscisin-like inhibitors increase to a maximum, and there is also considerable gibberellin-like activity. In the desiccated seed, growth is nil and there is little extractable hormone present. Eeuwens and Schwabe (1975) also observed a transient decline in growth-promoting substances at about 10 days of development that was followed by a lag in the growth rate

TABLE I

Hormone Levels during Pea Seed Development[a]

Stage	Growth rate	Hormone levels
Early	High	High gibberellin-like and auxin-like hormone levels
Middle	Rapid but decreasing	Little growth promoter present
Late	Low	Abscisin-like inhibitors and gibberelin-like inhibitors present
Desiccated	None	Little extractable hormone present

[a] From Eeuwens and Schwabe (1975).

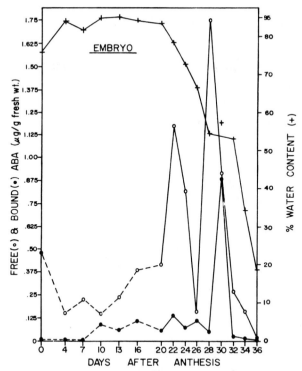

Fig. 12. Abscisic acid content of embryo of *Phaseolus vulgaris*. (Personal communication from F. Hsu and I. Sussex.)

measured as length increment per day. They suggested that this period from day 10 to day 12 coincides with embryo histodifferentiation, which is accompanied by little seed growth.

A more precise measurement of ABA levels in developing *Phaseolus* was recently made by F. Hsu and I. Sussex (personal communication). As shown in Fig. 12 ABA levels measured chemically are low until days 18–20, as predicted from the bioassay measurements in pea. After days 18–20 there is an eightfold increase in ABA content with the maximal level at day 28; during desiccation ABA content falls rapidly. Walton *et al.* (1973) reported that nonimbibed *P. vulgaris* seeds contain more ABA metabolites than ABA; the contents of ABA, phaseic acid, and dihydrophaseic acid were 0.06, 0.11, and 5.95 mg/kg dry weight, respectively.

In these two legumes it is apparent that the general hypothesis is true that levels of growth promoters are high during periods of rapid growth, and ABA and inhibitors appear during maturation. The relationship between hormone levels and growth and metabolism is less clear. The diauxic pattern in gibberellin content reported by Carr and Skene (1961) is matched in their study by a diauxic pattern of fresh and dry weight increase. However, the data presented in Fig. 5 show a continuous increase in weight of the seed and embryo. Starch, protein, and RNA also continuously increase until maturation. The first decrease in gibberellin in bean or in growth promoters in pea occurs before the accumulation of ABA in the embryo or seed. The second decrease in gibberellin and promoters occurs concomitant with the eight- to tenfold increase in ABA. These facts suggest that a decline in growth-promoting hormones per se is insufficient to affect the rate of RNA synthesis or growth in general since the fluctuations in growth promoter occur during the period of rapid accumulation of both fresh and dry weight. Data reported on an increment basis as was done by Carr and Skene and by Eeuwens and Schwabe tend to mask the important observation that the amount of fresh and dry weight increase and of accumulation of particular metabolites is very large on a per day basis through the period of declining growth promoter substance that occurs midway through embryogeny. In conclusion it appears that a decline in growth-promoting hormones is insufficient to modulate the continued growth of the embryo, and it is therefore the appearance and accumulation of the growth inhibitor ABA that regulates the rate of growth.

4. HORMONAL REGULATION OF MACROMOLECULAR SYNTHESIS

The increase in ABA content of the developing bean axis after day 18 (Fig. 12) coincides with a rapid decrease in the rate of RNA synthesis (Fig. 10), although the amount of RNA accumulated per axis continues to increase until day 24. By day 24 ABA content has increased fourfold, and the rate of RNA synthesis is only 10% of the maximum on day 18. Abscisic acid disappears rapidly from the embryo axis after day 28, and, although ABA is detectable in the dry seed, levels of metabolites of ABA are 100-fold higher than those of ABA itself (Walton et al., 1973). However, RNA synthesis is not restored during desiccation despite the decline in ABA content.

To determine whether the rate of RNA synthesis in the embryonic axis can be controlled by ABA, Walbot et al. (1975b) used excised embryonic axes germinated in the presence of ABA as a model for in vivo events; this study assumed that germinating and developing axes shared physiological properties. These experiments test the role of ABA in regulating growth and RNA synthesis and determine whether the embryonic axis retains sensitivity to ABA inhibition of RNA synthesis after desiccation. When axes are removed from the seed they imbibe water, elongate, initiate RNA synthesis, and show a marked increase in respiration rate and ATP pool size indicative of active metabolism (Walbot *et al.*, 1975b). The maximal rate of RNA synthesis occurs at the eighteenth hour of germination. If axes are pretreated with ABA for 15 min (from 17:45 to 18:00 hr) the rate of RNA synthesis is depressed about 40% during the next hour (total time in ABA, 75 min). With pretreatments of 2 or 4 hr the rate of RNA synthesis is depressed about 80% by 10^{-5} M ABA. Abscisic acid inhibition of RNA synthesis is concentration dependent with the maximal effect attained at 10^{-5} M, a level close to that found in maturing *P. vulgaris* embryos (Fig. 12). Physiological concentrations of ABA are therefore capable of quickly suppressing RNA synthesis in germinating bean axes. Abscisic acid has little effect on tissue respiration or ATP pool size during 5-hr experimental periods.

When ABA content increases dramatically after day 18, there is no cessation in the linear increase in fresh and dry weight by the whole embryo, nor is phytin, starch, or protein accumulation affected. The rapid effect of ABA on RNA synthesis seems to be specific for this macromolecule provided that germinating axes are an appropriate model for *in vivo* events. In the excised-embryo experiments, ABA also depressed the rate of wet weight increase after imbibition by approximately 40%. Since growth up to the eighteenth hour is due solely to cell expansion (Walton and Soofi, 1969), this effect of ABA is suppression of water uptake required for elongation prevents axis growth. This effect of ABA *in vitro* on water uptake may be important in understanding the inhibition of precocious germination required to prevent the premature elongation of the radicle when maximal embryo size is attained. The role of ABA in water relations is discussed in more detail in Section III,C.

Abscisic acid inhibition of RNA synthesis *in vitro* can be completely reversed by benzyladenine, a cytokinin, but not by auxin or gibberellins (Sussex *et al.*, 1975). This observation matches the *in vivo* measurements of growth in which decreases in growth rate are correlated with lack of growth promoter rather than solely the presence of growth inhibitor (Table I).

B. Cotton Cotyledons

1. STORED mRNA METABOLISM

Stored mRNA has often been implicated in the rapid development of both plants and animals following a quiescent period, e.g., postfertilization zygote development and seed germination. This information reserve is synthesized during one developmental stage and used to support protein synthesis required for a subsequent stage. It is of interest to determine when the mRNA's are synthesized, how they are selectively stored, and when and how they are activated. Preformed mRNA metabolism in embryonic cotton cotyledons has been described in detail, although explanation of the regulation of metabolism is still incomplete.

Waters and Dure (1966) noted that the early stages of germination of cotton cotyledons were relatively insensitive to actinomycin D. During seed maturation various polymeric nutrient reserves are synthesized and deposited in the cotyledons; these reserves are hydrolyzed during germination to provide nutrients to the growing seedling. Thus, there is a temporal separation of the anabolism and catabolism of the stored reserves in the cotyledonary tissue. Ihle and Dure (1969, 1972a, b) demonstrated that the synthesis of carboxypeptidase C and isocitratase, two enzymes required to hydrolyze the reserve materials, was not inhibited by actinomycin D after a particular stage of embryo development (two-thirds final weight) in a precocious germination test for enzyme appearance. The enzymes were synthesized *de novo* during germination. These data suggested that there is also a temporal separation of the transcription and translation of these mRNA's.

How are the mRNA's that are used during germination stored during embryogeny? Walbot *et al.* (1974) tested for the appearance of carboxypeptidase activity in embryos precociously germinated in a variety of inhibitors of nucleic acid synthesis. Actinomycin D has no effect on enzyme appearance, whereas cordycepin (3'-deoxyadenosine) treatment prevents enzyme appearance. Cordycepin probably acts as a chain terminator if inserted into a growing polynucleotide chain. However, the mRNA for carboxypeptidase is preformed, so the cordycepin effect is not due to inhibition of mRNA synthesis but suggests a requirement for polyadenylation of preformed mRNA during germination. Cordycepin-treated embryos showed no visible signs of germination, a clue that a multitude of activities are inhibited in a system that is relatively independent of additional RNA synthesis (i.e., insensitive to actinomycin D). By applying or removing cordycepin at different times during germination, it was also shown that the cordycepin-sensitive processing step of the stored mRNA occurred between the sixth and thirtieth hour after imbibition began. The kinetics of enzyme appearance following transfer to or from cordycepin solutions suggested that the polyadenylation of stored carboxypeptidase mRNA was essentially linear from hour 6 to hour 30.

As an additional control, precociously germinating embryos were also treated with 3'-deoxycytidine, which was assumed to inhibit RNA synthesis as effectively as cordycepin (data not shown) but which would presumably not interfere with preformed mRNA polyadenylation. Carboxypeptidase C appearance and total activity were normal in 3'-deoxycytidine-treated tissue; this result was interpreted as further evidence for the preformed nature of the carboxypeptidase mRNA. However, measurements of the actual effect of each nucleic acid inhibitor on 85 mg seeds used in these experiments by fractionation of total RNA into poly(A)+RNA and poly(A)–RNA and subsequent fractionation of these two classes of RNA by size to resolve each major species of RNA proved that RNA species synthesized by RNA polymerases I and III, ribosomal and transfer RNA's, were completely sensitive to both 3'-deoxycytidine and 3'-deoxyadenosine (Harris and Dure, 1974). However, poly(A)+RNA, the presumed mRNA fraction synthesized by RNA polymerase II, was insensitive to 3'-deoxycytidine. Cordycepin prevents polyadenylation, precluding a direct test of its effectiveness as an inhibitor of mRNA production in determinations such as this which depend on the selective recovery of mRNA by virtue of its poly(A) component. Therefore, the experiments with these base analogs do not allow the unambiguous conclusion that preformed message exists since by analogy to the 3'-deoxycytidine data cordycepin does not inhibit *de novo* mRNA synthesis.

2. POLYADENYLATION OF STORED mRNA DURING GERMINATION

The dilemma posed by these experimental results could be resolved by two approaches: either by following the processing of a specific mRNA for which DNA hybridization probes exist or by determining the processing of the class of stored mRNAs. The latter approach was adopted by Harris (1976) in a series of precise measurements to determine the proportion of mRNA in mRNA–poly(A) molecules that is preformed RNA.

In this analysis precociously germinating embryos were pulsed from the eighth to twentieth hour or thirty-second to fortieth hour with both [³H] adenosine and ³²P; poly(A)-containing RNA was selectively retained on oligo-d(T) columns; and a comparison of ³²P/[³H]adenosine incorporation into poly(A) and covalently attached RNA (presumed mRNA) was made after selective RNase treatment to hydrolyze the mRNA portion of mRNA–poly(A). Table II summarizes results from experiments in which embryos were germinated with and without actinomycin D from the eighth to twentieth hour compared to "delayed-pulse" embryos, which were incubated from the thirty-second to fortieth hour.

Cotyledons from seeds germinated in water were compared with seeds germinated in actinomycin D, since this antibiotic should suppress the bulk of new mRNA synthesis and thus reduce the apparent mRNA chain length. From the base composition of the mRNA and poly(A), average poly(A) chain length, and the proportion of ³²P and [³H]adenosine moieties in each portion of mRNA–poly(A) molecules, it is possible to calculate the apparent average length of mRNA. Separate determinations of mRNA length (Walbot and Dure, 1976; Harris, 1976) by electrophoresis of mRNA

TABLE II

Calculation of mRNA Chain Lengths from Experimentally Derived Data for Control, Actinomycin D, and Delayed-Pulse Treatments[a]

	Theoretical	Control (8th–20th hr)	Actinomycin D (8th–20th hr)	Delayed pulse (32nd–40th hr)
Average poly(A) chain length	110	110	110	110
Percent ^{32}P-labeled AMP in mRNA~poly(A)	29.7[a]	37.0	43.2	33.4
Percent ^{32}P-labeled AMP in mRNA	22.0	22.0	22.0	28.4
Percent [^{32}P]phosphate of mRNA~poly(A) in poly(A)	9.9[a]	19.0	35.7	7.5
Percent [2-^{3}H]adenosine of mRNA~poly(A) in poly(A)	33.0[a]	45.0	58.0	19.1
Calculated chain length of mRNA[b]				
Eq. (2)	1000	470	200	1360
Eq. (6)	1000	611	362	1642
Eq. (8)	1000	462	295	1466

[a] Based on a theoretical mRNA chain length of 1000 nucleotides, percent mRNA as ^{32}P-labeled AMP (22%), and poly(A) chain length of 110 nucleotides using Eqs. (2), (6), and (8).

[b] Calculation of mRNA chain length. The apparent chain length of mRNA is calculated in three different ways. The first method is based on the fraction of [^{32}P]phosphate found in the poly(A) portion of mRNA~poly(A), and the poly(A) chain length. Let L_M represent the mRNA chain length, L_A represent the poly(A) chain length, and $L_{M \sim A}$ represent the mRNA~poly(A) chain length. Then, by dividing L_A by the fraction of [^{32}P]phosphate of mRNA~poly(A) found in poly(A), $L_{M \sim A}$ is derived:

$$L_{M \sim A} = \frac{L_A}{\left(\dfrac{[^{32}\text{P}]\text{phosphate in poly(A)}}{[^{32}\text{P}]\text{phosphate in mRNA} \sim \text{poly(A)}} \right)} \tag{1}$$

Knowing that $L_{M \sim A} = L_M + L_A$, then

$$L_M = \frac{L_A}{\left(\dfrac{[^{32}\text{P}]\text{phosphate in poly(A)}}{[^{32}\text{P}]\text{phosphate in mRNA} \sim \text{poly(A)}} \right)} - L_A \tag{2}$$

The second method is based on the percentage of [2-^{3}H]adenosine found in the poly(A) portion of mRNA~poly(A) and the contribution of AMP to the total base composition of mRNA and poly(A). From this information the number of AMP moieties in poly(A) can be calculated as the product of L_A and the mole% of AMP in poly(A):

$$\text{No. AMP's in poly(A)} = (L_A) \, [\text{mole}\% \text{ AMP in poly(A)}] \qquad (3)$$

The number of AMP moieties in mRNA~poly(A) is determined by dividing Eq. (3) by the fraction of [2-³H]adenosine found in the poly(A) portion of mRNA~poly(A):

$$\frac{(L_A) \, [\text{mole}\% \text{ AMP in poly(A)}]}{\left(\dfrac{[\text{2-}^3\text{H]adenosine in poly(A)}}{[\text{2-}^3\text{H]adenosine in mRNA\~poly(A)}}\right)} = \text{No. AMP's in mRNA\~poly(A)} \qquad (4)$$

The number of AMP moieties in mRNA~poly(A) can also be derived from the summation of the AMP moieties in poly(A) and mRNA:

$$[\text{mole}\% \text{ AMP in poly(A)}] \, (L_A) + [\text{mole}\% \text{ AMP in mRNA}] \, (L_M) = \text{No. AMP's in mRNA\~poly(A)} \qquad (5)$$

By substituting Eq. (4) into Eq. (5) and rearranging, L_M can be calculated:

$$\frac{(L_A) \, [\text{mole}\% \text{ AMP in poly(A)}]}{\left(\dfrac{[\text{2-}^3\text{H]adenosine in poly(A)}}{[\text{2-}^3\text{H]adenosine in mRNA\~poly(A)}}\right)} - \left[(L_A) \left(\frac{\text{mole}\% \text{ AMP}}{\text{poly(A)}} \right) + \left(\frac{\text{mole}\% \text{ AMP}}{\text{in mRNA}} \right) \right] = L_M \qquad (6)$$

The third method of calculating L_M utilizes the same experimental data in Eq. (6) with the additional knowledge of the mole% of AMP in mRNA~poly(A). The mole% of AMP in mRNA~poly(A) represents the sum of AMP moieties in mRNA~poly(A) [Eq. (5)] divided by the total number of nucleotides in mRNA~poly(A) or $L_{M\sim A}$:

$$\frac{(\text{mole}\% \text{ AMP in mRNA}) \, (L_M) + [\text{mole}\% \text{ AMP in poly(A)}] \, (L_A)}{L_M + L_A} = \text{mole}\% \text{ AMP in mRNA\~poly(A)} \qquad (7)$$

By rearranging Eq. (7) L_M can be calculated using Eq. (8).

$$\frac{L_A \, [(\text{mole}\% \text{ AMP in poly(A)}) - (\text{mole}\% \text{ AMP in mRNA\~poly(A)})]}{(\text{mole}\% \text{ AMP in mRNA\~poly(A)}) - (\text{mole}\% \text{ AMP in mRNA})} = L_M \qquad (8)$$

If poly(A) is added to some preformed mRNA, then these calculations should yield mRNA chain lengths shorter than the length observed by polyacrylamide gel electrophoresis in 99% formamide. In fact, by comparing the calculated mRNA chain length with the lengths observed in electrophoresis it is possible to estimate the proportion of preformed mRNA to newly synthesized mRNA:

$$\% \text{ preformed mRNA in mRNA\~poly(A)} = 1 - \frac{L_M \text{ (calculated)}}{L_M \text{ (observed)}} \times 100 \qquad (9)$$

From Eq. (9) it is obvious that the percentage of preformed mRNA in mRNA~poly(A) approaches 100 as the calculated mRNA chain length approaches 0.

poly(A) using 99% formamide gels under fully denaturing conditions indicated that the average mRNA–poly(A) molecule is 1800 nucleotides, of which 100 nucleotides are in poly(A). The apparent mRNA chain in cotton tissue excised after the initial actinomycin D-insensitive period of germination ("delayed pulse" of Table II) is approximately 1500 nucleotides, in good agreement with the physical determination. However, "control" tissue of cotyledons during the actinomycin D-insensitive period of germination has an apparent mRNA chain length of only 500 nucleotides, indicating, if we assume that average mRNA chain length does not vary, that approximately two-thirds of the RNA that is polyadenylated during this period is performed and therefore is not labeled by [³H]adenosine or ³²P.

Incubation of cotyledonary tissue in actinomycin D during early germination to suppress *de novo* RNA synthesis reduces the apparent mRNA chain length further, indicating an enhancement of the proportion of preformed versus newly synthesized (labeled) mRNA polyadenylated during the incubation period.

3. ROLE OF ABSCISIC ACID

A corollary of the original observation of stored mRNA in cotton cotyledons was the observation that an ovular extract could prevent the precocious appearance of the enzyme activities diagnostic for germination; abscisic acid, a component of the ovular extract, can completely mimic ovular extract (Ihle and Dure, 1970). Abscisic acid prevented the translation of stored mRNA during the time when these mRNA's were synthesized and the developing cotyledons had an active protein synthetic machinery. Surprisingly the effect of ABA was itself inhibited by actinomycin D, suggesting that it was dependent on *de novo* RNA synthesis. This observation coupled with more recent findings indicating that polyadenylation of preformed RNA was the critical step in processing during germination led to the hypothesis that ABA might control, via a newly synthesized RNA or protein product, the special processing of the "to be stored" mRNA's of embryogeny. Specifically, it was predicted that ABA treatment might prevent the polyadenylation of preformed mRNA. Harris (1976) investigated the effects of ABA directly on RNA synthesis and polyadenylation. Since ABA has no effect on germination of mature cotton seeds (Ihle and Dure, 1969), developing embryos still active in the synthesis of the stored mRNAs were used. These embryos were precociously germinated in $5 \times 10^{-5} M$ ABA, ³²P, and [³H]adenosine. If ABA *in vivo* prevents precocious germination by inhibiting polyadenylation of mRNA, in the *in vitro* test ABA would be expected to inhibit incorporation of both radioisotopes into mRNA–poly(A). If the ABA effect is specific for the stored mRNA fraction, then other mRNA's should be normally synthesized and processed; if this occurs, calculation of the mRNA–poly(A) fraction would indicate the true *in vivo* mRNA chain length (1500–1700 nucleotides).

Abscisic acid treatment had no appreciable effect on incorporation into structural or the mRNA–poly(A) fractions of precociously germinated embryos (Harris, 1976). The calculation of mRNA chain length (Table III) in the presence of ABA gives a result very similar to the control. This result indicates that, *in vitro*, both preformed and newly synthesized mRNA's are polyadenylated in the presence of ABA, and no specific role for ABA can be assigned in the processing of mRNA. Even if exogenous ABA does not inhibit the processing of this class of stored mRNAs the appearance of carboxypeptidase and isocitratase is inhibited by ABA. It is possible that ABA acts, via a newly synthesized protein or RNA species, at the level of translation. Incorporation of radioactively labeled amino acids is enhanced in the presence of ABA, but without a correction for pool size fluctuations the rate of protein synthesis cannot be calculated. However, when the total protein of cotton cotyledons precociously germinated with ABA in the presence of ^{14}C-labeled amino acids is compared by acrylamide gel electrophoresis to control embryos there are qualitative differences in the profiles of newly synthesized, ^{14}C-labeled proteins. Several protein species not present in the ABA-treated embryos were present in the treatment without ABA (Harris, 1976).

Further support for the hypothesis that different proteins are synthesized in the presence of ABA is found in a comparison of the base composition of

TABLE III

Calculation of mRNA Chain Length for Control and Abscisic Acid-Treated Preparations

	Control	Abscisic acid
Average poly(A) chain length	100	120
Percent ^{32}P-labeled AMP in mRNA~poly(A)	37.5	39.4
Percent ^{32}P-labeled AMP in mRNA	21.2	23.4
Percent [^{32}P]phosphate of mRNA~poly(A) in poly(A)	16.0	18.9
Percent [2-^{3}H]adenosine of mRNA~poly(A) in poly(A)	37.3	38.3
Calculated chain length of mRNA[a]		
Eq. (2)	525	515
Eq. (6)	825	822
Eq. (8)	485	500

[a] See footnote *b*, Table II, for explanation of calculation.

TABLE IV

Molar Base Ratios of 0.1M NaCl Eluate and mRNA Poly(A) Fractions from Precociously Germinated Embryos in the Presence and Absence of Abscisic Acid

Fraction[a]	Control				Abscisic acid ($5 \times 10^{-5} M$)			
	G	A	C	U	G	A	C	U
0.1 M NaCl eluate	29.9	26.1	21.7	22.3	29.3	26.6	23.2	20.9
mRNA poly(A)	21.6	37.5	16.3	24.6	21.3	39.4	16.9	22.4
mRNA	27.3	21.2	21.4	30.1	25.8	23.4	21.8	29.0
Poly(A)		99+				99+		

[a] Low-salt (0.1 M NaCl) fraction from oligo-d(T) column is a poly(A)− fraction. The mRNA + poly(A) fraction binds to the oligo-d(T) column and is eluted with 0.4 M NaCl; the mRNA and poly(A) portions of this fraction are reported separately following RNase A and T_2 digestion to specifically remove the mRNA covalently linked to poly(A). (From Harris, 1976.)

mRNA–poly(A) isolated from control and ABA-treated embryos. A reproducible difference in the molar base ratios of these two preparations was found (Harris, 1976) (Table IV).

If preformed mRNA's are indeed stored as an informational reserve to support germination, how are they selectively distinguished from the normally metabolically active mRNA's of the cell? Harris' work indicates that a large body of stored mRNA exists; perhaps two-thirds of the messages utilized during early germination are preformed but newly adenylated during germination. Some evidence (Walbot et al., 1975a) has been offered to indicate that a ribonucleoprotein (RNP) particle lacking poly(A) is accumulated during cotyledonary development. However, definitive hybridization reactions have not been performed to prove that these RNP's contain RNA sequences that become polyadenylated during germination and function as mRNA's. This might prove to be an extremely important piece of information since, if true, it would be possible to ask whether this group of mRNA's coordinately transcribed, stored, adenylated, and translated shares some common recognition feature. The mRNA sequence itself is an unlikely candidate since mRNA's are presumably transcribed from unique structural gene elements. It would be interesting to test the Britten–Davidson model for gene regulation in eukaryotes (Davidson and Britten, 1973), in which repetitive sequences contiguous to structural genes serve as regulatory sequences, as it might pertain to the transcription and usage of a developmental, stage-specific body of mRNA's. For example, the simplest hypothesis would be that there is a single common repetitive sequence adjacent to each of the many genes coding for stored mRNA. This sequence would then be the means by which the diverse mRNA sequences are recognized as equivalent during processing, storage, adenylation, and translation. If one or a few of the approximately 1100 repetitive sequences in the cotton genome (Walbot and Dure, 1976) are adjacent to the stored mRNA structural genes, this would provide experimental evidence for the key hypothesis of the Britten–Davidson model. Seeds of cotton or other plants provide a convenient tissue from which to isolate sufficient quantities of both stored mRNA and functioning adenylated mRNA, and thus enable one to describe the transcription and processing sequence of a class of mRNA's.

C. Regulation of α-Amylase Production in Aleurone Cells of Barley

Although ABA is implicated in the cessation of RNA synthesis in developing bean axes and in the regulation of the processing of stored mRNA in cotton, the mechanism by which ABA acts is only incompletely described. Ihle and Dure (1970) demonstrated that the maintenance of translation inhibition of preformed carboxypeptidase C mRNA by ABA was inhibited by actinomycin D; this result was interpreted to indicate that ABA action required continued RNA and/or protein synthesis. However, there is a considerable delay between the time of inhibitor addition and the appearance of new enzyme, so that it is impossible to describe the initial ABA effects in this system.

If ABA is a selective inhibitor of a class of newly transcribed mRNA's, it is critical to determine the mechanism of inhibition. Such a determination will require a system exhibiting the same key features as cotton cotyledons but in which response to inhibitors is measurable in minutes. Ho and Varner (1974, 1976) demonstrated that the GA_3-enhanced synthesis of α-amylase by isolated barley aleurone tissue is regulated in an analogous manner to the cotton system. The important features of α-amylase synthesis are shown in Table V. After the twelfth hour there is a linear rate of α-amylase synthesis. If ABA is added, α-amylase synthesis ceases, although other cellular activities such as respiration continue at a normal rate. The effect of ABA on α-amylase mRNA translation is specific since other proteins continue to be synthesized in the presence of ABA. When actinomycin D or cordycepin are added to ABA-treated aleurone layers, there is resumption of α-amylase production at the pre-ABA treatment rate within 15 min (Fig. 13). Ho and Varner postulate that a short-lived RNA or protein species is required to inhibit α-amylase translation and that this regulator molecule is synthesized only when ABA is present. Since the regulator is short-lived, inhibition of RNA synthesis quickly releases the inhibition of α-amylase production.

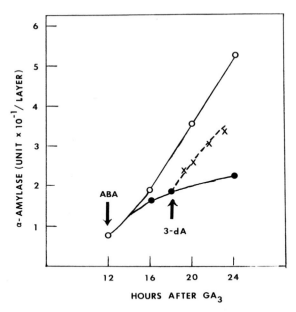

Fig. 13. Appearance of α-amylase in the medium in which barley aleurone cells are incubated: (O) linear increase in α-amylase activity in the absence of inhibitors; (●) addition of ABA at hour 12 results in the rapid inhibition of α-amylase appearance; (×) addition of 3'-deoxyadenosine results in a restoration of the linear rate of amylase appearance even in the continued presence of ABA. (Adapted from Ho and Varner, 1974, 1976.)

TABLE V

Steps in α-Amylase Production by Barley Aleurone Cells[a]

Time (hr)	Event	Sensitivity of α-amylase production to metabolic inhibitors
0	GA₃ applied	Sensitive to protein and RNA synthesis inhibitors
0–8	Synthesis of α-amylase mRNA	Insensitive to actinomycin D; sensitive to cordycepin and cycloheximide
8–12	Polyadenylation of α-amylase mRNA	Insensitive to actinomycin D and cordycepin; sensitive to protein synthesis inhibitors
12–24	Production and secretion of α-amylase by aleurone	α-Amylase production insensitive to actinomycin D or cordycepin; ABA regulation of α-amylase appearance sensitive to both actinomycin and cordycepin

[a] Adapted from Ho and Varner (1976).

D. Viviparous Embryo Development

Vivipary, precocious embryo germination within the fruit, is the normal developmental pattern of mangrove embryos; the embryo germinates and the seedling grows while the fruit is still attached to the parent plant. Vivipary is also observed in a variety of cultivated and wild species including cotton, various citrus species, tomato, corn, and jackfruit (V. Walbot, personal observation; Habib, 1973; Limberk and Ulrychova, 1972). In corn, vivipary is controlled by nine different loci (Robertson, 1955), suggesting that the suppression of the normal developmental pattern ending in developmental arrest is a complex process.

By studying the sequence of events and physiological responses of naturally viviparous embryos such as mangrove some of the criteria for genetic dissection of the regulation of arrest in corn and other plants may become apparent. Sussex (1975) described the direct, uninterrupted transition of *Rhizophora mangle* embryos from growth as embryos to growth as seedlings. Unlike embryos entering developmental arrest, *R. mangle* embryo–seedlings undergo no desiccation. *In vitro* these embryos show no sensitivity to inhibition by ABA at concentrations inhibiting RNA synthesis and growth in *Phaseolus* (Walbot *et al.*, 1975b). The ability of the mangrove embryos to synthesize ABA under stress was not measured. Sussex presents a specific hypothesis comparing vivipary with developmental arrest. He suggests that changes in growth-promoting and inhibitory substances (factor models; see Section III) are secondary effects since no specific triggering event has been identified that results in changes in hormonal balance. He further postulates that the trigger initiating the metabolic events leading to arrest is the development of water stress in the seed. The sequence of events leading to dormancy/arrest would be as follows: water stress, increased ABA content, cessation of RNA and protein synthesis, growth, and physiological arrest. This hypothesis is discussed in more detail in Section III,C.

Vivipary in terrestrial species is a lethal trait since aerial embryos have little chance of survival except in specialized aquatic environments as found for mangrove. The lethality of the viviparous trait in germinating embryos probably explains the dirth of literature on this important developmental mutant since embryos must be rescued and cultured from the fruit. However, I. Sussex (personal communication) examined the response to exogenous ABA of *Zea* embryos homozygous for vivipary (vp/vp). In *Phaseolus*, cotton, and barley, maturing or early germination stage embryos are highly sensitive to ABA inhibition of growth and germination at $10^{-5}\ M$ or less. This sensitivity is expressed both as suppression of overall growth and depression of rates of macromolecular synthesis. In normal corn embryos germination and growth are also suppressed by moderate levels of ABA; there is a 50% inhibition of germination at 5×10^{-5} M ABA. In the normally viviparous mangrove no effect of ABA is observed. In vp/vp Zea embryos germination is 100-fold less sensitive to ABA than normal embryo germination. These results suggest that even if ABA content is high in viviparous Zea embryos or surrounding tissue the embryo is incapable of a normal response to this growth antagonist. The heterozygous (vp/+) and homozygous plants do respond to ABA; the vp mutation is therefore specific to events of embryo development.

These observations suggest interesting experiments to explore the cause–effect relationship between water stress and ABA synthesis using the known loci in corn controlling vivipary. Do some viviparous corn embryos synthesize or respond to ABA? The number of loci influencing vivipary suggests that a number of different genetically determined viviparous physiological conditions might exist in which ABA metabolism is altered.

E. Embryo Development in Fraxinus and "Deeply" Dormant Embryos

$Fraxinus$ $excelsior$ embryo development has been studied by Wareing and his colleagues (Wareing, 1965; Villiers and Wareing, 1965; Wareing and Saunders, 1971) in light of their model of seed dormancy, in which the balance of growth-promoting/growth-inhibiting substances controls development. The ash embryo is typical of embryos considered to be "deeply" dormant. The ash embryo requires 2 years to develop. During the first year the embryo grows within the hydrated fruit; growth is most rapid during periods of warm temperature. Biochemical measurements indicate that growth-promoting substances are present during this period. Embryo maturation is delayed during the first winter by the indehiscent pericarp, which is impermeable to oxygen; during this winter the embryo remains hydrated. During the second growth season the embryo completes maturation, desiccates, and subsequently requires stratification to overcome dormancy. Reimbibed embryos that have not undergone stratification release germination inhibitors thought to include ABA. During the stratification treatment germination and growth-promoting substances are synthesized, and these substances apparently overcome the high levels of inhibitor found even in stratified seeds.

One additional observation concerning the development of $F.$ $excelsior$ is that seeds picked "green" before desiccation and sown directly may germinate without any intervening period of dormancy. This observation suggests that the development of ash up to this point is similar to the pattern described for *Phaseolus* and other nondormant embryos, in which physiological embryo dormancy is not imposed on the system. In *Fraxinus* the development of the stratification requirement occurs after the embryo has reached a condition in which germination is possible. The accumulation of ABA and other germination inhibitors before desiccation is similar to accumulation in bean and cotton. However, the ash embryo retains both the inhibitors and sensitivity to the inhibitors during desiccation. This contrasts with bean, in which the level of ABA is drastically reduced during desiccation although the embryo remains sensitive to inhibition of germination by ABA, and with cotton, in which the embryo becomes insensitive to ABA during desiccation. *Fraxinus* embryos, therefore, represent a case in which the embryo retains both a high level of inhibitors and sensitivity to these substances. In this embryo the low-temperature after-ripening provides a time period for the synthesis of growth-promoting substances required to overcome the inhibitory substances. In other "deeply" dormant seeds, after-ripening may provide a time period for the decay or leaching of such inhibitors without the concomitant synthesis of promoters.

In *F. ornus*, a nondormant species, and *F. americana*, a species requiring stratification to overcome dormancy, exogenously supplied ABA prevents germination (Sondheimer and Galson, 1966). The concentration of ABA in *F. ornus* seed and pericarp is considerably lower than in *F. americana* before stratification. During low-temperature after-ripening the ABA content of *F. americana* decreases to levels typical of *F. ornus* (Sondheimer *et al.*, 1968). Sondheimer *et al.* (1974) demonstrated that both dormant and stratified embryos of *F. americana* rapidly metabolize (S)-[2-^{14}C]abscisic acid. Phaseic and dehydrophaseic acids are the major initial metabolites as is true for *Phaseolus;* however, two additional compounds accumulated which have not been identified. The rate of ABA metabolism is approximately equal in dormant and stratified embryos ($10\mu g/g$ fresh weight per day). The ABA content in the dormant seed is only about 0.2 $\mu g/g$ fresh weight (Sondheimer *et al.*, 1968), so that endogenous ABA should be rapidly metabolized during stratification.

The ABA level per se seems unlikely, therefore, to maintain *Fraxinus* embryos in a dormant condition. Stratification is not required to develop the capacity to metabolize ABA (Sondheimer *et al.*, 1974), so that the role of ABA in maintaining dormancy is unresolved. The germination inhibitors present in *F. excelsior* may be metabolites of ABA or other unidentified compounds. That such inhibitors are unable to maintain seed dormancy suggests that growth-promoting factors play a more important role as postulated by Wareing and Saunders (1971).

III. REGULATION OF EMBRYO DEVELOPMENT

A. Factor Models

Wareing (Villiers and Wareing, 1965; Wareing and Saunders, 1971) outlined a general scheme to explain embryo development on the basis of changing levels of growth inhibitors and promoters during embryogeny. In this scheme promoter levels are high during early development and decrease as cell division and growth cease; inhibitors are accumulated during maturation, which further reduces the growth of the embryo. High levels of inhibitors in mature seed are the basis for dormancy induction and maintenance. However, measurements of hormone levels during development do not support such a simple view of regulation, although high levels of growth promoters are correlated with periods of rapid embryo growth. The correlation of hormone physiology and growth would be more secure if data on the chemical identity as well as activity in bioassay of various hormones were available for several species during the entire developmental period. For example, the ability of young embryos to synthesize hormones has not been documented *in vivo*, nor is much known of hormone metabolism by embryos except in studies of ABA catabolism (Walton *et al.*, 1973; Sonheimer *et al.*, 1974). The overall hormonal balance of the vegetative plant is also an important but neglected aspect of embryo hormone physiology.

What proportion of the extractable hormones found in embryo and fruit tissue is synthesized *in situ*? To what extent do changes in whole-plant hormone levels affect embryo hormone content and metabolism? Because our knowledge of the mechanisms of synthesis and actions of plant hormones is so rudimentary it is difficult to assign specific roles to each hormone or to understand why the concentration of a hormone changes.

One aspect of hormone balance that is a promising lead and amenable to experimental investigation derives from the fact that gibberellins and abscisic acid share a common biosynthetic pathway from mevalonic acid to farnesyl pyrophosphate (Milborrow, 1974; Hall, 1973; Lang, 1970). Farnesyl pyrophosphate is an important metabolite of intermediary metabolism in the biosynthesis of sterols, carotenoids, ubiquinones, some plant hormones, and diterpenes. Large increases in a particular hormone are not uncommon; for example, ABA synthesis increases to 40 times normal levels within 4 hr in wilted wheat leaves (Wright and Hiron, 1969). A thorough investigation of the control of partition of farnesyl pyrophosphate into different biosynthetic pathways is required to determine how the level of each hormone can be regulated by *de novo* synthesis. The potential feedback regulation of hormone level and/or antagonism between hormones at the biosynthesis stage should also be investigated.

B. Endogenous versus Exogenous Controls

The extent to which the developing embryo controls and is controlled by the surrounding maternal tissue and the vegetative plant is also an area of future research. The angiosperm embryo develops entirely within the confines of the ovary wall and is clearly dependent on the vegetative plant for water and nutrients and on the female gametophytic tissue for a structure in which to develop. Less clear than the anatomical relation of the embryo to the rest of the plant is the regulation of embryo development at the biochemical and biophysical levels. In addition, the seed system consists of three genetically distinct parts: the diploid embryo developing from the zygote, the typically triploid endosperm with two maternal and one paternal set of chromosomes, and the diploid tissue of the vegetative plant. The genetic capability of these three organs may be quite distinct as in the extreme case of a viviparous corn embryo growing in a seed and fruit of a nonviviparous constitution.

The resolution of endogenous versus exogenous control of embryo development should theoretically be amenable to experimental and genetic dissection. Embryo-lethal mutations have received little attention; rescue and culture of such tissue may provide autotrophic as well as developmental pattern mutants. The viviparous mutants of corn may also be of great value in the determination of ABA metabolism.

Although our knowledge is incomplete a distinction can be made between the exogenous versus endogenous control of embryo development. *Exogenous* controls of development include all information supplied by nonembryonic tissue. Since the vegetative plant is in more direct contact with the environment than is the embryo, it is likely that the embryo indirectly receives information regarding photoperiod, water stress, photosynthetic output, mineral nutrition, etc. Dure (1975) postulated that the hormonal environment and hence development of the cotton embryo is determined primarily by growth promoters supplied through the vascular tissue during early development and by ABA secreted by the degenerating ovule wall late in development. This view of development clearly places the embryo in a relatively passive condition as a system responding primarily to exogenous controls.

In pea the embryo regulates to some extent the growth of surrounding fruit tissue as Eeuwens and Schwabe (1975) demonstrated in their *in situ* embryo abortion experiments. The ability of embryos to synthesize and respond to exogenous and endogenous hormones must also change during development since seedlings are hormonally independent of the vegetative plant whereas young embryos depend on maternal tissue for a growth-promoting environment. For example, are the diauxic patterns of gibberellin content due to a change in the source of GA? Does the embryo undergo histodifferentiation under the hormonal regulation of the maternal/vegetative plant and then switch to endogenous regulation of subsequent growth and seed maturation?

The biophysical constraints placed on the embryo potentially include mechanical restraint, limitation of gas exchange, limitations of water, or changes in the osmotic environment of the seed system. Mechanical restraint and insufficient water potential are often invoked together to explain the inability of certain seeds to germinate (Nabors and Lang, 1971). Low gas exchange is also invoked as an explanation for persistent dormancy (Amen, 1968). These restraints may play a significant role during embryogeny as well. The major changes in macromolecular synthesis and metabolite concentration reported in *Phaseolus* (Section II) occur over a several-day period, from day 16 to day 20, the same stage at which the seed grows to fill the length of the pod cavity. The rapid influx of solutes and water from day 16 until day 24 may provide the embryo with enough hydrostatic potential to expand laterally against the pod and seed wall to provide sufficient embryo volume for the deposition of nutrient reserves. Continued expansion may also provide a means for continuing the rapid uptake of precursor molecules accompanied by water uptake to maintain a constant water pressure and availability.

Less is known about gas exchange by developing embryos. To the extent that respiration has been measured it is correlated with growth rate (Loewenberg, 1955) and hydration (Öpik and Simon, 1963). However, *in situ* measurements of respiration and gas partial pressue in developing embryos have not been made. More information is needed on the barrier provided by the pod and seed coats to gas exchange by the embryo since this restraint on development may be highly significant in regulating the development of embryos embedded in large fruits.

That quiescence or dormancy is controlled exogenously by biophysical or hormonal factors is clearly the case in many examples discussed in this review. There are four lines of evidence to support the hypothesis that developmental arrest is imposed on the embryo rather than a natural consequence of embryogeny. First, we know that lower plants such as mosses and ferns can proceed directly through embryogeny to germination without developmental or metabolic arrest. Second, when angiosperm embryos are removed from the ovule and placed in culture, they invariably attempt to germinate (if they grow at all) and do not continue growth as embryos. This can be seen in artificially induced embryos or embryoids as well, in which there is direct germination following histodifferentiation. Even in embryos that typically become dormant the young embryo will precociously germinate in culture if removed from the ovular tissue. Another line of evidence is found in the natural occurrence of a direct transition from embryo growth to germination in the mangroves (Sussex, 1975). In these plants the embryo germinates while still encased in the seed, and the hypocotyl undergoes considerable elongation before the embryo–seedling is shed from the vegetative plant. There is no cessation of growth during the transition from growth as an embryo to growth as a seedling. Therefore, it appears that there is no absolute requirement for a quiescent period between embryo development and germination. A fourth line of evidence can be drawn from examination of viviparous embryo mutants in various angiosperms such as corn. This genotype results in the germination of embryos before the dehydration and full maturation of the seeds.

Distinguishing between the roles of biophysical and hormonal (biochemical) restraints on embryo developmental pattern and growth is more difficult since there is insufficient information in any one system in either category. However, in Section III, C, 2 the role of water availability and ABA synthesis will be discussed in greater detail as a case study of the potential for examining changes in developmental pattern as consequences of biophysical factors. Perception of changes in biophysical restraints may be amplified in a physiologically significant way through the production of hormones. If this is true then the regulation of developmental arrest can be compared to the control of a process such as phototropism in which the perception of a physical parameter (light) clearly results in a change in growth pattern mediated by the plant hormones.

C. What Is the Signal for Developmental Arrest?

I. THEORETICAL CONSIDERATIONS

Why do embryos become quiescent? Seed quiescence and dormancy must be viewed as part of the overall strategy for survival of the species. Embryo development must lead to a seed system in which dispersal is facilitated and a high capacity for germination and growth is retained under optimal environmental conditions. Developmental descriptions as well as knowledge of the ecological conditions usually encountered by the species would seem to be essential to understand maturation and early germination development. Unfortunately, it is difficult to study development in angiosperm embryos, which must be dissected out of the ovarian tissue, nor has the strategy of plant survival received as much attention from ecologists as have strategies of animal survival. Consequently, embryo development is often explained by analogy to various physiological experiments. This approach is used to describe the action of plant hormones, photoperiod, drought, etc., on developmental and survival processes.

However, there are fundamental difficulties with the assumption that physiological experiments can explain developmental or survival phenomena. Plant growth regulator action on differentiated tissues, for example, may be quite different than the effects during differentiation. Physiological experiments often rank treatments by *order of effectiveness* in achieving a result, but during development the *order of appearance* of regulator molecules and signals may be more important than the strength of a particular signal. The *ability to perceive* signals and the *ability to respond to signals* may both vary during development, so that a comparison of different developmental stages with a standard bioassay using seedling tissue may not determine whether a particular signal is important. Another limitation of analogy to physiological experiments is that some treatment may not reflect realistic *in vivo* conditions. An exogenous agent that affects development *in vitro* may not even be present during normal ontogeny. The cybernetic model of seed dormancy proposed by Amen (1968) is an approach in which the whole system is taken into account. Further work on modeling seed dormancy and survival is clearly required.

Given the limitations of existing descriptions of embryo development and of physiological experiments in describing the role of potential regulators, it is difficult to define the trigger for developmental arrest. The signal for arrest must be sent, received, and processed, however, between the time the embryo attains the capacity for *in vitro* germination and the time when it is of sufficient size and contains sufficient nutrient reserves to germinate *in situ*. Before maturation embryos grow by expansion and cell division as the plantlet will grow during germination. During maturation—as the embryo approaches developmental arrest—growth is primarily by accumulation of dry weight with only a low frequency of cell division and expansion. Another major difference between growth state and the maturing embryo is that growing tissue retains a high water content and sufficient water potential to exert hydrostatic pressure against the cell wall, whereas the maturing embryo cells accumulate dry matter at the expense of percent water content. The maturing embryo develops a large negative water potential, which in other tissue would result in rapid hydration and growth by expansion. This paradox of growth (dry weight increase) without expansion despite a water potential favoring rapid hydration and expansion seems to be the key aspect of developmental arrest. Clearly, maturation is not the simple simultaneous cessation of metabolic activity, but rather the full-throttle growth of the system in the absence of a size increase. Furthermore, the low negative water potential developed during maturation and quiescence is not a barrier to germination, so that the factors that provide an environment in which growth without size increase is possible during maturation constitute no barrier to the hydration and expansion of the embryo during germination.

In the next section the relationship between ABA content and water metabolism is examined to determine whether ABA could serve the dual role of preventing germination while allowing growth to proceed.

2. ABSCISIC ACID METABOLISM AND WATER RELATIONS

The water relations of the developing embryo and germinating seed are extraordinarily important in determining the success of the species. In *Phaseolus* water content per cell and per organ continues to increase until day 28, although the rapid deposition of dry matter makes water a progressively smaller component of total weight from day 18 onward. Water pressure (ψ) is constant at -5 to -6 bars until day 18 and then decreases at approximately 0.5 bar/day until day 28. After day 28 there is progressive actual water loss from the seed system including the embryo, and water potential decreases rapidly to less than -100 bars. During germination seeds and embryos rapidly imbibe water. This initial non-growth-related imbibition is usually considered a passive process that results from the high negative water potential of the seed. During and immediately after imbibition many seeds retain the ability to reenter developmental arrest and to withstand desiccation.

What prevents the developing embryo from undergoing imbibition during maturation? The water relations of the developing embryo would seem to favor precocious germination: a decreasing water potential with high water availability. If imbibition of water during germination to increase ψ to -5 bars is sufficient, what prevents embryo imbibition? First, it might be argued that the barrier provided by the seed coat or other surrounding tissues to radicle penetration changes during development. The force required to penetrate this barrier in lettuce is 0.2–0.4 M (-3 to -8 bars, approximately), so that embryos with a water potential less than -3 bars will germinate (Nabors and Lang, 1971). Seed coats and other surrounding tissues may be a more substantial barrier during maturation, so that a much higher water potential is required for radicle elongation. The existence of viviparous germination in Mangrove and various mutants however, argues that the force required to penetrate developing tissues is not high. This suggests that an additional brake on germination must be applied.

In this section I would like to propose that ABA acts as the brake on germination while promoting a physiological state in which hydration and synthesis are maximized. The evidence that ABA, a growth inhibitor, is capable of regulating water uptake and loss to maximize hydration will be discussed. This paradoxical action of ABA explains how the embryo can remain hydrated and active metabolically but not germinate. An extension of this hypothesis is that ABA is actually required for maturation to maintain a high water content and sugar and amino acid concentration.

Hsiao (1973) summarized the literature on water stress, concluding that water loss can affect (1) the chemical availability of water, (2) hydrostatic pressure, and (3) the concentration of cell solutes. At moderate water stress (-10 bars) the chemical availability of water is reduced only 1%, so it is unlikely that the metabolic consequences of water stress are due to a change in the availability of water. The concentration of cellular solutes is affected by water stress more dramatically due to the decrease in solvent but more importantly to an accumulation of osmotically active molecules such as sugars and amino acids during stress. This accumulation decreases water potential further than the actual loss of water. These phenomena result also in a loss of hydrostatic pressure, so that growth by elongation is unlikely.

Under conditions of mild water stress, hormones may regulate cell metabolism by means of their effect on water relations. Arad and Richmond (1976) measured leaf RNase in water-stressed barley seedlings after hormone application. Kinetin promotes water loss and an increase in leaf RNase; ABA application, however, results in leaf hydration and a reduced level of RNase. The effect of ABA is most likely due to its ability to promote stomatal closure, preventing water loss (Raschke, 1975). This phenomenon of ABA promoting water retention is also demonstrated by the **flacca** mutant of tomato. Imber and Tal (1970) showed that the permanent wilted condition of the plant can be changed to a high turgor condition by the daily foliar application of ABA.

Mild water stress is a condition in which ABA is rapidly synthesized *de novo*. This observation was first made by Wright and Hiron (1969), who found that wilting wheat leaves synthesized ABA within several hours after water stress was evident. Milborrow and Noodle (1970) showed that ABA concentration actually increased 40 times within 8 hr in water-stressed wheat leaves. More recently Zabadal (1974) demonstrated a water potential threshold in the leaves of several plants for the synthesis of ABA. Under conditions in which water potential decreases -1 atm/hr ABA synthesis is initiated at -12 bars; the *de novo* synthesis of ABA under Zabadal's experiment conditions is sensitive to -1 atm changes. Synthesis of ABA in response to water stress is not confined to leaf tissue. Loveys *et al.* (1975) demonstrated that grape pericarp tissue in suspension culture responds to osmotic stress in a similar manner. Pericarp tissue placed in a -720 mOsm solution of mannitol secreted 30 times the control level of ABA into the medium within 8 hr.

The application of ABA in the **flacca** mutant is clearly related to the ability of that genotype to retain water. An experimental demonstration of the accommodation of leaf tissue to water stress resulting in recovery from wilt to normal turgor status accompanied by ABA production was provided by Hiron and Wright (1973). They used a continuous 38°C airstream to wilt *P. vulgaris* seedlings in which only low levels of ABA were detectable. Within 10 min of starting the airstream, there was a 50% increase in ABA concentration as the plants began to wilt; by 20 min ABA concentration had doubled, and it had increased 15 times at 4 hr. When ABA concentration had reached about 4 times the basal level, the plants that had been wilted began to regain turgor and actually recovered a normal appearance after 90 min in the continuous hot airstream. Clearly, increases in ABA are correlated with the ability of leaves to retain and regain water during water stress.

The hypothesis that ABA can regulate water relations can also be approached from the view of the growth-inhibiting effects of this hormone. Rehm and Cline (1973) showed that ABA prevents within a few minutes the auxin-induced cell elongation in *Avena* coleoptiles. Walbot *et al.* (1975b) demonstrated that ABA depresses the water content increases of isolated axes of *P. vulgaris* after imbibition is complete. Durley *et al.* (1976) reported a similar phenomenon in lettuce seed germination, in which ABA application inhibits radicle elongation and hypocotyl expansion. At normal, −5 bar water potential of elongating tissue, ABA actually prevents water uptake and increases in hydrostatic pressure.

Dhindsa and Cleland (1975) attempted to separate the effects of water stress, hydrostatic pressure deficits, and ABA on cellular metabolism in *Avena* coleoptiles. A decrease in water potential results in a decrease in hydrostatic pressure and an increase in ABA concentration in this tissue. Water stress also reduces the rate and changes the pattern of protein synthesis in this tissue. The overall effect of water stress is not reversible by GA. The two most obvious components of stress, decreased hydrostatic pressure and increased ABA, cannot singly match the overall effect of water stress. Exogenous ABA does prevent growth (Rehm and Cline, 1973) and changes the rate but not the pattern of protein synthesis within 1.5 hr (Dhindsa and Cleland, 1975); the depression of protein synthesis is reversible by GA. When a hydrostatic pressure of 8.2 atm was applied without water loss, there was a change in the pattern of proteins synthesized within 1.5 hr; this effect was not modified by GA. From these experiments it is clear that the ABA and hydrostatic pressure effects on metabolism are separable parts of the water potential equation. Water stress, although affecting both components, cannot be duplicated by either used singly.

Hydrostatic pressure effects on plant tissue are less well known than some effects of ABA on cell metabolism. Abscisic acid is rapidly synthesized *de novo* during water stress from a precursor pool shared by GA and many more plant metabolites. There is some evidence that elevated levels of ABA can disrupt the metabolism of the other products. For example, Durley *et al.* (1976) showed that exogenous ABA prevents the normal metabolism of Ga during lettuce seed germination including an inhibition of conversion of GA_4 to GA_1. In water-stressed shoots in which ABA is presumed to increase, cytokinin level fails (Itai and Vaadia, 1971). During barley germination ABA prevents the Ga-dependent secretion of two preformed enzymes (Ben-Tal and Varner, 1974). In a number of other studies in which ABA effects on metabolic events have been measured, there is an antagonism between either GA–ABA or cytokinin–ABA. Unfortunately, these studies have not examined the role of the exogenous hormone in regulating the production of the antagonist. The flux of precursors to GA and ABA should be measured under a variety of steady state conditions; after application of antagonist a change in synthesis rate and total flux of each could be measured.

In two genetically defined situations ABA metabolism is clearly linked to other metabolic processes. Five of the known viviparous corn embryo types also show lesions in anthocyanin or carotenoid production (Robertson, 1955). The **rin,** nonripening mutant of tomato produces no ethylene, although the endogenous levels of ABA are the same as in wild-type plants (Mizrahi *et al.*, 1975). Large doses of exogenous ABA accelerate ripening so that there may be a lesion in ABA perception or metabolism.

Experimental evidence suggests that ABA has differential effects on the water relations of plant tissue as a function of water stress. Under hydrated conditions (-5 bars), exogenous ABA decreases the growth by elongation dependent on fresh weight increase. Under conditions of mild water stress (-6 to -15 bars) ABA promotes water retention and solute accumulation. To the extent that ABA prevents severe water loss, normal metabolic activity can continue since the chemical activity of water will not be rate limiting. (In leaves stomatal closure may limit photosynthesis; however, in nonphotosynthetic tissues gas exchange rate would be expected to be less important.)

The effects of ABA can be described by reference to the water potential equation. Water potential (ψ) is the sum of hydrostatic pressure (ψ_p), matrix pressure (ψ_m), and solute/osmotic pressure (ψ_s). Under normal hydration conditons ψ_p determines growth potential; hydrostatic pressure is more sensitive to changes in water potential than is matrix or solute pressure in most circumstances (Hsiao, 1973). If we assume that ABA prevents increases in hydrostatic pressure but can promote solute accumulation (ψ_p), then several descriptions of water relations as affected by ABA during embryo maturation are possible. For example, if in developing embryos of *P. vulgaris* water potential is decreasing, solute concentration is constant, and matrix pressure remains constant, i.e., the strength of the cell walls and surrounding tissues, it is clear that hydrostatic pressure must decline at a rate exceeding the fall in water potential for the equation to balance. This decline in hydrostatic pressure will prevent the germination of the seedling despite high water availability. To the extent that ABA can promote the accumulation of solutes, ABA will stabilize hydrostatic pressure; germination will be prevented as long as hydrostatic pressure is insufficient to force the embryo through the surrounding tissues. Abscisic acid might also prevent actual water loss from seed and embryo tissue despite decreasing water potential if solute concentration decreases in parallel with the fall in water potential. This possibility seems at first paradoxical to rapid accumulation of reserves. However, it is the flux of metabolites, not concentration within reasonable limits, that determines the amount of nutrient reserves deposited.

These possibilities and others that might be proposed suggest a number of interesting experiments to determine the water relations of both the embryo and the surrounding tissues. To use *P. vulgaris* axes as an example, the following are potential experiments:

1. Using the axis as a Cartesian diver in D_2O columns or osmotic columns (Nabors and Lang, 1971), will water-stressed material find its ψ and then as a response to stress produce ABA and rise to a lower ψ in the column?

2. In the same experimental setup described above, will 10^{-5} M ABA included in the column promote a more rapid adjustment of axis ψ?

3. If ^{3}H-labeled water and ^{14}C-labeled solutes are included in the medium or in prelabeled axes, can changes in ψ_p and ψ_s be distinguished? How does ABA affect each component of ψ?

Similar investigation of water–ABA metabolism in the seed coats would also be of great interest. Without this experimental evidence the hypothesis that ABA promotes maintenance of nutrient reserve anabolism despite declining water potential remains unproved.

A. Genetic and Environmental Diversity in Embryo Development and Dormancy

Both the rate and extent of embryo development appear to be genetically determined variables. Embryos such as *Fraxinus* (Villiers, 1972) require 2 years to mature. Some mature seeds contain rudimentary embryos arrested at the heart-shaped stage (*Ilex;* Hu, 1975) or at other stages (reviewed by Maheshwari, 1950). Mangrove and other viviparous embryos proceed directly from growth as embryos to growth as seedlings (Sussex, 1975). Within the dicots that form mature embryos there are also differences in the extent to which the nutrients of the endosperm and nucellar tissues are transferred to the embryonic cotyledons. In *Phaseolus* and cotton, nutrient transfer is complete, whereas in fenugreek (Reid and Meier, 1970), castor bean (Marré, 1967), and lettuce (Bewley and Fountain, 1972) large endosperm reserves are characteristic of the seed. In monocots the embryo is often small and the cotyledon contains few storage reserves; the carbohydrate reserves are found in the endosperm, and storage protein is deposited in the aleurone layer (reviewed by Dure, 1975). It is clear that there is considerable diversity in the pattern of development *in vivo;* this variation is presumably under genetic control.

In vitro embryo culture and embryoid induction from single cells also follows species-specific developmental stages. Tissue clumps derived from single cells of tobacco (Vasil and Hildebrant, 1965) and of carrot (Steward *et al.,* 1958) are capable of organizing at least a root meristem in suspension culture. When these rooted clumps are transferred to agar-solidified medium a shoot is sometimes organized, creating a plantlet. This *in vitro* regeneration bypasses the normal order of differentiation of organs and meristems found in embryo development of the same species. However, later work with carrot (Halperin, 1966; Steward *et al.,* 1964) indicated that embryoids would develop in suspension culture that mimic the later stages of histodifferentiation from the globular or heart-shaped stage. Embryoids from a particular species retain some species-specific differences, although there are characteristic departures in embryoids from the normal *in vivo* pattern such as the relative size of organs and rate of differentiation. Embryoid formation is even possible with initially haploid tissue as demonstrated by Nitsch and Nitsch (1969), who developed methods for inducing embryoid formation from anther cultures in tobacco. Adventitious plantlet production is common in many members of the Crassulaceae, although the initial development of these plantlets is quite distinct from that of an embryoid or embryo (reviewed by Steeves and Sussex, 1972).

Another experimental approach to examine the universality of the developmental programs governing seed development is to switch from comparison among species to comparison within species. Not only are the individual elements of a developmental program such as a new enzyme activity or morphological feature specified genetically but the developmental program itself must be encoded in the genome. We can gain some information on the genetic basis for the developmental program by comparing developmental rate and steps within a species. If variation exists, we may be able to describe the particular genetic elements responsible. First, we must distinguish between developmental rate and developmental steps. Developmental rate is defined as the length of time required to complete a given developmental program; for example, *Phaseolus vulgaris* and *P. coccineus* have the same developmental rate from zygote formation through cotyledon development (Fig. 1), but *P. vulgaris* completes the developmental program through embryo maturation about twice as fast as *P. coccineus*. Developmental steps are the biochemical, ultrastructural, anatomical, and morphological criteria used to measure the progress of the embryo. For example, the appearance of ABA in significant quanitites in *P. vulgaris* embryos is a criterion used to mark the end of embryo expansion by cell division and the beginning of embryo maturation. The order of some developmental steps may be rigidly fixed so that step 1 always precedes step 2, as is found in a biochemical pathway. Development is too complex a process, however, to be explained by simple cascade model of regulation because, as the steps are more finely divided, variations in order and extent become apparent.

Some preliminary evidence of Radin and Trehalase (1976, and unpublished data) has major implications for our understanding of the order in which developmental steps occur. In repeating Ihle's and Dure's (Ihle and Dure, 1969) experiments on precociously germinated cotton embryos, they found that moderate doses of actinomycin D prevented the appearance of isocitratase and carboxypeptidase C in certain seed lots. Differences in growing conditions and in delinting and seed processing procedures of seed lots of the same variety resulted in various degrees of susceptibility of the seed lot to inhibition by actinomycin D. Smith *et al.* (1974) also demonstrated a different response of cotton germination to inhibition by actinomycin D. Halloin (1976) has evidence that some cotton seeds retain sensitivity to ABA during germination in contrast to the insensitivity reported by Ihle and Dure (1969). To some extent these differences reflect variation in experimental procedure; however, the exciting possibility remains that differences in regulation of enzyme appearance during cotton seed development have been discovered.

On a theoretical level it is possible that different genotypes of cotton respond to different environmental cues or that growing conditions vary significantly for the various agronomic varieties used in these investigations. If the expression of the genes coding for the germination enzymes is regulated by environmental conditions, it is possible that in particular environments these genes will not be activated or activated only very late in maturation. This will result in the storage of very little or no mRNA in the seeds. Another possibility is that the order of transcription of the stored mRNA's is regulated, so that during a long growing season all of the germination enzyme mRNA's are transcribed whereas in a shorter growing season only the mRNA's transcribed early in maturation are found in the seed.

Both environmental cues and regulatory gene variation could result in a variety of seed types with regard to a particular developmental step since these two elements introduce a "sliding scale" for measuring *developmental time*. Simplistic development models place developmental steps in a linear order so that, as in the domino theory of foreign policy, the completion of the first step inevitably creates conditions in which the second step must occur, etc. This notion, also referred to as cascade regulation, works best for the macroscopic aspects of seed development such as morphological characters. Biochemical events peripheral to the elaboration of the embryo body may, however, occur in a much less rigidly ordered manner. As discussed above a developmental step as important as the production of stored mRNA's coding for enzymes required for successful germination may be complete, incomplete, or nonexistent at a particular time (dry seed stage). All cotton seed genotypes were, however, able to germinate successfully, so that the enzymes required for germination are produced when required. Thus, particular biochemical steps are not constrained to occur only at a certain developmental time.

Nor would a slower developmental rate (longer real time) necessarily increase the number of developmental steps completed before dormancy (i.e., more stored mRNA's present) since developmental time may be quite independent of rate. It may be useful to think of developmental rate as speed of travel and developmental time as the direction of travel. Developmental steps are diagnostic events for either or both rate and direction. For example, embryo axis length measurements early in development would allow us to conveniently describe the rate of embryo growth as an embryo. In a precociously germinating embryo, however, axis elongation is a signal of a change in developmental program, a change in direction to a new developmental time. Similarly the presence of a particular enzyme may not be an accurate reflection of the developmental stage of the organism since the enzyme may be induced precociously, late, or on time.

Conditions such as vivipary represent an extreme inversion of the order in which development normally proceeds; germination occurs clearly independent of the biochemical steps usually required to make a mature seed. Dormancy is also a highly variable trait in genera such as *Fraxinus* in which "deeply" dormant and nondormant species exist. A more dramatic variation in dormancy is found in *Aegilops kotchyi* Boiss (Wurzburger *et al.*, 1976), in which caryopses from different positions within the same spikelet differ greatly in the degree of dormancy. The upper floret is least dormant and the basal floret is most dormant in *in vitro* germination tests. If the upper floret is removed the subapical floret germinates readily. Environ-mental conditions can also modulate the imposition of dormancy. Junttila (1973) showed that the extent of seed dormancy in *Syringa vulgaris* is affected by the environment of the vegetative plant during embryo develop-ment.

Genetic and environmental regulation of embryo development and con-trol of dormancy clearly require and deserve more study. Agronomic species in which high germination percentage is a desirable trait may represent only a fraction of the potential genetic diversity in extent of and cues for developmental arrest.

IV. CONCLUSIONS

A. Hypotheses to Explain Embryogeny and Arrest

It is perhaps overly ambitious to title this section "Conclusions" since there is not yet a complete description of the development of a single embryo or seed system. However, the descriptions of embryo development presented here logically point to a set of hypotheses for future investigation. My personal benefit from reviewing the experimental evidence summarized here has been a realization of the central importance of water relations in embryo development. All plant growth depends on water availability; during evolution responses to water stress must have preceded the production of and regulation of growth and development by hormones. The effects of plant hormones on water relations as far as described here depend on the water potential of the tissue. At water potentials near -5 bars growth is possible, and the growth-promoting hormones promote increases in hydro-static pressure whereas ABA prevents growth. Under mild water stress growth-promoting hormones can exacerbate the condition of the plant or tissue, whereas ABA application or production promotes homeostasis. To argue on a teleogical basis, the action of ABA is to promote the maintenance of the tissue under stress in the absence of growth until condi-tions are ameliorated. This view is supported by the evidence that ABA is rapidly produced under conditions of water stress and that ABA rapidly promotes tissue response to stress as demonstrated by stomatal closure and leaf-wilting experiments.

Despite the centrality of water relations to an explanation of embryo development and arrest no complete description of the water potential equa-tion is available for the embryo and seed of a species. Of the several ways of expressing water availability, organ water potential or cellular water content appears to reflect most accurately the availability of water for metabolic and growth processes. Before the precipitous loss of water from seeds during the final stages of maturation, deposition of nutrients and water availability are high despite decreasing water potential and percent water content. On the basis of this general outline and the data specifically reviewed on *Phaseolus* embryo development, I propose the following sequence of events and regulatory process controlling development.

Following histodifferentiation the embryo grows rapidly, indicating a high water availability and hydrostatic pressure. Growth-promoting hormones are also in high concentration. When the embryo has filled the seed cavity within the pod, a mechanism for reducing hydrostatic pressure must be found to prevent precocious germination. Although the embryo is capable of germination and growth as a seedling, at this developmental stage only a fraction of the nutrient reserves of the mature seed have been synthesized. Thus, a mechanism for suppressing growth while allowing the accumulation of dry matter would optimize the success of the seed during germination. An environment in which intermediary metabolism and storage product synthesis continue while growth is suppressed is created by an increase in the osmolarity of the embryo cytoplasm; osmotically active amino acids and sugars are rapidly accumulated and polymerized into storage reserves. Accompanying the uptake of solutes is an increase in the uptake of water to maintain ψ at a nearly normal value. This water uptake can be demonstrated both by the rapid increase in tissue fresh weight from day 18 to day 28 and by the substantial dilution of some cellular solutes such as phosphate and nucleotides from day 16 to day 20.

Only a slight water stress is initially imposed on the embryo; this stress, however, is sufficient to trigger the synthesis of ABA. The presence of ABA enhances solute uptake by the embryo and thereby prevents an increase in hydrostatic pressure. Abscisic acid synthesized by maternal tissues under water stress might also contribute to the regulation of water and solute uptake by cotton and other embryos. Absorption of ABA might be the trigger for the 1–2 bar decrease in ψ sufficient to prevent precocious germination. The continued presence of ABA is required to maintain the embryo in a nongrowing condition. Viviparous mutant and natural embryos that do not respond to ABA show no developmental arrest; these embryos germinate when embryo size and differentiation are sufficient to produce a seedling. Embryos maturing normally in the presence of ABA, i.e., cotton and *Fraxinus*, can be washed free of the hormone and subsequently precociously germinate.

Under the no-growth conditions imposed by insufficient hydrostatic pressure for radicle elongation, the embryo ψ also gradually falls. The embryo water content and, of greater physiological importance, water availability progressively decrease over the day 18 to day 28 period of rapid nutrient reserve accumulation. At ψ -10 to -15 bars the unavailability of water limits metabolism as well as growth. Concomitant with the rapid decrease in ψ after day 28 the ABA content of *Phaseolus* also rapidly decreases.

The fall in ABA content is most likely due to a cessation of ABA synthesis. The rapid conversion of ABA to catabolic derivatives by germinating bean axes suggests that ABA content would decrease rapidly without continued synthesis. However, the level of enzymes required for ABA breakdown has not been determined during *Phaseolus* embryogeny. It would be interesting to know, for example, whether the enzymes involved in ABA metabolism are synthesized *de novo* during maturation or whether they are constitutive functions of the embryo. The production of ABA as a function of water stress should also be examined. A cybernetic model for ABA metabolism is one intriguing possibility: ABA is synthesized in response to mild water stress and serves to prevent rapid water loss, but when water potential falls below some critical value ABA can no longer be synthesized, and ABA level and water potential drop precipitously. From this model one interesting deduction is that ABA prolongs the maturation phase of embryogeny by maintaining water potential greater than -15 bars. This prediction could be tested in a species with varieties of differing lengths of maturation and/or final seed weight by determining whether the maturation phase is accompanied by an ABA level sufficient to suppress growth and maintain a reasonable water potential for active nutrient reserve synthesis.

During maturation the embryo may also become more physically restrained by the seed coats and other ovular tissues which typically dehydrate and toughen prior to embryo desiccation. The deposition of insoluble material in embryonic storage tissue coupled with dehydration of these tissues may also prevent embryo elongation. The most severe hindrance to embryo growth is, however, desiccation of the embryo. There is little evidence on the water relations of mature embryos and fruits. Vascular tissue surrounding the embryo does not degenerate since imbibing bean or cotton seeds take up dye through the micropyle or funiculus and the dye is distributed through the extensive veination of the ovular tissue, (I. Sussex and V. Walbot, unpublished data). The quality of the vascular system between each seed and the fruit and between the fruit and the remainder of the plant has not been examined in detail.

B. New Experimental Approaches

The acceptance of a unitary concept of developmental arrest in plant embryo development is a first step in the elucidation of the molecular and structural basis for this developmental fate. Cursory description of a wide variety of developing seeds and the factors responsible for dormancy and/or promoting germination is already available as a background on the diversity of the species-specific details. It would behoove developmental botanists interested in this problem to concentrate their efforts on a few species in which genetic tools are available. The major points to be elucidated include (1) the genetic basis for vivipary, (2) the relationship between ABA and water potential, (3) the mechanism of rapid water loss during the final maturation of an embryo, (4) the importance of biophysical constraints to embryo growth when seeds reach nearly full size, (5) the metabolic consequences of rapid and slow changes in water potential in the range of -5 to -15 bars, and (6) the metabolic consequences of low (< -15 bars) water potential.

I believe that the answers to these questions will lead to a more complete understanding of developmental arrest. A moratorium on incorporation studies using radioactive precursors of protein and nucleic acid synthesis would provide a work force for experimental rather than descriptive explanation of arrest. The plant materials for which the most descriptive data exist are corn, beans, barley, ash, and lettuce. Corn viviparous mutants should provide an especially precise definition of dehydration and ABA metabolism.

ACKNOWLEDGMENTS

I would like to thank Keith Walker, Ian Sussex, Mary Clutter, and Tom Brady, who have discussed the concepts and data presented in this review with me on numerous occasions and who may find their own ideas restated here in a modified form. The framework within which the literature was evaluated was certainly a product of the discussions held over the years concerning the regulation of embryo development and germination in *Phaseolus*. Special thanks are due to Keith Walker for providing a thoughtful critique of the manuscript.

REFERENCES

Amen, R. D. (1968).*Bot. Rev.* **34,** 1.
Arad, S., and Richmond, A. E. (1976). *Plant Physiol.* **57,** 656.
Barton, L. V. (1965a). *In* "Handbuch der Pflanzenphysiologie" (W. Ruhland, ed.) Vol. 15, Part 2, p. 699. Springer-Verlag, Berlin and New York.
Barton, L. V. (1965b). *In* "Handbuch der Pflanzenphysiologie" (W. Ruhland, ed.), Vol. 15, Part 2, p. 727. Springer Verlag, Berlin and New York.
Ben-Tal, Y., and Varner, J. E. (1974). *Plant Physiol.* **54,** 813.
Berlyn, G. P. (1972). *In* "Seed Biology" (T. T. Kozlowski, Ed.), Vol. 1, pp. 223–312. Academic Press, New York.
Bewley, J. D., and Fountain, D. W. (1972). *Planta* **102,** 368.

Bhatnager, S. P., and Johri, B. M. (1972). *In* "Seed Biology" (T. T. Kozlowski, ed.), Vol. 1, pp. 78–149. Academic Press, New York.
Briarty, L. G., Coult, D. A., and Boulter, D. (1969). *J. Exp. Bot.* **20,** 358.
Carr, D. J., and Skene, K. G. M. (1961). *Aust. J. Biol. Sci.* **14,** 1.
Clutter, M., Brady, T., Walbot, V., and Sussex, I. (1974). *J. Cell. Biol.* **63,** 1097.
Davidson, E. H., and Britten, R. J. (1973). *Q. Rev. Biol.* **46,** 565.
Dhindsa, R. S., and Cleland, R. E. (1975). *Plant Physiol.* **55,** 782.
Dure, L. S. (1975). *Annu. Rev. Plant Physiol.* **26,** 259.
Durley, R. C., Bewley, J. D., Railton, I. D., and Pharis, R. P. (1976). *Plant Physiol.* **57,** 699.
Eeuwens, C. J., and Schwabe, W. W. (1975). *J. Exp. Bot.* **26,** 1.
Flinn, A. M., and Pate, J. S. (1968). *Ann. Bot. (London)* **32,** 479.
Habib, A. F. (1973). *Mysore J. Agric. Sci.* **7,** 120.
Hall, R. H. (1973). *Annu. Rev. Plant Physiol.* **24,** 415.
Halloin, J. M. (1976). *Plant Physiol.* **57,** 454.
Halperin, W. (1966). *Am. J. Bot.* **53,** 443.
Harris, B. (1976). Ph. D. Thesis, University of Georgia, Athens.
Harris, B., and Dure, L. S. (1974). *Biochemistry* **13,** 5463.
Hiron, R. W. P., and Wright, S. T. C. (1973). *J. Exp. Bot.* **24,** 769.
Ho, D. T. -H., and Varner, J. E. (1974). *Proc. Natl. Acad. Sci. U.S.A.* **71,** 4783.
Ho, D. T. -H., and Varner, J. E. (1976). *Plant Physiol.* **57,** 175.
Hofstein, A. (1973). *Physiol. Plant.* **29,** 76.
Hsiao, R. C. (1973). *Annu. Rev. Plant Physiol.* **24,** 519.
Hu, C. Y. (1975). *J. Am. Soc. Hortic. Sci.* **100,** 221.
Ihle, J. N., and Dure, L. S., III. (1969). *Biochem. Biophys. Res. Commun.* **36,** 705.
Ihle, J. N., and Dure, L. S., III. (1970). *Biochem. Biophys. Res. Commun.* **38,** 995.
Ihle, J. N., and Dure, L. S., III. (1972a). *J. Biol. Chem.* **247,** 5034.
Ihle, J. N., and Dure, L. S., III (1972b). *J. Biol. Chem.* **247,** 5048.
Imber, D., and Tal, M. (1970). *Science* **169,** 592.
Itai, C., and Vaadia, Y. (1971). *Plant Physiol.* **47,** 87.
Junttila, O. (1973). *Physiol. Plant.* **29,** 264.
Klein, S., and Pollock. B. M. (1968). *Am. J. Bot.* **55,** 658.
Kloz, J., Turkova, L., and Klozova, E. (1966). *Biol. Plant.* **8,** 164.
Koller, D., Mayer, A. M., Poljakoff-Mayber, A., and Klein, S. (1962). *Annu. Rev. Plant Physiol.* **13,** 437.
Lang, A. (1970). *Annu. Rev. Plant Physiol.* **21,** 537.
Limberk, J., and Ulrychova, M. (1972). *Phytopathol. Z.* **73,** 227.
Loveys, B. R., Brien, C. J., and Kriedemann, P. E. (1975). *Physiol. Plant.* **33,** 166.
Lowenberg, J. R. (1955). *Plant Physiol.* **30,** 244.
Maheshwari, P. (1950). "An Introduction to the Embryology of the Angiosperms." McGraw-Hill, New York.
Manohar, M. S. (1966). *J. Exp. Bot.* **17,** 231.
Marré, E. (1967). *Curr. Top. Dev. Biol.* **2,** 76.
Mayer, A. M., and Shain, Y. (1974). *Annu. Rev. Plant Physiol.* **25,** 167.
Milborrow, B. V. (1974). *Annu. Rev. Plant Physiol.* **25,** 259.
Milborrow, B. V., and Noodle, R. C. (1970). *Biochem. J.* **119,** 727.
Millerd, A., and Whitfeld, P. R. (1973). *Plant Physiol.* **51,** 1005.
Mizrahi, Y., Dostal, H. C., McGlasson, W. B., and Cherry, J. H. (1975). *Plant Physiol.* **56,** 544.
Nabors, M. W. and Lang, A. 1971. *Planta* **101,** 1.
Nitsch, J. P., and Nitsch, C. (1969). *Science* **163,** 85.

Öpik, H. (1968). *J. Exp. Bot.* **19**, 64.
Öpik, H., and Simon, E. W. (1963). *J. Exp. Bot.* **14**, 299.
Ory, R. L., and Henningsen, K. (1969). *Plant Physiol.* **44**, 1488.
Radin, J. W., and Trelease, R. N. (1976). *Plant Physiol.* **57**, 902.
Raghavan, V. and Torrey, J. G. (1963). *Amer. J. Bot.* **50**, 540.
Raschke, K. (1975). *Annu. Rev. Plant Physiol.* **26**, 309.
Rehm, M., and Cline, M. G. (1973). *Plant Physiol.* **51**, 93.
Reid, J. S. G., and Meier, H. (1970). *Phytochemistry* **9**, 513.
Robertson, D. S. (1955). *Genetics* **40**, 745.
Skene, K. G. M., and Carr, D. J. (1961). *Aust. J. Biol. Sci.* **14**, 13.
Smith, D. L. (1974). *Protoplasma* **79**, 41.
Smith, R. H., Schubert, A. M., and Benedict, C. R. (1974). *Plant Physiol.* **54**, 197.
Sondheimer, E., and Galson, E. C. (1966). *Plant Physiol.* **41**, 1397.
Sondheimer, E., Tzou, D. S., and Galson, E. C. (1968). *Plant Physiol.* **43**, 1443.
Sondheimer, E., Galson, E. C., Tinelli, E., and Walton, D. C. (1974). *Plant Physiol.* **54**, 803.
Steeves, T. A., and Sussex, I. M. (1972). "Patterns in Plant Development." Prentice-Hall, Englewood Cliffs, New Jersey.
Steward, F. C., Mapes, M. O., and Mears, K. (1958). *Am. J. Bot.* **45**, 705.
Steward, F. C., Mapes, M. O., Kent, A. E., and Holsten, R. D. (1964). *Science* **143**, 20.
Sussex, I., Clutter, M., Walbot, V., and Brady, T. (1973). *Caryologia* **25**, 261.
Sussex, I. M. (1975). *Amer. J. Bot.* **62** 948.
Sussex, I., Clutter, M., and Walbot, V. (1975). *Plant Physiol.* **56**, 575.
Tronier, B., Ory, R. L., and Henningsen, K. (1971). *Phytochemistry* **10**, 1207.
Vasil, V., and Hildebrandt, A. C. (1965). *Science* **150**, 889.
Villiers, T. A. (1972). *In* "Seed Biology" (T. T. Kozlowski, ed.), Vol. 2, p. 220–281. Academic Press, New York.
Villiers, T. A., and Wareing, P. F. (1965). *J. Exp. Bot.* **16**, 533.
Walbot, V. (1971). *Dev. Biol.* **26**, 369.
Walbot, V. (1972). *Planta* **108**, 161.
Walbot, V. (1973a). *New Phytol.* **72**, 479.
Walbot, V. (1973b). *Caryologia* **25**, 273.
Walbot, V., and Dure, L. S. III (1976). *J. Mol. Biol.* **101**, 503.
Walbot, V., Clutter, M., and Sussex, I. M. (1972a). *Phytomorphology* **22**, 59.
Walbot, V., Brady, T., Clutter, M., and Sussex, I. (1972b). *Dev. Biol.* **29**, 104.
Walbot, V., Capdevila, A., and Dure, L. S., III. (1974). *Biochem. Biophys. Res. Commun.* **60**, 103.
Walbot, V., Harris, B., and Dure, L. S., III. (1975a). *In* "Developmental Biology of Reproduction" (C. Markert, ed.), pp. 165–187. Academic Press, New York.
Walbot, V., Clutter, M., and Sussex, I. M. (1975b). *Plant Physiol.* **56**, 570.
Walker, K. A. (1974). Ph.D. Thesis, Yale University, New Haven, Connecticut.
Walton, D. C. and Soofi, G. S. (1969). *Plant Cell Physiol* **10**, 307.
Walton, D. C., Dorn, B., and Fey, J. (1973). *Planta* **112**, 87.
Wareing, P. F. (1965). *In* "Handbuch der Pflanzenphysiologie" (W. Ruhland, ed.), Vol. 15, Part 2, pp. 909–924. Springer-Verlag, Berlin and New York.
Wareing, P. F., and Saunders, P. F. (1971). *Annu. Rev. Plant Physiol.* **22**, 261.
Waters, L. C. and Dure, L. S. III (1966). *J. Mol. Biol.* **19**, 1.
Wright, S. T. C. and Huron, R. W. P. (1969). *Nature, Lond.* **224**, 719.
Wurzburger, J., Lesham, Y., and Koller, D. (1976). *Plant Physiol.* **57**, 670.
Zabadal, T. J. (1974). *Plant Physiol.* **53**, 125.
Zacharius, R. M. (1970). *Phytochemistry* **9**, 2047.

*In a study of Avena it is necessary to gather as
much data as possible from a variety of external
factors, and, recognizing the uncertainty to which
any one line of enquiry alone leads, to judge the
situation from the combined results.*

Atwood, 1914

4

Metabolic Regulation of Dormancy in Seeds—A Case History of the Wild Oat (*Avena fatua*)

GRAHAM M. SIMPSON

DORMANCY AND DEVELOPMENTAL ARREST

I. INTRODUCTION

A. General Considerations

There are wide ranges within the plant kingdom of both structure and germination behavior in seeds. These variations can be seen whether classification is approached taxonomically or through arbitrary divisions into starch, protein, and oilseeds. Within these ranges differences in the type and degree of seed dormancy are considerable (Villiers, 1972; Nikolaeva, 1969). The approach used in this chapter is to consider one species, the wild oat (*Avena fatua* L.), and to attempt an explanation of seed dormancy based primarily on the evidence from this species. In places reference may be made to the common oat (*Avena sativa* L.) or to the winter wild oat (*Avena ludoviciana* L.), which are both closely related taxonomically and also considered to be derived from *Avena fatua* (Aamodt *et al.*, 1934). The reason for this narrow approach is to avoid the temptation of constructing a general explanation of the mechanism of dormancy which might be applicable, for example, to at least all cereals by combining evidence obtained largely from nondormant domesticated cereals, such as wheat and barley, which provide us with the bulk of our physiological and biochemical knowledge of normal germination processes in cereals. Domestication of the cereals has involved genetic selection against natural long-term dormancy, which probably existed in ancestral wild species.

B. Definition of Dormancy and Metabolism

Various definitions of seed dormancy have been given in the literature (Amen, 1968; Villiers, 1972), and the final definition will be given only when seed–environment systems have been completely elucidated. As a starting point for a definition in its most simplistic and general sense, seed dormancy can be described as that state in which no macro changes occur in the visible appearance of the seed following abscission from the parent plant and in which viability is retained. Growth is suspended, in some cases for periods of many years, as long as the seed remains desiccated. The initiation of this resting stage of the sporophyte generation is presumably located in the genetic control exerted by the female parent plant, suitably modified by some specific set of environmental factors. From a practical viewpoint the cessation of dormancy is measured by the onset of visible growth (germination), which is generally in the form of protrusion of the root or shoot from the protecting testa shortly after the seed has been rehydrated.

Most frequently the term "seed dormancy" is used to describe a specific modification of the general desiccated state described above and, in this more restricted sense, dormancy is that state in which a seed fails to resume growth when it is rehydrated in an environment that will support normal germination and seedling growth of apparently identical, but nondormant, seeds from the same species or even the same parent plant. Usually the elapse of some period of time or exposure to a set of environmental changes will eventually entrain internal changes in either the desiccated or hydrated seed such that it can germinate and grow at a later time in the same environment that initially failed to support growth. This extension of the resting state beyond the limits enforced by desiccation ensures greater variation in the time of germination and thus increases the chance of survival of at least some members of the population within a species. This more specialized form of dormancy represents a subtle adaptation of the seed to certain specific sets of environmental factors and confers a competitive advantage on those species that have become important weeds in crops. These crops generally have little or no dormancy in the hydrated seed.

For the purpose of reviewing literature pertaining to the mechanism of seed dormancy in the specific case of *A. fatua*, it is sufficient to use the loose definition that a dormant seed is one that fails to germinate when it is rehydrated in an environment that supports normal germination of rehydrated nondormant seeds from the same population. Although this is deliberately vague and ambiguous, it is the intention to demonstrate by the later discussion and review of the literature that a precise definition of dormancy cannot be used in the general sense to apply to all seeds but can only be given for each individual seed considered in the context of a precisely defined set of environmental conditions.

The term "metabolism" will be used in the broadest sense (Street and Cockburn, 1972) as the "sum total of chemical reactions of the living seed," which permits discussion of all levels of change occurring in the germinating or nongerminating seed.

C. Conceptual and Semantic Problems

The experimental approach used to determine the nature of dormancy as a general phenomenon in seeds has been based for the most part on the reductionist approach summarized by Koestler (1967), which emphasizes the systematic division of the whole into its constituent parts through successively lower levels of organization. Carried to the current limits of biological and biochemical technology, analysis of a seed down to the level of its constituent molecules will reveal, at that level, the biochemical (metabolic) relationships. On the other hand, modern systems analysis applied to biological organisms (Koestler, 1967; von Bertalanffy, 1968; Whyte *et al.*, 1969) stresses that all parts of a system are organized as a hierarchy of levels, which are either structurally, functionally, or taxonomically interrelated (Fig. 1). The requirements on one level act as the constraints of the level beneath. Thus, in the case of a functional hierarchy the proper functioning on a given level requires that all the levels below that level function normally. Systems analysis stresses that the rules governing function at one level are not the same rules that govern the function of higher or lower levels in the system. Similarly, in a structural sense, the whole is made up of parts that themselves have attributes of wholeness and autonomy. We recognize this when we start at the level of the cell to describe structure and proceed upward in the hierarchy through tissues, organs, organisms, and populations of organisms or downward through organelles, plastids, macromolecules, molecules, atoms, etc. The semiautonomous subwholes

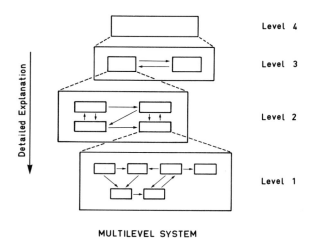

MULTILEVEL SYSTEM

Fig. 1.   A multilevel hierarchical system. (After Whyte *et al.*, 1969.)

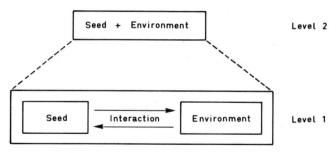

Fig. 2. A seed–environment system.

that are the stable units of any level in a hierarchy have been called holons (Koestler, 1967). At any one level there may be any number of holons, which can function autonomously but which must all function normally for the proper expression of the next level in the hierarchical system. Ideally, a systems approach to understanding seed dormancy would require that structural levels be separated from functional levels and that all levels be connected logically. For example, growth regulators, which are molecules, presumably exert their effects directly on other molecules and not nonspecifically at tissue or organ levels. In reality it is often extremely difficult to separate function from structure.

It is often assumed that the most important description of dormancy is at the biochemical level. However, if the rules that govern higher levels in the same system are not the same as those for lower levels, there is an obvious weakness in looking for an explanation for dormancy solely in biochemical terms. The most satisfying description is not necessarily the most detailed description at the lowest level of the hierarchical structure but rather is the holistic view of the entire system, which brings an understanding of the relationships among levels as well as within levels. A brief conceptual analysis of the question "Why does a dormant wild oat seed fail to germinate following rehydration?" can illustrate the complexities and technical difficulties that face a plant physiologist who examines the problem of seed dormancy by a systems approach. The initial problem is to define the extent of the system. An extensive literature on seeds, produced over the last 100 years, indicates that seeds cannot be considered in the absence of environment. Thus, in the simplest sense, the whole system can logically be described as the seed and environment (Fig. 2). By subdividing the environment at level 2 into all of its constituent parts the common observation that dry seeds do not germinate becomes less significant and in perspective beside the many other factors (light, temperature, oxygen, ethylene, CO_2, ions, soil matrix, etc.) that also make up the natural environment of a seed. Each of the environmental factors, if considered alone, may have very significant effects on germination and dormancy.

Given any specific set of environmental factors, further understanding of the system can be attained through anatomical subdivision of the seed (Fig. 3). It is difficult to subdivide on a functional basis without making some kind of anatomical subdivision. Interactions may occur between parts within the seed subsystem and between the environmental factors within the environmental subsystem. Interactions may also occur between each of the components of each subsystem so that complexities become multiplied rapidly, particularly if the interactions are reciprocal effects in the sense that a seed component may be affected by the environment but may itself affect the environment.

Another dimension that must be added to the system is the dynamic nature of the changes that occur during the arbitrary time frame chosen to measure rate of change. The time frame that has commonly been used to determine whether dormancy is present in a seed is from that point at which the seed begins imbibing water to the appearance of an emerging radicle that can be seen with the naked eye (Fig. 4). This, of necessity, can be measured only by using a nondormant seed since the dormant seed, however else we characterize it, does not generally have an emerging radicle. The vast majority of investigations of seed dormancy, which would include many of the more than 20,000 papers cited in one text (Barton, 1967), use the emergence of the radicle as the key indicator of dynamic change in the system. Nevertheless, it is only a small part of the seed subsystem, and there are many cases in which normal growth and development do not continue after radicle emergence (Haber, 1962). Failure of one part of the system (for example, radicle emergence) does not necessarily imply that the remainder

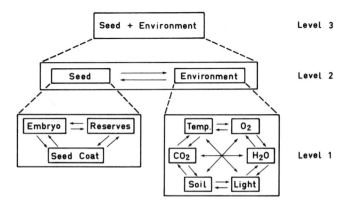

Fig. 3.   Three levels of a seed–environment system.

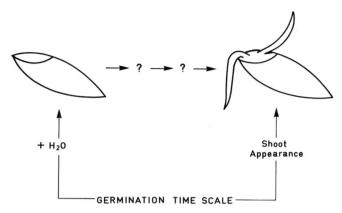

Fig. 4.   A time frame for germination based on shoot appearance.

of the system is inoperative. Analysis (by experimental techniques) of the full nature of a seed–environment system that exhibits dormancy is very difficult. Nevertheless, the concept of a systems approach for comprehending dormancy is useful because it emphasizes the importance of investigating the connecting links between levels of the system and of conceptually describing the system in a logical manner so that the confusion caused by misrelating levels and sublevels can be avoided.

The general approach used in this chapter is to examine the system of *Avena fatua* seed and its environment and to determine the relative importance of metabolism, as against other structural and functional aspects of the system, as the source of the mechanism of dormancy. Since the experimental approaches used by the different investigators of *A. fatua* have been quite varied and focused on different levels of the system, any overall picture of the total system is of necessity sketchy. In addition, some parts of the system have never been investigated at all, whereas others have been studied in considerable detail.

D. Individuals and Populations

Observations on dormancy, or for that matter on nondormant seeds, are difficult to make on single seeds, and for this reason most studies are made by comparing populations of seeds. Although this may not pose a problem when genetic differences are easily discernible, for example, as distinct morphological differences, so that populations can be separated accurately, it can be a potentially serious problem as biochemical aspects are investigated.

At this level the presence of a few nondormant seeds in a population of otherwise dormant seeds can produce qualitative and quantitative changes in enzymes, metabolites, and growth regulators that can obscure the real differences between dormant and nondormant seeds. Although *Avena fatua* is self-pollinating (Aamodt *et al.*, 1934) and the proportion of outcrossing is very low, there is evidence for considerable genetic variation in both the presence or absence and depth of dormancy in natural populations (Aamodt *et al.*, 1934; Baker and Leighty, 1958; Chancellor and Peters, 1972; Coffman and Stanton, 1938, 1940; Johnson, 1935a; Lindsay, 1956; Lute, 1938; Marshall and Jain, 1970; Thurston, 1957, 1963; Toole and Coffman, 1940). For these reasons it is desirable, as a general principle, to work with uniform, inbred populations when investigating the mechanism of seed dormancy. In the specific case of *Avena fatua* it is probable that of the more than 100 laboratories that have investigated dormancy in this species not more than 5 (Andrews and Burrows, 1972; Chen and Varner, 1969; Johnson, 1935b; Naylor, 1966; Simpson, 1965) have worked with inbred populations. For somewhat similar reasons it is an important practical point to keep a constant check on the germination percentage of even an inbred population which will always undergo after-ripening in storage and thus shift with time to become either a partially or completely nondormant population. A way around this difficulty is to grow populations of inbred plants continuously so that freshly harvested and completely dormant populations of seeds are always available which can be contrasted with genetically identical material that has lost dormancy completely through a lengthy period of after-ripening (Naylor, 1966; Simpson, 1965).

II. DORMANCY AND THE WEED PROBLEM

A. Seed Survival

Seed dormancy, which leads to the persistence of the seed in a viable state for periods of many years, is the main reason for the survival of wild oat populations (Banting, 1966a,b). Seed dormancy, combined with other attributes of the plant such as competitive growth of the seedling and mature plant and the hairy nature of the seed and suckermouth, which cause the seed to be trapped in agricultural machinery, makes wild oats one of the most serious weeds of cereal and oil crops in the temperate zone (Evans, 1958; Thurston, 1962a, 1964).

The combination of dormancy in the caryopsis and a protective hull (lemma and palea) ensures survival of the seed under a wide range of environmental conditions. Seeds can persist and remain viable under high temperatures (Bibbey, 1935; Hopkins, 1936; Sexsmith, 1969; Williams and Thurston, 1964), low temperatures (Andrews and Burrows, 1972, 1974), burying to a considerable depth in soil (Thurston, 1961), various forms of soil cultivation (Banting, 1966a; Leggett, 1950; McDonald, 1949; Tingey, 1961; Wilson and Cussans, 1975), under pasture leys (Thurston, 1966), under various forms of fertilizer application (Rademacher and Kiewnick, 1964), on the soil surface under burning straw (Wilson and Cussans, 1975) and under a range of light conditions (Cumming and Hay, 1958; Thurston, 1963; Whittington *et al.*, 1970), in barnyard manure up to 4 months, and after passage through the intestine of bovine ruminants (Kirk and Courtney, 1972).

B. Genotype and Environment

There is considerable genotypic variation in the expression of seed dormancy in natural populations of *Avena fatua* (Section III,E). In addition there are also differences in depth of dormancy between seeds from the same spikelet; secondary seeds are generally more dormant than primary seeds (Coffman and Stanton, 1938; Johnson, 1935a). In certain environmental circumstances secondary dormancy can be induced both in genotypically dormant, but after-ripened, seeds that have lost primary dormancy and in nondormant seeds from a nondormant strain of *Avena fatua*. Seeds with this induced secondary dormancy fail to germinate when restored to an environment that previous to the induction could support normal germination. The most documented situation for induction of natural secondary dormancy is an excessively wet situation, which leads to partial restriction of the oxygen supply due to the presence of hulls and adhering water (Banting, 1966a,b; Hay, 1962, 1967; Hsiao and Simpson, 1971; Kiseleva, 1956; Kommedahl *et al.*, 1958; Simmonds, 1971; Thornton, 1945). Light (Cumming and Hay, 1958; Hart and Berrie, 1966; Hay and Cumming, 1959; Hsiao and Simpson, 1971) and CO_2 (Hart and Berrie, 1966) can interact with oxygen status for the induction of secondary dormancy. Soaking dormant seeds in water followed by drying prolongs the period of primary dormancy (Banting, 1966b; Johnson, 1935a).

Under the influence of either primary or secondary states of dormancy wild oat seeds have been shown to survive under natural conditions in the soil for as long as 5 years under a pasture (Thurston, 1966), 6 years under cultivated conditions (Cates, 1917), and 7 years in a light-textured, cultivated soil (Banting, 1966a).

III. ROLE OF SEED STRUCTURE IN
DORMANCY OF WILD OATS

A. Outline of Seed Morphology

No detailed description of the seed of *Avena fatua* has yet been published. The flower and early developing embryo have been described (Cannon, 1900) as has the mature embryo of the closely similar *Avena sativa* (Avery, 1930). For the purposes of the approach used in this chapter a reproduction of the caryopsis (fruit), with or without the enclosing lemma and palea

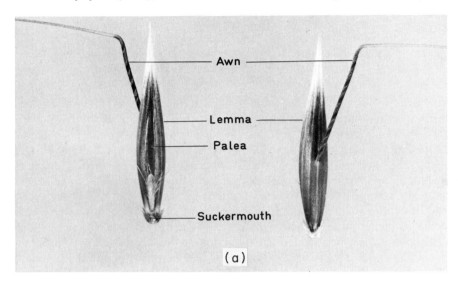

(a)

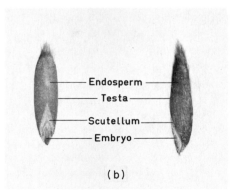

(b)

Fig. 5.   Seed (a) and dehulled fruit (b) of *Avena fatua*.

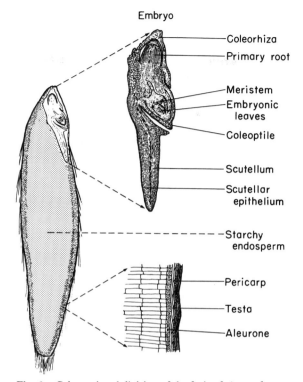

Embryo

Coleorhiza

Primary root

Meristem
Embryonic
leaves
Coleoptile

Scutellum

Scutellar
epithelium

Starchy
endosperm

Pericarp

Testa

Aleurone

Fig. 6. Schematic subdivision of the fruit of *Avena fatua*.

(hulls), is shown (Fig. 5). A schematic subdivision of the fruit into the components discussed under structure is also shown (Fig. 6).

B. Lemma and Palea

The earliest observations on dormancy in *Avena fatua* seeds demonstrated that removal of the hull promotes germination of the fruit (Zade, 1909; Atwood, 1914; Kommedahl *et al.*, 1958). Several explanations for the inhibitory role of hulls have been put forward. Germination inhibitors that can be leached from the hulls by water have been demonstrated (Helgeson and Green, 1957a; Naylor and Christie, 1957; Hay, 1962). Hulls of a dormant, as compared with a nondormant, strain contained a growth inhibitor that permitted germination of *Avena fatua* seeds but stopped subsequent seedling growth (Pawlowski, 1959). Several substances, which could be separated by paper chromatography and which inhibited germination of lettuce seeds, were extracted from hulls of semidormant *Avena fatua* seeds by Black (1959). He also concluded that hulls prevented the leaching out of germination inhibitors present in the fruit of *Avena fatua*.

Hulls may interfere with gas exchange. One study (Baker and Leighty, 1958) concluded that hulls act as a barrier to the uptake of both oxygen and water and that after imbibition of water by the fruit the hulls also confine inhibitory products of respiration within the fruit, thus delaying germination. An increase in oxygen uptake of fruits accompanied by an increase in germination was noted after hulls were removed (Hay and Cumming, 1959), but it was not established whether the increased oxygen uptake preceded or followed the increased germination. A 40% reduction of germination was noted in seeds with hulls compared with hulls removed even in the presence of an atmosphere of 20% oxygen. The authors (Hart and Berrie, 1966) concluded that hulls interfere with oxygen uptake by the fruit.

While hull and seed coat color have often been implicated as factors associated with dormancy, a comprehensive study (Toole and Coffman, 1940) showed no relationship between dormancy, seed coat color, hull color, or a number of other hull characteristics for a wide range of types of *Avena fatua* selected from the western United States. A later study (Kommedahl *et al.*, 1958) also indicated no relationship between dormancy and either seed coat or hull color.

Although one study indicated that hulls do not interfere with water uptake by the fruit (Black, 1959), there is some indication that hulls have a role in mediating an interaction between light and water which affects germination. Whittington *et al.* (1970) demonstrated that removal of hulls promotes germination in both light and darkness. Nevertheless, germination was higher in light than in darkness when hulls were kept on. In a similar kind of study with hulled and dehulled seed (Hsiao and Simpson, 1971) an opposite effect was noted. In the presence of hulls germination was inhibited in white light compared to darkness, and this difference was even greater when hulls were removed. The same study indicated that the volume of water available to the seed determines whether light inhibits germination (very low or very high volumes) or promotes germination (intermediate volumes). The presence or absence of hulls modifies this relationship.

Another aspect of the light response that is modified by hulls has been noted (Hart and Berrie, 1966). If CO_2 concentration is zero in the atmosphere in which seeds are germinated, light inhibits both intact and dehulled seed. When CO_2 is added to the atmosphere, dehulled seeds are no longer inhibited by light. This suggests a role for the hull in modifying the rate of diffusion of CO_2 either into or out of the fruit.

Although hulls undoubtedly play a role in modifying germination in seeds exhibiting various degrees of dormancy, their presence cannot account for the total expression of dormancy since it has been shown (Atwood, 1914; Johnson, 1935a; Naylor and Simpson, 1961a,b) that a fruit can be dormant in the absence of lemma and palea.

C. Pericarp and Testa

The pericarp, consisting of remnants of the original ovary wall, integuments, and epidermis of the nucellus (Cannon, 1900), together with the thin testa, makes up the outer coat of the fruit. Atwood (1914) was the first to show that pricking the fruit coat after the hulls were removed increased germination from about 40 to 100% when fruits were given normal conditions for germination. The effect disappeared as fruits were after-ripened over a period of several months. He also observed that in immature fruit the CO_2/O_2 ratio changed from 0.800 in air to 0.649 when coats were punctured and from 0.800 to 0.557 when intact fruits were placed in 93% oxygen. He concluded that the coat is a barrier to the diffusion of oxygen. The same author showed that puncturing the coat of immature fruits (milky dough stage) promoted germination from 30 to 80%.

In another series of experiments (Hay and Cumming, 1957, 1959; Hay, 1962) it was shown that puncturing the coat increased germination of fruits with primary dormancy and also fruits with induced (secondary) dormancy. The latter fruits, when punctured, germinated even when the wound was covered with lanolin or agar. Puncturing the fruit coat also increased oxygen uptake. Puncturing fruits, with or without the hulls removed, increased germination from 15 to 85% (Baker and Leighty, 1958), and a similar effect was noted when coats were punctured by pricking through the hulls (Drennan and Berrie, 1961).

The above evidence indicates that the coat has some effect on germination by interfering with either gas exchange or imbibition of water. Nevertheless, the presence of an intact seed coat, as with the presence of hulls, cannot account for the total expression of dormancy because it has been demonstrated that puncturing the seed coat or excising the embryo completely from the fruit does not promote germination of embryos from highly dormant populations of *Avena fatua* (Naylor and Simpson, 1961a; Simpson, 1965).

D. Endosperm and Aleurone

The endosperm, which constitutes the bulk (70–80%) of the fruit, is a store of inert food reserves, mainly starch, to support the postgerminal growth of the embryo. The breakdown of these reserves is mediated by enzymatic activity of both the surrounding aleurone layer and the digestive organ of the embryo, the scutellum. The triploid endosperm is genetically distinct from the diploid tissues of the embryo and scutellum, and after maturation of the fruit on the parent plant there is no further growth by cell division of either the aleurone layer or the scutellum.

Unlike barley, which has been studied in great detail because of its significance to the brewing industry (Briggs, 1973), the structure and function of the endosperm and aleurone of *A. fatua* have not been well investigated. Nevertheless, there is sufficient evidence to suggest a close similarity of *A. fatua* to barley in the roles of endosperm and aleurone in germination. Aleurone cells release hydrolytic enzymes, which degrade endosperm starch to simple sugars. Other hydrolytic enzymes that are produced in aleurone cells and are specific for reserve proteins or minerals degrade these reserves to amino acids, inorganic phosphorus, etc., which, together with the simple sugars, are taken up by the embryo via the scutellum.

In both dormant and nondormant seeds of *Avena fatua* the aleurone layer and scutellum produce α-amylase and aleurone produces β-amylase (Simpson and Naylor, 1962). They are generally considered to be the two principal enzymes necessary for the degradation of starch to glucose and maltose (Briggs, 1973). Nevertheless, the study in *Avena fatua* (Simpson and Naylor, 1962) demonstrated that when dormant seeds are imbibed on water there was no hydrolysis of starch in the endosperm despite the presence of these two enzymes. Quantitative analysis of α-amylase present in either the scutellum or endosperm, and of β-amylase in the endosperm, indicated similar levels in both dormant and nondormant fruits at various times after imbibition of water by the fruit. A detailed *in vitro* study of the actions of α- and β-amylase, alone or in combination, on various forms of starch indicated that these enzymes could not hydrolyze intact native starch granules from the endosperm of either dormant or nondormant *Avena fatua* unless the enzyme maltase (α-glucosidase) was present. In the same study it was found that maltase activity was very low, or absent, in endosperm of dormant fruits when compared to nondormant fruits. Addition of an exogenous supply of maltase promoted embryo germination if low concentrations of the hormone gibberellic acid (GA_3) were present. The GA_3 was found to be essential for the embryo to utilize the glucose produced by the degradation of starch. High concentrations of GA_3, without added maltase, broke dormancy and brought about a large increase in the production of maltase. The general conclusion from this study was that the utilization of endosperm starch reserves is dependent on the secretion of gibberellin-like substance(s) by the embryo. These substances stimulate the production of maltase, which then acts synergistically with the α-amylase secreted by the scutellum and aleurone and the β-amylase of the aleurone to degrade starch.

Other studies have indicated that degradation of endosperm reserves is not a prerequisite for germination of the embryo but rather a postgermination phenomenon that is dependent on hormones for triggering the process. In dormant seeds of both *Avena fatua* and *Avena ludoviciana* there was some restriction in the amount of α-amylase produced after imbibition of the seeds in water when compared to nondormant seeds (Drennan and Berrie, 1961). Nevertheless, significant amounts of α-amylase were produced in both types by 22 hr, when embryos had either germinated or were dormant. The authors concluded that α-amylase activity does not control dormancy and that endosperm degradation is a postgerminal phenomenon. In another study (Andrews and Simpson, 1969) endosperm excised from either dormant or nondormant fruits supported normal germination of excised embryos from nondormant fruits, whereas neither kind of endosperm supported the germination of excised embryos from dormant fruits.

It is not clear to what extent the degradation of endosperm reserves is totally dependent on the hormones secreted by the embryo. There is reason to believe that endosperm of nondormant strains of *Avena fatua* and also of the domesticated oat (*Avena sativa*), which has no dormancy, is autonomous with respect to hormonal supply. When crosses were made between dormant and nondormant strains of *Avena fatua* and nondormant *Avena sativa*, it was noted (Garber and Quisenberry, 1923) that every seed that failed to germinate because of dormancy nevertheless became soft from degradation of the endosperm. A study (Naylor, 1969) of isolated aleurone tissue from an inbred dormant strain of *Avena fatua* indicated that α-amylase production was more dependent on an exogenous supply of GA_3 and amino acids than was α-amylase production in aleurone tissue excised from nondormant *Avena sativa*. The author concluded that there is genetic variability in the degree of hormonal control over endosperm degradation within wild oat populations. Isolated aleurone tissue from dormant strains of seed after-ripened for 2 years, so that embryo dormancy was lost, still had a dependency on an exogenous supply of GA_3 for the production of α-amylase and thus starch degradation. Amino acids could partly substitute for GA_3, and from this and other evidence it was concluded that the controlling factor in the degradation of endosperm is the absence of protease. Other papers from the same laboratory (Naylor, 1966; Maherchandani and Naylor, 1971, 1972) showed that RNA and mRNA are produced in dormant aleurone and are independent of exogenous GA_3. The action of exogenously supplied GA_3 merely coincides with the onset of the independent RNA synthesis. The conclusion from these four studies was that aleurone tissues of dormant fruits lack GA_3 and that this hormone is needed to produce protease, which in turn produces the products, such as amino acids from the breakdown of reserve protein, necessary for synthesis of sufficient hydrolytic enzymes to support rapid degradation of the starchy endosperm. There is some evidence for the presence of the inhibitor *cis*-abscisic acid (ABA) in aleurone of *Avena fatua* (Quail, 1968). Abscisic acid inhibits the production of α-amylase in barley (Khan and Downing, 1968), but it has not yet been shown to have a similar role in *Avena fatua*.

The endosperm and surrounding aleurone tissue clearly have a role in supporting early seedling growth. It is unlikely that they have a direct effect in controlling dormancy of *Avena fatua* seeds, at least in highly dormant fruits, because of the dependency of the aleurone on the embryo for the hormone(s) that initiate(s) the hydrolysis of endosperm. As shown in Section III,D an isolated embryo may exhibit dormancy and lose dormancy in the complete absence of aleurone and endosperm tissue.

E. Embryo and Scutellum

Seed dormancy in *Avena fatua* is in fact a function of the physiological state of the embryo. The first attempt (Atwood, 1914) to isolate mature embryos from surrounding endosperm tissues and hulls showed that germination increased from 41 to 87%, a result that could be interpreted to mean that the surrounding tissues imposed some degree of dormancy and that true embryo dormancy (13%) was not very significant. Nevertheless, subsequent investigations have all confirmed that embryos of highly dormant fruits are completely dormant when excised from their surrounding tissues.

Embryos excised from freshly matured fruits with a small covering of pericarp and testa and a thin layer of endosperm adhered to the scutellum failed to germinate when imbibed on water (Naylor and Simpson, 1961a). If the intact fruits were held in dry storage for various periods up to 3 years, both the rate of germination and final proportion of germinable embryos increased with period of storage. This indicated that embryos gradually recover the capacity to germinate through an after-ripening process. The onset of this embryo dormancy can be prevented by administering GA_3 to the parent plant during seed maturation (Black and Naylor, 1959), and a somewhat similar effect of GA_3, that of breaking dormancy of partially after-ripened excised embryos with some endosperm attached, has been used as a bioassay technique for gibberellins (Naylor and Simpson, 1961b).

Further studies were made in the same laboratory (Simpson, 1965) using excised embryos completely freed of surrounding pericarp, testa, and starchy endosperm and with the scutellum undamaged (naked embryos). Naked embryos from freshly matured seed of a highly dormant inbred strain of *Avena fatua* failed to germinate on a liquid nutrient medium containing sugars, amino acids, and vitamins which supported normal and rapid germination of naked embryos from nondormant (after-ripened) fruits of the same strain (Simpson, 1965). Embryos of genetically nondormant strains also germinated on the same liquid culture medium. A major conclusion from this study was that dormant excised embryos lacked the ability to synthesize a gibberellin-like substance necessary for promoting the synthesis of enzymes in turn needed for the utilization of sugars and amino acids supplied by either degraded endosperm or a liquid nutrient solution. The gibberellin-like substance, demonstrated by a diffusion technique and identified by its ability to promote degradation of endosperm in a similar manner to GA_3, was absent in freshly harvested, mature, excised embryos. It increased with period of after-ripening up to about 18 months, by which time embryos had lost dormancy when they were excised. When excised embryos from highly dormant fruits were placed in a nutrient medium containing low concentrations of GA_3, dormancy was overcome. On the other hand, excised embryos from either a highly dormant strain after-ripened for some time or a nondormant strain not after-ripened had no requirement for exogenous GA_3, indicating that nondormant embryos produce sufficient endogenous GA_3 (or gibberellin-like factor) to permit normal germination. Furthermore, addition of the gibberellin synthesis inhibitor (2-chloroethyl)trimethylammonium chloride to nondormant embryos prevented the secretion of the endogenous gibberellin-like substance (Simpson, 1966).

Further evidence that dormancy is a property of the embryo was provided in a study of the germination and growth of embryos of *Avena fatua* excised at various stages of development, beginning 10 days after anthesis (Andrews and Simpson, 1969). Embryos from a highly dormant strain excised at various stages of development failed to germinate on a liquid nutrient medium that supported normal germination and growth of similar embryonic stages excised from a nondormant strain. Addition of GA_3 to either the liquid nutrient medium or intact developing fruits overcame dormancy of the different stages of embryo from a dormant strain. It was concluded that the requirement for gibberellin by the embryo was not absolute since germination of dormant embryos occurred if the nutrient medium, without gibberellin, was solidified with agar.

Studies on *Avena ludoviciana* corroborate the early observations made on *Avena fatua* (Zade, 1909; Atwood, 1914; Lute, 1938), which indicated that embryos show dormancy at all stages of maturity after anthesis. Developing fruits of *Avena ludoviciana* in which the fruit coat was pricked showed embryo dormancy soon after anthesis, and this was expressed completely by the late dough stage (Morgan and Berrie, 1970). Drying the fruits at different stages of maturity deepened the degree of embryo dormancy, and the overall conclusion was that the embryo is the true site of dormancy. A similar study on *Avena ludoviciana* (Quail, 1968) indicated that 60% of the fruits were dormant by 1 week after anthesis and 100% were dormant at maturity.

Indirect proof that the embryo is the site of the main physiological expression of dormancy in *Avena fatua* has accumulated from studies of the inheritance of seed dormancy in this species. Crosses of *Avena fatua* with the nondormant species *Avena sativa* show that dormancy is a recessive character (Garber and Quisenberry, 1923; Coffman and Stanton, 1938, 1940; Pawloski, 1959). When either *Avena fatua* or *Avena sativa* is the maternal parent in forming F_1 hybrids in reciprocal crosses, all the resultant fruits are nondormant (Johnson, 1935b). From this study Johnson concluded that it must be the *Avena sativa* contribution to the zygote that overcomes embryo dormancy and that the maternal contribution to the environment of the fruit is not responsible for expressing dormancy in the zygote. He also concluded that the observed variation in dormancy could be explained on a triple allelomorph basis with three factors of more or less equal potency, one of which was linked with the factor for grain type.

IV. ROLE OF THE ENVIRONMENT IN DORMANCY OF WILD OATS

A. After-ripening

Early use of the term "after-ripening" inferred the significance of structural modification of the seed with the passage of time. The first use of the term in connection with *Avena fatua* (Atwood, 1914) drew upon the definition given by Eckerson (1913), i.e., "increased viability in succeeding weeks or months . . . due to alterations in the structure enclosing the seed, or to modifications in the enclosed members themselves." Atwood modified the term to "all changes in the seed subsequent to harvest, as a result of which greater germination percentages may be obtained." Since that time the term has been used even more generally as a kind of descriptor for any interactions of the seed and environment taking place over a time period that led to loss of dormancy and therefore increased germination. Thus, a more useful modern definition for after-ripening that would place more emphasis on the causality of environmental factors and less on structural changes could be as follows. After-ripening is the loss of dormancy caused by exposure of the seed to varying environmental conditions over some period of time. In a humorous sense the old adage that "time conquers everything" can be applied to dormancy.

The reason for discussing the term here is that most investigators of seed dormancy in *Avena fatua* have observed that dormancy is lost when seeds are stored, generally under dry conditions, for some period of time. Thus, observations about after-ripening can provide insight into two important parameters: (1) crude estimates of the time period during which the dormancy mechanism can exert control over the germination process and (2) rough estimates of the specific environmental factors that are likely to be important in prolonging or terminating dormancy.

Zade (1909) noted that dormancy in *Avena fatua* seeds stored in the laboratory was lost more rapidly than dormancy in seeds left in the field. He attributed loss of dormancy to a variety of external factors, particularly high temperature. Johnson (1935a) concluded that the period of after-ripening varied with genotype; the most dormant seeds needed at least 18 months. On the same plant there was considerable variability: Seeds in the upper part of the panicle (earlier maturing whorls) required less after-ripening than the lower ones, and even on the same spikelet, after 18 months, secondary seeds were still more dormant than primary seeds.

In a highly dormant inbred strain (Naylor and Simpson, 1961a) dormancy was lost completely after 30 months when seeds were after-ripened by dry storage under laboratory conditions. Seeds left on the soil surface for 3 months, followed by after-ripening in a range of temperatures under laboratory storage (Thurston, 1962a), required 5–6 months from maturity to change from 10 to 50% germination. Quail (1968) demonstrated that after-ripening in storage was faster with increasing temperatures up to 30°C together with increasing relative humidities over the range 16–100% and also with continuous light rather than darkness.

A study of after-ripening by storage in air, oxygen, or nitrogen (Simmonds and Simpson, 1971) showed that nitrogen delays after-ripening and induces even deeper dormancy, whereas oxygen accelerates after-ripening compared to air. All of the above evidence supports the argument that loss of dormancy can be accelerated by changes in the environment. Furthermore, not one of the six papers quoted above, in which the term "after-ripening" was used, presented any evidence for structural changes in the seed accompanying loss of dormancy with the passage of time. On the contrary, the evidence in each case emphasized that the environment initiates changes in the state of dormancy. There seems good reason, therefore, for using the term "after-ripening" as an environmental rather than as a seed structure descriptor for the causal agency in loss of dormancy.

Dormancy is lost within 18–36 months when dormant seeds are stored dry under laboratory conditions, whereas field observations (Section II,B) indicate that some seeds remain dormant as long as 7 years when left in soil.

This indicates that the moist and more variable field environment may prolong the period of after-ripening for primary dormancy and in some cases induce a state of secondary dormancy.

B. Water

While water is essential for the normal germination of wild oat seeds, variation in the volume of water in which seeds are imbibed has significant effects on the expression of seed dormancy. As with other cereal seeds two phases of water uptake can be distinquished. There is an initial rapid uptake to about 30% of total weight within several hours, due primarily to absorption of water by the hulls and testa (Atwood, 1914; Green and Helgeson, 1957a). This is followed by a slower phase, taking 40 hr to reach 40% moisture content (Green and Helgeson, 1957a) and about 170 hr to the maximal uptake of about 75% moisture content, which varies with different types (Atwood, 1914). Since evidence of germination, expressed by emergence of the coleorhiza from the fruit coat of nondormant dehulled seeds, can generally be seen with the naked eye at 12 hr, it can be assumed that major effects of water on the expression of dormancy are likely to take place during this initial phase of water uptake.

Detailed experiments (Hsiao and Simpson, 1971) with hulled and dehulled seeds germinated in a range of volumes of water under various bands of light (white, blue, red, far red) indicated that small or large volumes of water, when compared to the medium range of volumes, markedly restricted germination. Furthermore, small volumes of water interacted with light to further inhibit germination, compared to darkness. Large volumes of water together with white light promoted germination, compared to darkness. The authors demonstrated that by merely choosing specific combinations of light and volume of water it was possible to obtain any percentage of germination from 0 to 100% with a population of seeds that would be considered to be nondormant on the basis of showing germination in an optimal amount of water in darkness.

A partially after-ripened population of dormoats (*A. sativa* × *A. fatua* cross exhibiting dormancy) gave higher germination percentages in a small volume of water (3 ml) than a larger volume (5 ml) (Andrews and Burrows, 1974). From these and other observations on seeds in moist soil they concluded that germination was enhanced by conditions of low moisture. On the other hand, altering volumes of equimolar KNO_3 in which seeds were germinated (Johnson, 1935a) resulted in small volumes giving low germination percentages and large volumes giving high germination percentages.

If nondormant seeds were completely immersed in still as against running water they become secondarily dormant (Hay, 1962, 1967; Hay and Cumming, 1957, 1959). The explanation given for this effect was that respiration becomes rate limited because of low diffusion of oxygen into nonagitated water. That the induction of secondary dormancy may be due to more than simply a temporary lack of oxygen is indicated by observations on excised embryos of nondormant *Avena fatua* (Simmonds and Simpson, 1972). When embryos were immersed for 24 hr in circulating deoxygenated water saturated with nitrogen gas, they did not germinate but remained viable Embryos leached in the same manner in oxygenated water germinated during the treatment, indicating that embryos have an absolute requirement for oxygen to germinate. Nevertheless, when embryos leached under anaerobic conditions were transferred to aerobic conditions, germination occurred normally. This suggests that, although lack of oxygen can prevent germination, the absence of oxygen in a large volume of water is not an essential causal factor in the induction of secondary dormancy. Leaching excised dormant embryos with large volumes of water enhanced the gibberellin-promoted loss of dormancy (Naylor and Simpson, 1961a). The authors attributed the effect to a loss of an inhibitory substance in large volumes of water.

Interrupting the germination of seeds in water, by drying, caused the induction of secondary dormancy to the extent that 7 months of after-ripening in dry storage could not remove the dormancy (Johnson, 1935a). This manner of induction of secondary dormancy by soaking and drying occurs only when hulls are present (Kommedahl *et al.*, 1958).

Unpublished observations by the author (G. M. Simpson) have shown that nondormant dehulled fruits of *Avena fatua* germinate within 24 hr if left dry on the bottom of a petri plate in which the atmosphere is saturated with water vapor by suspending a wet filter paper in the lid. This indicates that dry seeds have high enough imbibition potentials to germinate normally in the presence of only water vapor. All of the above observations, taken together, suggest that variability in germination (expression of either primary or secondary dormancy) associated with variation in the volume of liquid water available for germination is not due to lack of uptake of water by the embryo. Rather, the observations suggest an interaction of water volumes with hulls, and possibly the pericarp and testa, which in turn modifies the effects of light and movements of gases, metabolites, and growth regulators that are essential for normal germination.

C. Gases

The first observations on the role of oxygen in relation to dormancy in *Avena fatua* (Atwood, 1914) were experiments to determine why puncturing the coat of the fruit promoted some degree of germination of dormant seeds. Atwood began with the assumption that the fruit coat was a barrier to the diffusion of oxygen from the surrounding air to the embryo. He found that pricking the seed coat greatly promoted germination in freshly matured (dormant) seed at any oxygen level in the atmosphere between 4 and 20%. Following a period of after-ripening the differences between pricked and nonpricked seeds disappeared. Nevertheless, there was still a marked germination promotion with increase in oxygen concentration; germination percentage in either the after-ripened or non-after-ripened intact fruit was less than 5% in an atmosphere with less than 4% oxygen. A curious result not commented on by Atwood was that breaking the fruit coat of newly mature seeds gave nearly 100% germination at a relatively low oxygen concentration (8%), whereas breaking the coats of after-ripened seeds and germination in the same oxygen concentration gave only 50% germination. Ignoring this inconsistency, the data of Atwood indicate that above 8% oxygen level in the atmosphere (the level below which germination is rate limited in nondormant intact *Avena sativa* seeds) germination of intact fruits of *Avena fatua* was much lower in newly matured seed at all concentrations of oxygen up to 100% than in well after-ripened seeds. Taking into account the inconsistency stated above, Atwood's data do not provide conclusive evidence that the seed coat is a significant barrier to the uptake of oxygen by the embryo. If anything the data support the view that the requirement for oxygen by the dormant embryo is very high and that this requirement decreases with after-ripening.

Increasing the oxygen concentration in the atmosphere to either 50, 60, 70, or 80% stimulated germination by 17, 18, 21, and 30%, respectively, above that in air (Baker and Leighty, 1958) in fruits with hulls present. Nevertheless, the most dormant strain, which increased from 1 to 41% germination when hulls were removed and fruit coat was broken, failed to show an increase in germination over the same range of oxygen concentrations. This suggests that, although hulls and seed coat may act as a partial barrier to uptake of oxygen, there is still residual dormancy in the fruit. A somewhat similar effect was noted (Hart and Berrie, 1966) when both hulled and dehulled seeds were germinated in a range of concentrations of oxygen (10–80%). Removal of hulls increased germination percentage, indicating they can act as a partial barrier to oxygen uptake. Nevertheless, with hulls removed the optimal oxygen concentration (20%) only gave 75% germination; below 10% oxygen, germination was markedly decreased.

In a study of oxygen requirements for germination of intact *Avena fatua* seeds exposed to a range of temperatures (Müllverstedt, 1963) it was found that at temperatures around 20°C germination percentage increased with increasing oxygen partial pressure over the range 6–15% O_2; optimal germination was 75% at 15% O_2. At low temperatures seeds used very little oxygen.

In 1.8% oxygen, over a range of temperatures, some interesting effects were observed. At 20°C all the seeds died. If temperatures were alternated between 10° and 24°C at the same oxygen concentration, 65% germination occurred. Alternating temperatures between 6° and 15°C maintained 100% viability of seeds, whereas a steady temperature of 7°C permitted only 2% germination. These effects, particularly the alternating-temperature response, indicate the significance of metabolic processes in determining the amount of oxygen uptake by the fruit.

A similar observation was noted by Atwood (1914), who showed that the Van't Hoff coefficient (rate of oxygen uptake at 26°C/rate of oxygen uptake at 16.2°C) of fruits was 2.38, a value that might be expected if the oxygen uptake was dependent on metabolic processes and if the fruit coat was not a significant barrier to oxygen uptake.

Differences with respect to oxygen uptake of both excised embryos and the separated endosperm portion of the fruit can be shown between dormant and nondormant strains of *Avena fatua* (Table I). Nevertheless, comparison of oxygen uptakes at intervals over 60 months in excised embryos from a highly dormant strain showed that, although embryos after-ripened 60 months had a significantly higher rate of oxygen uptake than those after-ripened 5 months, there were no significant differences in uptake at the time when very significant changes occurred in the ability to germinate (Table II). Moreover, the same study (Simmonds and Simpson, 1971) demonstrated that when gibberellic acid overcame dormancy of excised embryos there was no alteration in the rate of oxygen uptake.

A number of observations indicate that when *Avena fatua* seeds are placed in an atmosphere of nitrogen or in a situation in which the oxygen level is much reduced, induction of secondary dormancy occurs in nondormant fruits (Christie, 1956; Black, 1959; Hay and Cumming, 1959; Banting, 1962; Hay, 1962; Simmonds and Simpson, 1972). Soaking seeds under a vacuum also decreases subsequent germination (Kommedahl *et al.*, 1958). An indication that these effects are directly on the embryo is given by a study of maturing embryos excised from a highly dormant strain (Andrews and Simpson, 1969). Excised embryos, which are otherwise dormant, can be induced to germinate slowly in either air or oxygen if they are placed on a nutrient medium solidified with agar. The same embryos fail to germinate on this medium under nitrogen gas. However, GA_3, which breaks primary dormancy, stimulates the germination of these embryos under nitrogen.

Although the earliest study of the effects of an increased CO_2 partial pressure indicated no effects on germination of *Avena fatua* (Bibbey, 1948), which may have been due to the completely nondormant strain examined, a later study (Hart and Berrie, 1966) demonstrated an important interaction of CO_2 with light in partially dormant seeds. These authors showed that at

TABLE I

Oxygen Uptake[a] of Embryo and Endosperm Tissues from Dormant (D) and Nondormant (ND) Strains of Avena fatua[b]

Duration of imbibition in water (hr)	Excised						Intact fruit	
	Embryo		Endosperm					
	D	ND	D	ND			D	ND
1	2.7 ± 0.5	3.4 ± 2.7	3.0 ± 2.1	2.3 ± 3.2			1.5 ± 1.4	4.5 ± 1.2
4	38.0 ± 0.7	48.6 ± 2.5	8.3 ± 2.2	8.3 ± 2.2			35.3 ± 3.2	60.0 ± 3.8
8	89.9 ± 3.2	120.4 ± 0.6	52.8 ± 3.0	78.3 ± 0.6			104.3 ± 8.8	151.0 ± 11.7

[a] Expressed as microliters per milligram dry weight at 30°C.
[b] Duplicate measurements.

190

TABLE II

Effect of After-ripening on Rates of Oxygen Uptake during the Period 0–10 hr of Imbibition of Water and on Germination of Excised Embryos of *Avena fatua*[a]

	Period of after-ripening (months)					
	2	3	5	12	24	60
O_2 uptake (μl/hr per 10 embryos $\pm$ SE)	8.71 $\pm$ 0.38 ab	9.33 $\pm$ 0.18 ab	8.40 $\pm$ 0.18 a	9.06 $\pm$ 0.48 ab	8.92 $\pm$ 0.08 ab	10.12 $\pm$ 0.30 b
Germination index	0.37	1.96	1.19	6.15	6.90	6.67

[a] After Simmonds and Simpson (1971).

191

zero CO_2 and various proportions of oxygen (0–80%), with inert argon as the background gas, white light inhibited germination of both entire and dehulled seeds. At 3% CO_2, again over a range of O_2 concentrations, white light inhibited entire seed but not dehulled seed. In 20% CO_2, over the same range of oxygen concentrations, there was no effect of light on either entire or dehulled seeds. At this concentration of CO_2 there was also considerable reduction of germination in the entire seeds but only a small reduction in dehulled seeds, implying that this concentration was above the optimum for normal germination. The CO_2 interaction with light disappeared completely in seed after-ripened 2 years when there was no residual dormancy. In all combinations of CO_2, germination was proportional to oxygen concentration over the range 0–20% oxygen with the exception of two cases in which seeds with hulls present, in 0 or 3% CO_2, showed an optimal oxygen requirement of 50%.

The explanation given by these authors for the light inhibition in the absence of carbon dioxide was that CO_2 is needed for metabolism in some type of condensing reaction such as the Wood–Werkman step, implying that light-promoted dormancy in this species is associated with a block in the respiratory system of the type, such as lack of dehydrogenase activity, suggested by Hay (1962). The authors also made reference to the possibility of light blocking the tricarboxylic acid cycle.

To summarize the above evidence the following conclusions can be made about the effects of gases on dormancy. Both oxygen and carbon dioxide have important effects on germination through their effects on metabolism, particularly in the embryo. Both gases are essential for germination. Structures such as the hulls and fruit coat can restrict gas exchange, but they are not the sole determinants of dormancy associated with variation in gas composition of the atmosphere. The level of metabolic activity of the embryo can also determine both oxygen and carbon dioxide requirement and thus uptake. The rate and type of metabolism are a function of the degree of dormancy present within the embryo. This dormancy can be lost independently of variations in gas supply but can in part be sustained or reintroduced by variations in gas supply. Induction of secondary dormancy in seeds placed under anaerobic conditions or in light can in part be attributed to effects on metabolism of the embryo.

D. Light

The first determinations of the effects of both sunlight and artificial light on germination of *Avena fatua* were, in the words of the author (Johnson, 1935a), "inconclusive because of conflicting results." Tests with newly ripened seeds indicated a 15% promotion by light of both hulled and dehulled seeds from four different strains. On the other hand, in well after-ripened seeds light was detrimental to germination, compared to darkness. In a comparison of light and dark effects on germination of 12 strains of *Avena fatua* the mean value for light was 51% germination and for dark 43% germination (Leighty, 1958), a difference of doubtful significance. Other observations (Cumming, 1957; Cumming and Hay, 1958) indicated that completely nondormant seeds (with hulls removed and seed coats pricked) failed to show any response to white, red, blue, or far-red bands of artificial light. However, partially dormant seeds were inhibited by white, blue, and far-red, but not red, bands. Sunlight was also inhibitory to partially dormant seeds (15% germination) compared to darkness (55%). A later publication by the same authors (Hay and Cumming, 1959) indicated that, in seeds with hulls on that had been induced to become secondarily dormant under anaerobic conditions, white light was inhibitory (zero germination). In the same seeds with hulls removed light reduced germination from 70 to 30% when compared with darkness.

   In their studies on the effects of CO_2 on germination Hart and Berrie (1966) confirmed that white light can inhibit germination of partially dormant seeds; the inhibition was less when hulls were removed. Their explanation for the light/dark response was that light and CO_2 interacted together in the germination response. Absence of CO_2 permitted a high degree of light inhibition in both hulled and dehulled seeds, whereas 3% CO_2 permitted germination of dehulled, but not hulled, seeds in the presence of light. High concentrations of CO_2 (20%) inhibited germination in both light and dark. Their observations suggest that light has effects on metabolism and these effects can be modified by the presence or absence of hulls. Further evidence for the interaction of light and metabolism was provided from a complex multifactorial experiment designed to compare temperature, light, and hull effects on germination of *Avena fatua* and *Avena ludoviciana* (Whittington et al., 1970). The authors concluded that light promotes germination of *Avena fatua* at both low (5°C) and higher temperatures (18°C), whereas in *Avena ludoviciana* it inhibits only at the higher temperature. They also concluded that the inhibitory effects of the hull in *Avena fatua* could be alleviated either by removal or through illumination.

   Changes in length of photoperiod have significant effects on germination of *Avena fatua* (Thurston, 1963). Seeds germinated for 33 days at 15°C in dim light (less than 50 fc from mixed fluorescent and tungsten lamps) showed 53% germination under 8-hr photoperiods, but only 18% under 16-hr days. When ungerminated seeds were dehulled, pricked, and returned to the same conditions the inhibitory effect of long days still persisted: 5% still failed to germinate in long days, whereas all seeds germinated in short days. The effect of long days was more marked in *Avena ludoviciana* than in

TABLE III

Effect of Light on Percent Germination of a Dormant Strain of Wild Oat Seeds[a]

Age of seeds (months)	Germination (%)	
	White light	Darkness
9	12.4	58.0
21	55.8	97.9
34	97.2	99.7
Mean	55.1	85.2

[a] Mean values from three-factorial experiments, which included hull treatments in the presence or absence of gibberellic acid (2 ml). After Hsiao and Simpson (1971).

Avena fatua. After-ripening under continuous light, compared to darkness, promoted loss of dormancy of stored seeds (Quail, 1968).

An explanation for some of the conflicting results reported in the literature, where in some cases light promotes and in others it inhibits germination, was suggested after a study of the effects of water volume and light on germination of *Avena fatua* (Hsiao and Simpson, 1971). White, red, far-red, and blue light were all found to have inhibitory effects if seeds were germinated in small volumes of water. However, in large volumes of water, white light promoted germination compared to darkness. An interesting observation was that white light induced a reduction in the rate of increase of water potential in seeds. Light inhibition was most marked in newly ripened seeds (Table III), the effect disappearing with after-ripening. Light was effective in both the presence and absence of hulls (Table IV). It was concluded that the site of perception of light is in the embryo and that the presence of hulls does not constitute a significant barrier to the passage of light to the embryo. In addition white light considerably reduced the promotory effect of GA_3 (small volumes) in overcoming dormancy (Table V), a further indication of the significance of light effects on metabolism, since GA_3 has been shown to affect enzyme reactions in both the embryo and endosperm (Section III,D and E).

E. Temperature

Effects of temperature variations on the state of dormancy in *Avena fatua* seeds should be distinguished from temperature variations that influence the normal processes of germination occurring in the absence of dormancy. Optimal germination of nondormant seeds has been shown to

TABLE IV

Inhibitory Effects of White Light on Percent Germination of a Dormant Strain of Wild Oat Seeds After-ripened 9 Months and Given Different Hull Treatments[a]

Hull treatment	White light		Darkness	
	I	II	I	II
Hulled	5.6 c	11.3 d	12.5 ab	21.9 cd
Hull cut at chalazal end	8.9 bc	12.3 d	20.7 a	32.4 bc
Hull cut at micropylar end	21.1 a	47.3 ab	71.7	77.7
Dehulled	49.6	49.6 a	98.7	98.7
Mean	21.3	30.9	50.9	57.7

[a] Mean of treatments with and without gibberellic acid (2 ml): I, percentage of radicles emerged through hulls at 120 hr; II, total percentage of emerged radicles seen when hulls are removed from the caryopsis at 120 hr. The test for significance of the factorial design is by a two-way layout within either I or II. After Hsiao and Simpson (1971).

occur around 21°C; temperatures above 27°C or below 16°C were found to be detrimental to germination (Friesen and Shebeski, 1961). A lower optimum (15°C) has been reported for seed grown in Australia (Quail, 1968).

The first report of a temperature effect on seed dormancy in *Avena fatua* (Zade, 1909) indicated that the higher temperatures of dry laboratory

TABLE V

Combined Effect of Light and Gibberellic Acid on Percent Germination of Dormant Wild Oat Seeds [a]

Conc. GA (ppm)	White light	Darkness
0	7.6 fg	22.6 de
1	5.4 g	38.9 cd
50	18.5 def	62.7 bc
100	21.6 def	54.1 bc
500	20.0 def	75.3 ab
1000	9.5 def	87.7 a
Mean	13.8	56.9

[a] The test for significance is by a two-way layout. Seeds were after-ripened 9 months and given different hull treatments and 2 ml of solution. After Hsiao and Simpson (1971).

storage favored more rapid loss of dormancy than the cooler environment of seeds left under field conditions. Other workers have made observations that support the general conclusion that high temperatures favor more rapid after-ripening than do low temperatures. For example, after-ripening of seeds at temperatures near freezing was slower than at high temperatures (Johnson, 1935a), and a similar conclusion, that warm, dry conditions promote after-ripening faster than do cold, humid conditions, was made by Banting (1966b). Cobb and Jones (1962) indicated that at low temperatures (5°C) after-ripening was faster in seeds that were dried than in moist seeds. Another study showed that when seeds were after-ripened in dry storage over a period of 5 months, dormancy was lost rapidly at higher temperatures (16° and 21°C) compared with negligible loss of dormancy at lower temperatures (4° and 7°C). The optimum temperature for loss of dormancy in dry storage was 16°C (Thurston, 1962b). Sexsmith (1969) reported that high temperatures during maturation of seeds on the parent plant reduced dormancy compared to cool temperatures.

A conclusion that has often been drawn from the above evidence is that either high temperatures promote loss of dormancy by accelerating an oxidative process that removes a factor inhibitory to germination, or enhanced respiration produces promotory factors necessary for germination. Studies on secondary dormancy indicate that temperature effects are centered on the metabolic system of the seed.

Whereas high temperatures appear to promote loss of primary dormancy in dry seeds, they also promote induction of secondary dormancy in moist seeds. Hay (1962) demonstrated that induction of secondary dormancy was greater at 25°C than at 7°C. Another detailed study of the induction of secondary dormancy in *Avena ludoviciana* and *Avena fatua* (Thurston, 1962b) demonstrated some striking effects of high temperatures. Moist seed of *Avena ludoviciana* showed no induction of dormancy at 23° and 24°C, but at 27°C there was a sharp threshold for the induction of deep dormancy. At 27°C more than 50% of the seeds of *Avena fatua* died, and 95% of the survivors showed secondary dormancy. A particularly interesting observation was that the critical temperature for induction (27°C) remained constant for over 3 years in dry stored seeds and that different strains showed different threshold temperatures. This suggests both genotypic variation for the response and a high degree of stability for the underlying response in the seed.

Evidence that secondary dormancy is probably not the same as primary dormancy has been obtained from a study of temperature effects on dormoats (*Avena fatua* × *Avena sativa* crosses) (Andrews and Burrows, 1972, 1974). Seeds with induced secondary dormancy (soaking in water) were only slightly promoted in germination by low incubation temperatures

(7°C), whereas seeds with partial primary dormancy were stimulated to germinate at the same low temperature, in comparison to seeds germinated at 20°C. Both primary and secondary dormancy of dormoats were overcome with GA_3, an indication of the association of metabolism with both these forms of dormancy. Evidence from a multifactorial study of primary dormancy (Whittington *et al.*, 1970) indicated that germination was better at 5°C than at 18°C, and a combination of low temperatures (4°–7°C) and 0.2% KNO_3 was found to be very effective in breaking primary dormancy (Baker and Leighty, 1958).

F. Growth Regulators

Evidence that regulatory compounds have a role in the initiation, maintenance, and termination of dormancy in *Avena fatua* is derived largely from observations on the effects of exogenously supplied substances. Plant growth regulators in low concentration (hormones) and other organic and inorganic compounds act in various ways to control dormancy and germination. Despite the fact that many hormones can either promote or inhibit growth according to the concentration present, the various substances reviewed here are discussed under two general categories: substances that terminate dormancy (promoters) and substances that induce dormancy (inhibitors).

1. PROMOTERS

The first observation that a plant hormone has significant effects on the termination of dormancy in *Avena fatua* was made with one of the family of gibberellins (Green and Helgeson, 1957b, Helgeson and Green, 1957b) and to this date, despite the many papers substantiating this first report, only the one gibberellin, GA_3, has been studied. The first report on GA_3 demonstrated that newly matured dormant seeds germinate in a range of concentrations of GA_3 (25–500 ppm). The optimal concentration was 50 ppm for newly ripened seed (4% in water; 50% in GA_3). The response to exogenous GA_3 decreased with length of after-ripening period since after 90 days the number of seeds germinating in GA_3 increased to 83%. Baker and Leighty (1958) confirmed this effect of GA_3 and indicated the possibility of genetic variability for the response since one of the strains examined failed to respond to GA_3 during the 13-day test. Fruits from excised panicles with the stems placed in a range of concentrations of GA_3 were germinable on water at maturity, whereas fruits from the water control were dormant (Black and Naylor, 1959). It was not clear whether the effect was due to prevention of the induction of dormancy or to the direct effect on breaking dormancy observed by Green and Helgeson (1957b).

A detailed study (Naylor and Simpson, 1961a) of the germination response of excised embryos to both gibberellin and sucrose provided an explanation for the distinctly bimodal response curve to GA_3 observed in a partially after-ripened population of seeds (Naylor and Simpson, 1961b). Two distinct concentration optima [(high (50 ppm) and low (5×10^{-1} ppm)], separated by a 100-fold difference in concentration, were noted in the semidormant population. Freshly harvested, highly dormant seeds showed only a single optimal response (50 ppm) to GA_3. If these highly dormant seeds were given sucrose along with GA_3, the optimal response to GA_3 shifted from the high value to the low (5×10^{-1} ppm) value. The other major finding reported in this paper was that a combination of sucrose with the low concentration optimum of GA_3 produced a synergistic response in breaking dormancy quickly; that is, either substance alone could not break dormancy. The explanation given for the role of the high level of exogenous GA_3 was that it stimulated production of sugar in the endosperm. The interpretation for the action of the low level was that it was needed for the utilization of sugar by the embryo. This hypothesis was strengthened by later observations (Simpson and Naylor, 1962) that, in dormant seed, starch granules stored in the endosperm could not be broken down to yield glucose, the principal sugar utilized by the embryo. The hydrolases needed to do this, α- and β-amylases, were both present but proved to be ineffective unless the enzyme maltase was also present. Maltase was absent in dormant seed and could be induced by the high level of GA_3. Alternatively, application of a low level of GA_3 together with the pure enzyme maltase broke dormancy through a synergistic action in the same manner as a low level of gibberellin and sugar together.

The above observations on the role of exogenous gibberellin suggested that endogenous gibberellins may play a role in the termination of dormancy. Helgeson and Green (1958) reported that newly germinated wild oats have an extractable gibberellin-like activity that breaks dormancy. Nondormant 2-year-old dry seeds contained a substance that cochromatographed on a two-dimensional paper chromatogram along with GA_3 and promoted germination of semidormant embryos of *Avena fatua* (Naylor and Simpson, 1961a). A comparison of dormant and nondormant strains of *Avena fatua* for an endogenous, diffusible, gibberellin-like substance proved that synthesis of the gibberellin-like substance(s) in the embryo was blocked in mature dormant seed (Simpson, 1965). The ability to synthesize this substance, assayed by its ability to trigger the hydrolysis of endosperm reserves, was gradually recovered after about 24 months of seed storage. An inhibitor of gibberellin synthesis, (2-chloroethyl)-trimethylammonium chloride), prevented production or secretion of this gibberellin-like factor by embryos (Simpson, 1966).

An investigation of three gibberellin-like factors present in developing embryos (identified by three different assays of extracts partitioned by paper chromatography) indicated that one factor, associated with embryo growth, was present at similar levels in both dormant and nondormant strains (Andrews, 1967). The other two substances, associated with germinability of the embryo, were present in much higher levels in the nondormant strain than in the dormant strain. One of the latter substances increased greatly during imbibition of the seed to a much higher level in nondormant embryos.

Evidence for one role of the endogenous gibberellin-like substances found in embryos has accumulated from studies of the effects of exogenously supplied GA_3 on metabolism. Simpson (1965) showed that the effect of a low concentration of GA_3 on dormant embryos was to overcome a block in the embryo that prevented utilization of sugars supplied from the hydrolysis of endosperm reserves. This effect of GA_3 was confirmed by experiments with exogenously supplied, labeled glucose (Chen and Park, 1973). The activity of the pathway for oxidative metabolism of glucose, as measured by the oxygen uptake of embryos, was apparently not altered when dormancy was broken by the addition of GA_3 (Simmonds and Simpson, 1971). However, observations at the same time of the C_6/C_1 ratio of respiratory CO_2 produced by both dormant and nondormant embryos fed radioactive glucose, in the presence or absence of GA_3, indicated a greater activity of the pentose phosphate pathway compared to the glycolytic pathway in the nondormant and GA_3-treated embryos. This was confirmed by observations on the relative activities of nine key enzymes of oxidative metabolism including the two key enzymes controlling the oxidative pentose phosphate pathway [glucose-6-phosphate dehydrogenase (G6PDH) and 6-P-gluconate dehydrogenase (6PGDH)] (Kovacs and Simpson, 1976). Comparison of the relative activities of these two enzymes in dormant and nondormant mature dry seed showed that initially the levels were similar, but following steeping in water there was a rapid increase in activities of both enzymes in nondormant seeds and a decrease in dormant seeds. Steeping dormant embryos in GA_3 increased activities of both enzymes; the increase in G6PDH, the first and therefore most important controlling step in the pathway, was much greater than that in 6PGDH.

Chen and Park (1973) also confirmed a role for GA_3 in influencing metabolism, by demonstrating enhanced biosynthesis of RNA and proteins in both the embryo and endosperm of dormant fruits of *Avena fatua* treated with GA_3.

Other chemical substances shown to have promotory effects on germination of dormant seeds are KNO_3 (Johnson, 1935a; Hay and Cumming, 1957, 1959; Baker and Leighty, 1958); NaCNS (Johnson, 1935a); sodium iodoacetate, *p*-chloromercuribenzoate, and coumarin (Morgan and Berrie, 1970); and malonic acid (Simmonds and Simpson, 1972). All of these substances are normally considered to be metabolic inhibitors, and the significance of their unexpected promotory effects on germination is discussed in Section V,B,2.

2. INHIBITORS

The first report of a substance in *Avena fatua* seeds inhibitory to germination was that of a toxic, dialyzable substance in hulls that could be adsorbed onto charcoal (Helegeson and Green, 1957a). However, a somewhat similar study (Baker and Leighty, 1958) indicated that there was no growth inhibitor in hulls that could stop the germination of nondormant seed. These authors also showed that if hulls were loosened but left under the fruit, as compared to over the fruit, there was an inhibitory action during steeping in water, which they attributed to a delay in leaching of an inhibitor from either the embryo or fruit. Black (1959) chromatographed aqueous extracts of both hulls and fruits and demonstrated two substances, identified by their inhibitory effects on lettuce seed germination, which was present in equal amounts in hulls and fruit on a dry weight basis. He found a decrease in the amount of inhibitors in moistened seeds placed under oxygen and an increase under nitrogen. The general conclusions reached were that hulls prevented leaching of inhibitors from the fruit and that during induction of secondary dormancy under anaerobic conditions there were marked increases in the total inhibitor levels in fruits.

Another study (Hay, 1962) of methanol extracts of dormant and nondormant fruits concluded that there were inhibitors of seedling growth, but not of germination, present in the extracts. However, an inhibitor found in relatively high concentration in aqueous extracts of hulls was shown to cause a reversible inhibition of germination. The inhibitor stopped growth completely without injuring embryos and could be reversed by GA_3. Because the inhibitor could stop germination in the presence of sucrose the author concluded that dormancy can be due to a block at a point other than the conversion of starch to sugar.

Further evidence for a germination inhibitor in embryos was obtained from leaching experiments (Naylor and Simpson, 1961a). Excised dormant embryos leached with low concentrations of GA_3 increased their germination response from 20 to 90% compared to nonleached seeds on the same concentration of GA_3. A germination inhibitor found in both primary and secondary seeds was identified through gas chromatography as *cis*-ABA (Quail, 1968). The author also concluded that the inhibitor could prevent the GA_3-promoted synthesis of α-amylase in barley. Another study (Andrews and Burrows, 1972) demonstrated that ABA can inhibit the germination of excised embryos of dormoats and that this inhibition can be reversed with GA_3.

Some of the discrepancies reported in the literature of *Avena fatua* studies on inhibitors of germination are undoubtedly due to differences in ages and strains of seeds investigated by different workers. An added difficulty in working with inhibitors is the fact that crude extracts of most plant material can be inhibitory to either germination or growth because of concentration, pH, and additive effects of a variety of substances.

V. METABOLIC SYSTEM OF WILD OATS

A. Endosperm

1. MOBILIZATION OF RESERVES

a. Enzymes. Hydrolysis of the starchy endosperm reserves of *Avena fatua* is dependent on enzymes secreted by the aleurone layer. Starch is hydrolyzed to low molecular weight sugars, which are then taken up through the scutellum and transported to the embryo. The first observations of α- and β-amylase (Drennan and Berrie, 1961) concluded that major changes in these two enzymes occurred only as a postgermination phenomenon related to early seedling growth. This was confirmed (Simpson and Naylor, 1962) when it was shown that both α- and β-amylase were present in mature dry seeds in sufficient amounts to support germination. The amounts present in both dormant and nondormant seed were similar during the pregermination phase. However, Simpson and Naylor (1962) established that intact starch granules from *Avena fatua* endosperm could not be hydrolyzed by either α- or β-amylase alone, or in combination, unless the enzyme maltase (α-glucosidase) was also present. This enzyme was absent in dormant fruits and present in nondormant fruits. Either addition of an exogenous supply of maltase or the induction of maltase by the addition of an exogenous supply of GA_3 permitted a synergistic response of this enzyme with the amylases, causing hydrolysis of the starch granules to glucose, maltose, maltotriose, and small dextrins.

Later studies (Chen and Chang, 1972; Chen and Park, 1973) confirmed that α-amylase is present in the endosperm of both dormant and nondormant mature seed. When dormancy was broken with GA_3, activity of α-amylase did not increase until 48 hr, well after germination had occurred. A linear increase in α-amylase activity from 48 hr to 5 days was noted.

A study of enzymes involved in breakdown of endosperm in *Avena sativa* (Sutcliffe and Baset, 1973) pointed to the presence of proteases and phytase in addition to the amylases. In another study the onset of protein synthesis was shown to be independent of newly formed RNA in freshly imbibed aleurone cells of *Avena fatua* (Maherchandani and Naylor, 1972). It was concluded that all the fractions of RNA essential for protein synthesis can survive for several years in dry stored fruits. A further study involving a comparison of α-amylase activities of dormant *Avena fatua* endosperm and *Avena sativa* endosperm indicated a much greater dependence on exogenous GA_3, or amino acids, for formation of this enzyme in wild oats (Naylor, 1966, 1969). An important conclusion in this study was that *Avena sativa* and *Avena fatua* differ in the genetic control of synthesis of one or more proteases that function in the degradation of aleurone reserve protein. This may be related in part to unusual DNA turnover, or amplification, during differentiation of aleurone cells in *Avena fatua* (Maherchandani and Naylor, 1971) since aleurone in *Avena fatua* showed a continuous range of DNA values from 2C to greater than 6C, with no accumulation of 3C and 6C as expected in a mature triploid tissue.

Inhibition of the formation of enzymes necessary for endosperm degradation cannot be a primary rate-limiting step that causes dormancy in embryos of *Avena fatua* because excised nondormant embryos germinate in the absence of endosperm. Rather, it must be the failure of dormant embryos to secrete GA_3 that provides the control mechanism for the embryo to prevent premature degradation of endosperm reserves to low molecular weight substrates before they can be utilized. Nevertheless, variations apparently exist in the degree of independence of endosperm from the hormones secreted by the embryo. With the domestic oat, *Avena sativa*, during domestication and breeding varieties, there has been no selective pressure to maintain ancestral genetic dormancy. Thus, *Avena sativa* has endosperm that is degraded rapidly to sugars in the absence of the embryo, after water is imbibed. Strains of *Avena fatua* show a range of variability in the degree of autonomy exhibited by excised endosperm with respect to degradation of reserves in the absence of hormones secreted by the embryo. Natural selection could be expected to maintain genotypes conferring strong hormonal control of this physiological system in wild oats, which must depend, for survival, on both embryo dormancy and preservation of endosperm reserves (Naylor, 1969).

b. Carbohydrates. Starch is the principal storage carbohydrate in the endosperm of *Avena fatua*, and there have been no reports in the literature indicating any differences in amounts between dormant and nondormant endosperm tissues. Hydrolysis of starch by endogenous enzyme action yielded mainly glucose and maltose (Simpson and Naylor, 1962; Chen and Varner, 1969) with some maltotriose, fructose, and low molecular weight dextrins (Simpson and Naylor, 1962). There was some indication that maltose acts as an end-product inhibitor of α-amylase activity and that the function of maltase is to remove this block (Simpson and Naylor, 1962). Glucose is the primary source of energy taken up by the scutellum, where it is converted to sucrose (Chen and Park, 1973; Edelman *et al.*, 1959).

c. Nitrogenous Compounds.   There have been few studies of the content or mobilization of nitrogenous compounds in either the endosperm or aleurone layer of *Avena fatua*. Endosperm from a dormant strain was unable to synthesize α-amylase until 8 days after imbibing water; addition of amino acids promoted a significant increase in the enzyme, suggesting that the breakdown of reserve protein to amino acids is inhibited (Naylor, 1966, 1969). Nevertheless, it is doubtful whether the absence of an amino acid pool in the endosperm is a limiting factor in the germination of the embryo since either nondormant or dormant excised embryos exposed to GA_3 begin germination in the complete absence of amino acids (Simpson, 1965). Amino acids are essential for subsequent embryo and seedling growth (Harris, 1954; Simpson, 1965).

d. Phosphorus.   A very comprehensive study of inorganic and organic phosphorous compounds related to germination in *Avena sativa* seeds showed that phytic acid (Ca, Mg, K, salt of inositolphosphoric acid) is the primary reserve phosphate of the endosperm (Hall and Hodges, 1966). Phytate was approximately 50% of the total phosphorus of the seed. In the first several days of germination the nearly stoichiometric losses of phosphorus from the endosperm and gains by the roots and shoot of nucleic acids, phospholipids, and phosphoproteins demonstrated the great significance of endosperm reserves of phosphorus in supporting postgerminal growth. A later study on *Avena sativa* (Sutcliffe and Baset, 1973) demonstrated a strong correlation between development of phytase activity and breakdown of phytic acid in the endosperm; maximal activity of the enzyme developed several days after germination. The above evidence indicates that major changes in phosphorus metabolism of endosperm are probably a postgermination phenomenon.

A study that indicated a possible relationship between metabolism of phosphorus in the endosperm and seed dormancy in *Avena fatua* was an analysis of free phosphate levels and acid phosphatase of fruits (Kovacs and Simpson, 1976). Free phosphate levels declined in nondormant seeds, whereas the level remained constant in dormant seeds imbibed for 24 hr on water. At the same time acid phosphatase activity increased up to 24 hr in nondormant seed, whereas it remained constant in dormant seed. A somewhat similar study of free inorganic phosphorus in excised embryos of the same strain of *Avena fatua* (Simpson, 1965) indicated that levels of phosphorus remained constant in dormant embryos but increased ninefold in nondormant embryos during the first 48 hr of imbibition on water. A parallel observation was that levels of the enzyme 3'-nucleotidase, which splits phosphorus from adenylic acid, declined in activity in dormant embryos but increased markedly in nondormant embryos. The activity of this enzyme increased when embryo dormancy was broken by treatment with GA_3.

There is no evidence yet to suggest that phosphorus levels of the endosperm control embryo germination. If anything, the evidence supports the view that embryo germination, and therefore dormancy, is independent of the endosperm. Utilization of endosperm reserves is governed by the rate of demand by the embryo for the products of reserve hydrolysis.

2. TRANSPORT

It is unlikely that transport of the substrates formed from hydrolysis of endosperm reserves is a significant factor in dormancy of *Avena fatua*. No detailed study of transport has yet been made in *Avena fatua* similar to that in either *Avena sativa* (Sutcliffe and Baset, 1973) or wheat (Edelman *et al.*, 1959). Both these studies indicated that there was a zone of hydrolysis, small at first and close to the scutellum, which then spread gradually through the remainder of the endosperm. The hydrolysis and translocation were regulated by the growth of the axis, which stimulated the activity of the various hydrolytic enzymes; these in turn catalyzed the macromolecules to smaller, more easily transported molecules that can move by diffusion (Sutcliffe and Baset, 1973). Thus, germination of the embryo must occur before the hydrolysis and translocation of endosperm reserves.

In a test with *Avena fatua* to see whether GA_3 could enhance translocation of sugar from the endosperm to the embryo (Chen and Park, 1973), radioactive glucose was fed through the distal end of the endosperm after the seeds were soaked in GA_3 for 12 hr. Glucose uptake by embryos was nearly three times greater in the GA_3-treated seeds compared with the water control. The authors were unable to conclude whether the effect was caused by increased translocation because of enhanced membrane permeability or enhanced sugar utilization by the embryo.

B. Embryo and Scutellum

1. UTILIZATION OF RESERVES

An early study suggested the possibility of germination being supported by the breakdown of fat reserves in the embryo (Hay, 1962), but this was rendered improbable when it was shown that the fat content of dormant and nondormant embryos was similar and did not change during the early phases of germination (Simmonds and Simpson, 1971). A study of sterol fractions in dormant *Avena fatua* and in *Avena sativa* indicated no relationship between these fractions and dormancy (Knights, 1968).

The first observation that dormant embryos were unable to utilize the products of endosperm hydrolysis was made during a study of the influence of exogenously supplied GA_3 on excised embryos of *Avena fatua* (Naylor and Simpson, 1961a). The role of exogenously supplied GA_3 in overcoming this block in the embryo was later confirmed (Simpson and Naylor, 1962), and proof for the presence of naturally occurring gibberellins in embryos of nondormant, but not in dormant, seeds was obtained in the same laboratory (Simpson, 1965). The latter study also demonstrated that germination of isolated embryos was dependent on both an endogenous source of gibberellin and an exogenous source of sugars and amino acids.

Chen and Varner (1969) studied embryonic utilization of maltose and glucose derived from endosperm starch hydrolysis and concluded that glucose is the key sugar taken up into the scutellum, where it is converted to sucrose by the enzyme sucrose synthetase. They also concluded that an important block in dormant embryos was the failure to synthesize sucrose because of the absence of sucrose synthetase. However, even if sucrose synthetase was not present in the scutellum of dormant embryos, some rate-limiting factor other than lack of sucrose must have been responsible for embryo dormancy since addition of sucrose to dormant embryos did not break dormancy unless a low concentration of GA_3 was also supplied (Naylor and Simpson, 1961a). This suggests that failure to utilize sugar is an important characteristic of dormant embryos.

2. RESPIRATION

a. EMP Pathway. The first report of a possible regulatory role of respiration in dormancy of *Avena fatua* was the observation that dormant fruits failed to reduce the dye 2,3,6-triphenyltetrazolium chloride (Ebell, 1957). This was confirmed by Hay (1962), who showed that the dye could be taken up through the intact seed coat but was not reduced unless the seed coat was punctured to allow gas exchange. He concluded that the inactivity of dehydrogenases in dormant seeds, as shown by the tetrazolium tests, was due to an inhibition of the electron transport system. He speculated that during the induction of dormancy electrons, instead of going to oxygen, are passed to another acceptor, which in its reduced form then blocks the electron transport system at some specific locus where there are no alternative pathways or, alternatively, general metabolism is stopped. He speculated further that the reversibility of dormancy might be due to an inhibitor that is active in its reduced form but is inactive when oxidized.

Observations on the respiratory quotient of dormant and nondormant seeds (Simmonds and Simpson, 1971) indicated no difference between excised embryos of dormant (0.97) and nondormant (0.95) fruits during the first 10 hr of imbibition of water. This established that oxidation of sugar is not blocked in dormant embryos. In the same study, observation of the C_6/C_1 ratio of CO_2 evolved during the first 10 hr indicated values of 0.76–0.91 for dormant embryos and 0.54–0.59 for nondormant embryos, establishing that glycolysis is the main pathway for glucose oxidation in dormant embryos and that the pentose phosphate (PP) pathway has a more important role in nondormant embryos. When dormancy was broken by GA_3 the overall rate of respiration was not altered, but the C_6/C_1 ratio shifted in a manner similar to the change in the ratio with increased after-ripening; i.e., the oxidation shifted from predominantly glycolysis to a greater participation of the PP pathway.

Measurements of activity of three key enzymes of glycolysis (fructose diphosphate aldolase, NAD-dependent glyceraldehyde-3-P dehydrogenase, and phosphoglycerate kinase) confirmed the activity of the glycolytic pathway in dormant fruits (Kovacs and Simpson, 1976). There were no significant differences in specific activity of the three enzymes between dormant and nondormant fruits during the first 48 hr of imbibition on water.

b. Pentose Phosphate Pathway. The first indication that glucose oxidation, via the PP pathway shunt, was of potential significance in embryo dormancy of *Avena fatua* was the observation that the enzyme 6PGDH, a key enzyme of the pathway, was present in dormant embryos (Simmonds and Simpson, 1971). Quantitative estimation (Simmonds and Simpson, 1971) of the relative activities of the glycolytic and PP pathways using the C_6/C_1 ratio method of Stetton and Bloom indicated that when dormancy was lost in the embryo, either by after-ripening or by application of GA_3, the C_6/C_1 ratio shifted toward a greater participation of the PP pathway.

When malonic acid, a competitive inhibitor of the Krebs cycle enzyme succinate dehydrogenase, was applied to dormant and nondormant embryos it decreased oxygen consumption and CO_2 evolution in dormant embryos; at the same time, it increased germination percentage and shifted the C_6/C_1 ratio toward a greater participation of the PP pathway (Simmonds and Simpson, 1972). The malonate-related inhibition of oxygen consumption decreased with increased after-ripening period. Treatment of nondormant embryos with malonic acid reduced their rate of germination, which indicated that a limited activity of the Krebs cycle was essential for germination. The authors concluded from this and other experimental evidence that the Krebs cycle in dormant embryos was supraoptimal for germination and that inhibition of the cycle reduced the overall requirement for oxygen, thereby allowing other oxygen-requiring processes, such as the PP pathway, to operate more readily. They noted that oxygen consumption continued, even in the presence of malonate, and that oxygen was essential for germination since there was no germination in its absence.

An analogous observation on *Avena sativa* indicated that stimulation of germination could be obtained not only with malonate but also with five inhibitors (KCN, $NaNO_3$, CO, H_2S, and NH_2OH) of cytochrome oxidase, which is the conventional terminal oxidase of glucose oxidation via the EMP pathway (Major and Roberts, 1968). An important conclusion reached by Major and Roberts (1968) was that respiratory inhibitors other than those that affect cytochrome oxidase can also stimulate germination by increasing the participation of the PP pathway. The central significance of the PP pathway in germination of many species of seeds, including *Avena fatua*, was emphasized in an extensive review of oxidation processes in germination (Roberts, 1973). He put forward the following important general hypothesis:

> In order to germinate, seeds need to operate the PP pathway during the initial stages of germination and this pathway is associated with an oxidase which is relatively insensitive to the conventional terminal-oxidase inhibitors—KCN, NaN_3, CO, H_2S and NH_2OH. Dormant seeds are less capable of operating this pathway than non-dormant seeds, although they show a high activity of conventional respiration (the EMP pathway) involving cytochrome oxidase. Any treatment which increases the activity of the PP pathway tends to alleviate dormancy. Since the condition within the seed is relatively anaerobic, removing the covering is such a treatment since it reduces the competition between the oxidase of the PP pathway and the highly competitive cytochrome oxidase. Alternatively the same effect can be obtained by increasing the oxygen pressure, or by reducing the EMP competition by applying respiratory inhibitors.

Further support for the significance of the PP pathway in dormancy of *Avena fatua* was obtained from a comparision of the activities of representative enzymes of glycolysis, the PP pathway, and the Krebs cycle (Kovacs and Simpson, 1976). The activities of G6PDH and 6PGDH were similar in dry, mature, dormant and nondormant fruits. However, the activity of both enzymes increased considerably in nondormant fruits following imbibition of water, whereas their activities in dormant fruits declined steadily to essentially zero by the end of a period of 48 hr. There were no major differences between dormant and nondormant fruits in the activities of the four representative enzymes from glycolysis and the Krebs cycle during the first 24 hr of imbibition of water. At the end of this period germination was completed in the nondormant population.

An analogous, unpublished study by S. A. Olusuyi (Roberts, 1973) of the above two enzymes in dormant and nondormant barley seeds also included a study of the oxidized and reduced forms of the two coenzymes NAD and $NADH_2$ (important for the EMP pathway) and NADP and $NADPH_2$

(essential for the PP pathway). The study suggested that the relative inactivity of the PP pathway in dormant seeds is not caused by lack of the appropriate dehydrogenases or coenzymes. Rather, the equilibrium between the oxidized and reduced forms of the enzymes suggested that, although the oxidation of $NADH_2$ is not limiting, the oxidation of $NADPH_2$ can well be because of the relative inactivity of the postulated, as yet unknown, $NADPH_2$-oxidase system or, alternatively, because of the superoptimal activity of the EMP pathway operating through the cytochrome system and competing with the unknown terminal oxidase for oxygen.

c. TCA Cycle. The first indication of the involvement of the TCA (tricarboxylic acid or Krebs) cycle in embryo dormancy was the observation (Atwood, 1914) that acidity was greater in dormant than in after-ripened embryos. Later, dry seeds of dormant *Avena fatua* were shown to contain a functional TCA system since these seeds were able to metabolize [^{14}C]ethanol to proteins, presumably via the pathway ethanol → acetaldehyde → acetate → α-keto acids → proteins (Chen, 1972; Chen and Varner, 1973). Most of the radioactivity was incorporated into sugars and amino acids, and only a small amount was fixed in organic acids. Chen and Varner (1973) were unable to provide an explanation for the large conversion of ethanol to sugars during after-ripening, which paralleled the conclusion from another study (Hart and Berrie, 1967) that labeling of sugars from uptake of $^{14}CO_2$ must have been due to reverse glycolysis. Possibly both of these observations are proof for gluconeogenesis. Hart and Berrie (1967) also assayed the levels of malic and succinic acids and showed that the level of malic acids was inversely correlated with the degree of seed dormancy; the level of succinic acid did not change. They also indicated that secondary dormancy may be qualitatively different from primary dormancy since malate decreased but succinate increased when secondary dormancy was induced by placing seeds under anaerobic conditions. They found that light was inhibitory to germination and that there was one-third the amount of fixation of $^{14}CO_2$ into malic acid in light than in darkness.

A further indication of the significance of the TCA cycle in dormancy was provided by the demonstration that GA_3-induced germination of embryos was enhanced by succinic acid (10^{-1}–5×10^{-2} M) (Simmonds and Simpson, 1972). The combination of a suboptimal concentration of GA_3 (1.0 ppm) and succinate was more effective than the optimal (50 ppm) concentration of GA_3 alone. When ADP was fed to dormant embryos as a means of enhancing generation of ATP, germination was promoted after a considerable lag period. Inhibition of germination of nondormant embryos by the uncoupler of oxidative phosphorylation 2,4-dinitrophenol (DNP) demonstrated that ATP production is essential for germination. The several observations that ADP, succinate, and malonate promote germination of dormant embryos and that DNP inhibits nondormant embryos were incorporated into a hypothesis (Simmonds and Simpson, 1972) that overproduction of ATP leads to a reduction in the rate of glycolysis and a concomitant activation of a gluconeogenic pathway. Thus, under conditions of high atmospheric oxygen tension the TCA cycle oxidative system would become saturated so that a greater proportion of the available oxygen could be used via the PP pathway.

In attempting to obtain an overall perspective on the contributions from the several pathways of oxidative metabolism in seeds from a range of different species, Roberts (1973) summarized the role of various dormancy-breaking agents. The data he presented for *Avena fatua* together with additions by this author are summarized in Table VI. To this information can also be added the effect of the important dormancy-breaking hormone GA_3, which stimulates the activity of at least two enzymes of the PP pathway (Kovacs and Simpson, 1976).

3. PROTEIN SYNTHESIS

The first study of protein synthesis in embryos of *Avena fatua* was a comparison of the soluble-protein content of dormant and nondormant embryos during the first 48 hr of imbibition on water (Simpson, 1965). In nondormant embryos the amount fell sharply during the first 24 hr, possibly due to secretion of enzymes, and was then built up by 48 hr to an amount greater than that of the mature dry embryo. In dormant embryos the soluble protein continued to fall throughout the 48-hr period unless GA_3 was present, in which case it increased again after the initial decline. The conclusion was made that the site of GA_3 stimuluation of protein synthesis was in the scutellum since the excised root–shoot axis, separated from the scutellum, was not blocked in the reaccumulation phase in either dormant or nondormant embryos, with or without GA_3. A further conclusion was that the phase of reaccumulation of soluble protein seen in nondormant embryos was under nuclear control since it was inhibited by actinomycin D, an inhibitor of RNA synthesis; GA_3 was able to reverse this effect. The activity of 3'-nucleotidase, an enzyme that splits phosphate from 3'-adenylic acid, was also monitored in both dormant and nondormant seeds. The activity increased by 350% in nondormant embryos during the first 48 hr but increased by only 30% in dormant embryos. Significant increases could be induced in dormant embryos by treatment with GA_3, suggesting that it has a role in the breakdown of nucleotides through the activation or synthesis of at least one specific enzyme.

An autoradiographic study of dormant *Avena fatua* seeds administered [14C]leucine showed that protein synthesis occurred in all parts of the

TABLE VI

Dormancy-Breaking Agents of *Avena fatua* and Their Role in Oxidative Metabolism[a]

Agent	Effect of germination	Reference[b]
Hydrogen acceptors		
O_2	+	1, 2, 5
NO_3	+	2, 4, 5
NO_2	+	7
Methylene blue	0	7
Reduced N compounds	0	7
Respiratory inhibitors		
Terminal oxidases	+	5, 7
CN^-	+	5, 7
N_3^-	+	7
Krebs cycle		
malonate	+	8
Glycolysis		
iodoacetate	+ (*Avena ludoviciana*)	6
NaF	0	7
Uncouplers		
DNP	0	7
	(Dormant)	8
Na_3ASO_4	− (Nondormant)	7
ADP and Succinate	0	8
CO_2	+	3
	+	

[a] After Roberts (1973).

[b] Key to references: (1) Atwood (1914);   (2) Baker and Leighty (1958);   (3) Hart and Berrie (1966);   (4) Hay and Cumming (1957, 1959);   (5) Johnson (1935a);   (6) Morgan and Berrie (1970);   (7) E. H. Roberts and A. Madden (unpublished); Roberts (1973);   (8) Simmonds and Simpson (1972).

embryo, including apical meristem, coleoptile, primary leaf, radicle, root cap, and scutellum (Chen and Varner, 1970). Dormant seeds were apparently capable of synthesising protein at a rate comparable to that of nondormant seeds during the first 24 hr of germination. When dormant seeds were exposed to actinomycin D during the first 12 hr of imbibition the incorporation of [14C]leucine was somewhat reduced, indicating both that preformed RNA was present in sufficient quantity to support most of the protein synthesis and that some DNA-dependent synthesis was required for maximal rate of protein synthesis.

The role of GA_3 in stimulating protein synthesis in dormant embryos was confirmed when it was found that GA_3 stimulated incorporation of

[³H]leucine and [³H]uridine into embryos during the first 14 hr of imbibition (Chen and Park, 1973). Cordycepin (3'-deoxyadenosine), an inhibitor of RNA synthesis, inhibited germination of the same seeds in the absence of GA_3.

The protein synthesis inhibitor chloramphenicol (CP) interacts with both exogenous sucrose and exogenous GA_3 on the germination of embryos (G. M. Simpson, unpublished results). Sucrose enhanced the stimulating effect of CP on germination. Together they reduced the time taken for a population of semidormant embryos to achieve 50% germination, from 150 hr in the water control to 42 hr in the treatment. The combination of CP and sucrose was nearly as effective as the optimal combination of GA_3 (10^{-1} ppm) and sucrose (5%). Addition of GA_3 to the combination of CP and sucrose did not further enhance the rate of germination. This evidence suggests that CP may prevent the synthesis of proteases early in germination or alternatively inhibit protein synthesis so that relatively scarce ATP is diverted to driving cell elongation in the radicle. Proteases have a role not only in the breakdown of reserve protein but also in the breakdown of such enzymes as 6GPDH and G6PDH, which catalyze sugar oxidation. While these two enzymes are present in mature dry dormant seeds they are broken down during the first 48 hr unless the dormant seed is exposed to GA_3 (Kovacs and Simpson, 1976).

VI. A SYSTEMS CONCEPT TO EXPLAIN CONTROL OF SEED DORMANCY

A. Interrelationship of Metabolism, Seed Structure, and Environment

The review (Sections III–V) of the seed–environment subsystems in *Avena fatua* indicates that without exception each one of the components has some role in either the initiation, maintenance, or termination of dormancy. Despite the fact that an excised embryo can exhibit dormancy independently of the presence of endosperm, pericarp–testa, or hulls, these structures clearly have an important interaction with the embryo, either through sustaining the initial embryo dormancy or by facilitating the induction of secondary dormancy under specific environmental conditions.

Several concepts for explaining the mechanism of dormancy in *Avena fatua* were previously put forward. The first of these was based on the obvious significance of the plant growth regulator GA_3 (Simpson, 1965). The role of GA_3 in both the mobilization and utilization of endosperm reserves by the embryo was linked with the observation that dormant

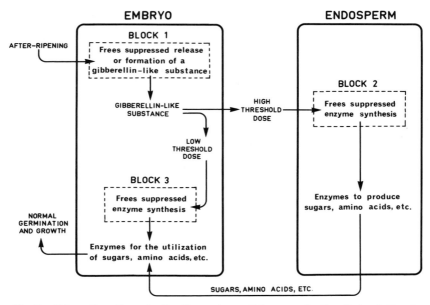

Fig. 7. Schematic pathway leading from release of dormancy by the effects of after-ripening to normal germination and growth in seeds of *Avena fatua*. (From Simpson, 1965.)

embryos failed to secrete gibberellin-like substances. Dormancy was thus envisaged as a single block in the production of GA_3 within the embryo. This block in turn imposed two other blocks due to the requirement for gibberellin as a trigger: first, for synthesis of the enzymes necessary for the mobilization of endosperm reserves and, second, for the synthesis of enzymes necessary for the utilization of these reserves by the embryo (Fig. 7). In some unknown way the embryo was freed from the block to gibberellin production by events occurring during the after-ripening period.

Essentially the same scheme as Fig. 7 was later modified to take account of the demonstrated or assumed effects of both seed structure and environment on the release or formation of the gibberellin-like substance in the embryo (Hsiao and Simpson, 1971). The effects of light (presumed to act via the phytochrome system), gases, growth inhibitors, and variations in water status were all incorporated into the scheme to account for either the loss of dormancy through an after-ripening period or the induction of secondary dormancy in nondormant seeds.

Shortly after the proposal of the second scheme, investigations on oxidative metabolism in dormant and nondormant embryos (Simmonds and Simpson, 1972) led to the discovery that oxygen supply, as well as GA_3, could modify the relative amounts of glucose oxidized through either the PP pathway or the Krebs cycle. Simmonds and Simpson (1972) proposed a scheme for the regulation of oxidative metabolism (Fig. 8) in which variation in the supply of oxygen moderated the activities of the two pathways. Restriction of the PP pathway, because of high activity of the Krebs pathway, would reduce the supply of NADPH, which is produced through activity of the PP pathway and which was considered to be essential for the synthetic processes of germination. The concept of mutually exclusive oxidative pathways regulating dormancy was considered to be a biochemical mechanism that in effect enabled a seed to measure time. That is, the period needed for after-ripening would have to be long enough to permit generation of an excess of ATP in the dry stored seed in sufficient amounts to restrict the Krebs cycle activity through feedback inhibition. Inhibition of the Krebs pathway would thus permit the PP pathway to operate more actively, thereby providing the substrates necessary for germination.

Gibberellic acid can alter the PP pathway activity by bringing about an increase in the activity of the first two enzymes of the pathway (G6PDH and 6PGDH). It can also increase the activity of at least three other

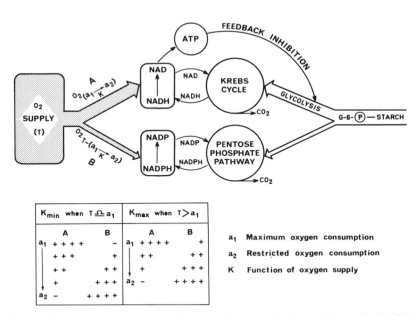

Fig. 8. Schematic representation of the regulation of oxidative metabolism involved in the control of germination of *Avena fatua*. Relative activities within each pathway are designated in the boxed table as: high $(++++)$, varying to low $(+)$, and zero $(-)$. (From Simmonds and Simpson, 1972.)

enzymes of the fruit (α-amylase, maltase, 3'-nucleotidase). The question therefore arises whether natural shifts in the ratio of the PP pathway to the Krebs pathway are dependent first on endogenous gibberellin production, which could be determined by environmental factors such as oxygen partial pressure or temperature, or whether these environmental factors directly affect the pathways in the absence of gibberellin. Considerable experimental effort will be needed to resolve this question since at present the site of gibberellin production (whether intracellular or within specific tissues) in the embryo of *Avena fatua* is unknown. In addition, although it is known that the enzymes of the PP pathway are usually found in the cytoplasm and are thus spatially separated from the Krebs enzymes in the mitochondria (Schnarrenberger *et al.*, 1973), there is no knowledge of their general distribution within the embryo of *Avena fatua*.

General models of seed dormancy (Amen, 1968; Villiers, 1972) have placed emphasis on an inhibitor–promoter complex as a general controlling system. In the case of *Avena fatua* the gibberellins undoubtedly have an important role as true hormones in initiating the breakdown of endosperm reserves and embryo growth. Experimental evidence suggests that there are inhibitors in the hulls and fruit, and abscisic acid has been identified; nevertheless, there is little evidence as yet for their physiological role. It is tempting to speculate that inhibitors might play a role in controlling the activity of the PP pathway—for example, by allosteric inhibition of enzymes. However, much more work must be done to prove whether an inhibitor–promoter balance is a determinant of dormancy in *Avena fatua*.

Roberts' (1973) review and resultant hypothesis of the central significance of the PP pathway in seed dormancy of a range of species gains considerable significance in view of other reports on the effects of different factors on the activity of this pathway. Hendricks and Taylorsen (1975) gave a schematic explanation for the germination-promoting effects of thiourea, sodium nitrate, and hydroxylamine salts on dormant lettuce seeds which could be similar in *Avena fatua* seeds. They showed that these substances inhibited the enzyme catalase. Thus, the metabolically derived hydrogen peroxide that was spared by the inhibition of catalase could oxidize reduced NADPH needed as the oxidant of the PP pathway. The oxidation was mediated by the pyridine nucleotide quinone oxidoreductase, which could be the unknown oxidase postulated by Roberts (1973). An important conclusion was that the reaction system controlling the activity of the PP pathway can also be limited by the $NADP^+$–NADPH level. Hendricks and Taylorsen (1975) cited evidence that the pigment phytochrome activates NAD-kinase, which controls the interconversion of NAD and NADP, thus establishing links between the effects of light, the PP pathway, and control of dormancy. A similar effect was described by Frosch and Wagner (1973).

Evidence for the role of the PP pathway in controlling protein synthesis was reported by Lange *et al.* (1970). They showed that, in potato tissue placed under conditions of restricted oxygen supply in water, protein synthesis was severely inhibited and the activities of both G6PDH and 6PGDH were both low compared with activities when potato tissue was exposed to air. The same two enzymes were associated with the rise and fall of soluble protein in a study of bud activity and dormancy (Sagisaka, 1974). It is probable that the "net energy charge of the adenylate pool" (Atkinson, 1968) has a regulatory role in dormancy of *Avena fatua* since both the PP pathway and glycolysis are likely to be sensitive to varying levels of ATP. For example, in other plant tissues G6PDH of the PP pathway is inhibited by both ATP and NADPH (Ashihara and Komamine, 1974; Grossman and McGowan, 1975; Muto and Uritani, 1972; Sagisaka, 1974), and both orthophosphate and ATP control the activity of glycolysis in germinating pea seeds (Givan, 1972). The subject of energy charge as a control in seed metabolism was reviewed (Ching, 1972), and it was shown that when the energy charge is less than 0.5, ATP-regenerating systems are dominantly active; when the energy charge is greater than 0.5, ATP-utilizing systems become active until the energy charge approaches one where ATP-utilizing systems are predominant. Thus, any environmental factors that can influence adenylate charge can also be expected to influence dormancy and germination.

B. A Concept of Control

At present the most attractive single concept for explaining dormancy of *Avena fatua* is that of the blocked, or restricted, oxidative pentose phosphate pathway proposed by Roberts (1973). Nevertheless, this may prove to be too narrow a view. The argument that the PP pathway limits the supply of NADPH needed for the synthetic biochemical reactions of growth may be valid, but so little is known at present about the interrelationships among the pathways of oxidative metabolism in *Avena fatua* and their connection to growth, by either cell enlargement or division, that it would be premature to attach too much significance to the central importance of the pathway. Nevertheless, the observations so far made on metabolism indicate that most, if not all, of the effects of the environment and seed structure on dormancy can be interpreted in terms of their ultimate effects on metabolism.

Looked at from a systems viewpoint, initiation of primary dormancy by the blocking of an important biochemical pathway can be considered as a temporary disconnection of some of the subsystems in the lower levels of the hierarchy of functional levels found in the normal germinating seed. It is

216 Graham M. Simpson

possible, but not necessarily essential, that this disconnection is brought about via control of the genetic activity in the zygote. It could also be due to the genetically imposed influence of the maternal tissue, suitably modified by the external environment of the plant, on the metabolism of the developing zygote. In the latter case the environment of the senescing maternal parent plant would be important in determining the timing and degree of induced dormancy permissible within the genetic system of the female plant. The possibility that the imposition of dormancy is via maternal tissue should be resolvable by reciprocal crosses between genetically dormant and nondormant parents.

The systems analogy implies that much of the seed system continues to operate normally but the steps of growth that begin germination cannot take place until the whole system, on every level, is operative. There is an analogy here between seed dormancy and the old-fashioned pocket watch. When the winding knob of a watch is pulled out, the wheels that turn the hour and minute hands (growth) become disconnected from the rest of the mechanism (metabolism), which continues to function as long as the spring (endosperm reserves) provides energy. Pushing the knob in entrains the gears, and the hands move again. In the simplest sense this seems to be the case with dormancy and growth in the wild oat seed. Growth of the embryonic axis of the imbibed fruit does not occur, despite continued respiratory activity and RNA and protein synthesis, until something entrains the total system. From the available evidence it seems logical to assume that it is the environment around the seed that initiates both the disengagement and reentrainment of the subsystems that are temporarily nonfunctional. Exactly how and where the environmental force is exerted is not yet understood. Experimental evidence suggests that secondary dormancy may be different from primary dormancy in *Avena fatua*. If this is confirmed it would demonstrate that more than one possibility exists for the expression of dormancy, according to whichever subsystems can be temporarily dislocated from the total system.

A systems concept of a hierarchy of relatively autonomous subsystems arranged in levels implies connecting links between each of the subsystems and levels. The substrates of metabolism, including gases, and plant growth regulators functioning either as inhibitors or promoters can be these links. In this respect the concept of adenylate charge controlling the subsystems or levels is attractive because it has already been shown in animal tissues that compartmentalization of AMP, ADP, or ATP within a cell or within different tissues exerts a powerful influence on metabolic activity. The fact that light can influence the magnitude of the adenylate charge through an influence on the phytochrome system and NAD-kinase suggests a way in which light can directly influence dormancy of wild oat seeds.

C. Proof for the Concept

Experimental proof for a systems concept of dormancy in which some of the subsystems are temporarily disengaged is scanty. Nevertheless, experimental proof is unlikely to be obtained unless a hypothetical framework is first established and then systematically tested. Much experimental work remains to be done to determine where the metabolic events that are involved in dormancy of *Avena fatua* are located within the embryo. Location of phytochrome, location of sites of hormone formation and action, and identification of sites of activity of the PP pathway and EMP–Krebs pathways are essential to any further understanding of the role of metabolism in dormancy. Protease activity and protein synthesis are important in the initiation of normal germination and may therefore be equally important in determining the state of dormancy. Atwood's (1914) approach of "judging the situation about dormancy from the combined results rather than from any one line of enquiry" still seems relevant more than 60 years later. A combination of holistic thinking together with an evaluation of each of the subsystems in any one level of the seed–environment system would seem to offer the most hope in further elucidating the nature of seed dormancy in *Avena fatua*.

ACKNOWLEDGMENTS

Thanks are due to Professor J. M. Naylor for reviewing the manuscript, to Mrs. J. McLean for typing, and to Mr. J. Diduck for preparing the figures. The author is indebted to the Canada Department of Agriculture and the National Research Council of Canada for financial support of his research on *Avena fatua*.

REFERENCES

Aamodt, C. S., Johnson, L. P. V., and Manson, J. M. (1934). *Can. J. Res.* **11**, 701.
Amen, R. D. (1968). *Bot. Rev.* **34**, 1.
Andrews, C. J. (1967). Ph.D. Thesis, University of Saskatchewan, Saskatoon.
Andrews, C. J., and Burrows, V. D. (1972). *Can. J. Plant Sci.* **52**, 295.
Andrews, C. J., and Burrows, V. D. (1974). *Can. J. Plant Sci.* **54**, 565.
Andrews, C. J., and Simpson, G. M. (1969). *Can. J. Bot.* **47**, 1841.
Ashihara, H., and Komamine, A. (1974). *Plant Sci. Lett.* **2**, 331.
Atkinson, D. E. (1968). *Biochemistry* **7**, 4030.
Atwood, W. M. (1914). *Bot. Gaz. (Chicago)* **57**, 386.
Avery, G. S. (1930). *Bot. Gaz. (Chicago)* **89**, 1.
Baker, L. O., and Leighty, D. H. (1958). *Proc. West. Weed Control Conf.* **16**, 69.
Banting, J. D. (1962). *Can. J. Plant Sci.* **42**, 22.
Banting, J. D. (1966a). *Can. J. Plant Sci.* **46**, 129.

Banting, J. D. (1966b). *Can. J. Plant Sci.* **46,** 469.
Barton, L. V. (1967). "Bibliography of Seeds." Columbia Univ. Press, New York.
Bibbey, R. O. (1935). *Sci. Agric.* **16,** 141.
Bibbey, R. O. (1948). *Plant Physiol.* **23,** 467.
Black, M. (1959). *Can. J. Bot.* **37,** 393.
Black, M., and Naylor, J. M. (1959). *Nature (London)* **184,** 468.
Briggs, D. E. (1973). *In* "Biosynthesis and Its Control in Plants" (B. V. Milborrow, ed.), pp. 219–277. Academic Press, New York.
Cannon, W. A. (1900). *Proc. Calif. Acad. Sci.* [3] **1,** 329.
Cates, H. R. (1917). *U.S., Dep. Agric., Farmers' Bull.* **833.**
Chancellor, R. J., and Peters, N. O. B. (1972). *Proc. Br. Weed Control Conf.* **1,** 218.
Chen, S. S. C. (1972). *Naturwissenschaften* **59,** 123.
Chen, S. S. C., and Chang, J. L. L. (1972). *Plant Physiol.* **49,** 441.
Chen, S. S. C., and Park, W. (1973). *Plant Physiol.* **52,** 174.
Chen. S. S. C., and Varner, J. E. (1969). *Plant Physiol.* **44,** 770.
Chen, S. S. C., and Varner, J. E. (1970). *Plant Physiol.* **46,** 108.
Chen, S. S. C., and Varner, J. E. (1973). *Seed Sci. Technol.* **1,** 325.
Ching, T. M. (1972). *In* "Seed Biology" (T. T. Kozlowski, ed.), Vol. 2, pp. 103–218. Academic Press, New York.
Christie, L. A. (1956). M.Sc. Thesis, University of Saskatchewan, Saskatoon.
Cobb, R. D., and Jones, L. G. (1962). *Am. J. Bot.* **49,** 658.
Coffman, F. A., and Stanton, T. R. (1938). *J. Agric. Res.* **57,** 57.
Coffman, F. A., and Stanton, T. R. (1940). *Am. Soc. Agron. J.* **32,** 459.
Cumming, B. G. (1957). *Res. Rep. Can. Agric. Nati. Weed Comm., West. Sect.* p. 133.
Cumming, B. G., and Hay, J. R. (1958). *Nature (London)* **182,** 609.
Drennan, D. S. H., and Berrie, A. M. M. (1961). *New Phytol.* **61,** 1.
Ebell, L. F. (1957). Ph.D. Thesis, University of Alberta, Edmonton.
Eckerson, S. A. (1913). *Bot. Gaz. (Chicago)* **55,** 286.
Edelman, J., Shibko, S. I., and Keys, A. J. (1959). *J. Exp. Bot.* **10,** 178.
Evans, S. A. (1958). *Agriculture* **64,** 502.
Friesen, G., and Shebeski, L. H. (1961). *Weeds* **9,** 634.
Frosch, S., and Wagner, E. (1973). *Can. J. Bot.* **51,** 1521.
Garber, R. J., and Quisenberry, K. S. (1923). *J. Hered.* **14,** 267.
Givan, C. V. (1972). *Planta* **108,** 29.
Green, G. J., and Helgeson, E. A. (1957a). *Proc., North Cent. Weed Control Conf.* **14,** 5.
Green, G. J., and Helgeson, E. A. (1957b). *Proc., North Cent. Weed Control Conf.* **14,** 39.
Grossman, A., and McGowan, R. E. (1975). *Plant Physiol.* **55,** 658.
Haber, A. H. (1962). *Plant Physiol.* **37,** 18.
Hall, J. R., and Hodges, T. K. (1966). *Plant Physiol.* **41,** 1459.
Harris, G. P. (1954). *New Phytol.* **55,** 253.
Hart, J. W., and Berrie, A. M. M. (1966). *Physiol. Plant.* **19,** 1020.
Hart, J. W., and Berrie, A. M. M. (1967). *Phytochemistry* **7,** 1257.
Hay, J. R. (1962). *Can. J. Bot.* **40,** 191.
Hay, J. R. (1967). *Physiol. Oekol. Biochem. Keimung, Mater. Int. Symp., 1963* Vol. I, pp. 319–323.
Hay, J. R., and Cumming, B. G. (1957). *Proc. North Cent. Weed Control Conf.* **14,** 39.
Hay, J. R., and Cumming, B. G. (1959). *Weeds* **7,** 34.
Helgeson, E. A., and Green, J. G. (1957a). *Proc. North Cent. Weed Control Conf.* **14,** 38.
Helgeson, E. A., and Green, J. G. (1957b). *N.D., Agric. Exp. Stn., Bull.* **455,** 121.
Helgeson, E. A., and Green, J. G. (1958). *N.D., Agric. Exp. Stn., Bull.* **B20,** 7.

Hendricks, S. G., and Taylorsen, R. B. (1957). *Proc. Natl. Acad. Sci. U.S.A.* **72,** 306.
Hopkins, C. Y. (1936). *Can. J. Res., Sects. C–D* **14,** 178.
Hsiao, A. I., and Simpson, G. M. (1971). *Can. J. Bot.* **49,** 1347.
Johnson, L. P. V. (1935a). *Can. J. Res., Sect. C* **13,** 283.
Johnson, L. P. V. (1935b). *Can. J. Res., Sect. C* **13,** 367.
Khan, A. A., and Downing, R. D. (1968). *Physiol. Plant.* **21,** 29.
Kirk, J., and Courtney, A. D. (1972). *Proc. Br. Weed Control Conf.* **1,** 226.
Kiseleva, N. A. (1956). *Agrobiologia* **2,** 130.
Knights, B. A. (1968). *Phytochemistry* **7,** 2067.
Koestler, A. (1967). "The Ghost in the Machine." Hutchinson, London.
Kommedahl, T., Devay, J. E., and Christensen, C. M. (1958). *Weeds* **6,** 12.
Kovacs, M. I. P., and Simpson, G. M. (1976). *Phytochemistry* **15,** 455.
Lange, H., Kahl, G., and Rosenstock, G. (1970). *Physiol. Plant.* **23,** 80.
Leggett, H. W. (1950). *Proc. West. Can. Weed Control Conf.* **4,** 139.
Leighty, D. H. (1958). M.Sc. Thesis, Montana State College, Bozeman.
Lindsay, D. R. (1956). *Weeds* **4,** 1.
Lute, A. M. (1938). *Proc. Assoc. Off. Seed Anal.* **33,** 70.
McDonald, B. K. (1949). *Proc. West. Can. Weed Control Conf.* **3,** 167.
Maherchandani, N. J., and Naylor, J. M. (1971). *Can. J. Genet. Cytol.* **13,** 578.
Maherchandani, N. J., and Naylor, J. M. (1972). *Can. J. Bot.* **50,** 305.
Major, W., and Roberts, E. H. (1968). *J. Exp. Bot.* **19,** 77.
Marshall, D. R., and Jain, S. K. (1970). *Ecology* **51,** 886.
Morgan, S. F., and Berrie, A. M. M. (1970). *Nature (London)* **228,** 1225.
Müllverstedt, R. (1963). *Weed Res.* **3,** 154.
Muto, S., and Uritani, I. (1972). *Plant Cell Physiol.* **13,** 377.
Naylor, J. M. (1966). *Can. J. Bot.* **44,** 19.
Naylor, J. M. (1966). *Can. J. Bot.* **47,** 2068.
Naylor, J. M., and Christie, L. A. (1957). *Proc. West. Can. Weed Control Conf.* **10,** 56.
Naylor, J. M., and Simpson, G. M. (1961a). *Can. J. Bot.* **39,** 281.
Naylor, J. M., and Simpson, G. M. (1961b). *Nature (London)* **39,** 281.
Nikolaeva, M. G. (1969). "Physiology of Deep Dormancy in Seeds." Nat. Sci. Found., Washington, D.C.
Pawlowski, S. H. (1959). M.Sc. Thesis, University of Alberta, Edmonton.
Quail, P. (1968). Ph.D. Thesis, University of Sydney, Sydney, Australia.
Rademacher, B., and Kiewnick, L. (1964). *Z. Acker- Pflanzenbau* **119,** 369.
Roberts, E. H. (1973). *In* "Seed Ecology" (W. Heydecker, ed.), pp. 189–218. Butterworth, London.
Sagisaka, S. (1974). *Plant Physiol.* **54,** 544.
Schnarrenberger, C., Oeser, A., and Tolbert, N. E. (1973). *Arch. Biochem. Biophys.* **154,** 438.
Sexsmith, J. J. (1969). *Weed Sci.* **17,** 405.
Simmonds, J. A. (1971). Ph.D. Thesis, University of Saskatchewan, Saskatoon.
Simmonds, J. A., and Simpson, G. M. (1971). *Can. J. Bot.* **49,** 1833.
Simmonds, J. A., and Simpson, G. M. (1972). *Can. J. Bot.* **50,** 1041.
Simpson, G. M. (1965). *Can. J. Bot.* **43,** 793.
Simpson, G. M. (1966). *Can. J. Bot.* **44,** 115.
Simpson, G. M., and Naylor, J. M. (1962). *Can. J. Bot.* **40,** 1659.
Street, H. E., and Cockburn, W. (1972). "Plant Metabolism." Pergamon, Oxford.
Sutcliffe, J. F., and Baset, Q. A. (1973). *Plant Sci. Lett.* **1,** 15.
Thornton, N. C. (1945). *Contrib. Boyce Thompson Inst.* **13,** 487.
Thurston, J. M. (1957). *J. Agric. Res.* **49,** 259.

Thurston, J. M. (1961). *Weed Res.* **1**, 19.

Thurston, J. M. (1962a). *Weed Res.* **2**, 192.

Thurston, J. M. (1962b). *Rothamsted Conf.* p. 236.

Thurston, J. M. (1963). *Rothamsted Conf.* p. 90.

Thurston, J. M. (1964). *NAAS Q. Rev.* **65**, 22.

Thurston, J. M. (1966). *Weed Res.* **6**, 67.

Tingey, D. C. (1961). *Weeds* **9**, 607.

Toole, E. H., and Coffman, F. A. (1940). *J. Am. Soc. Agron.* **32**, 631.

Villiers, T. A. (1972). *In* "Seed Biology" (T. T. Kozlowski, ed.), Vol. 2, pp. 219–281. Academic Press, New York.

von Bertalanffy, L. (1968). "General System Theory—Foundations, Development, Applications." George Braziller, New York.

Whittington, W. J., Millman, J., Gatenby, S. M., Hooper, B. E., and White, J. C. (1970). *Heredity* **25**, 641.

Whyte, L. W., Wilson, A. G., and Wilson, D. (1969). "Hierarchical Structures." Am. Elsevier, New York.

Williams, G. C., and Thurston, J. M. (1964). *Ann. Appl. Biol.* **53**, 29.

Wilson, B. J., and Cussans, G. W. (1975). *Weed Res.* **15**, 249.

Zade, A. (1909). Inaugural Dissertation, pp. 1–48. Jena.

5

Environmental and Hormonal Control of Dormancy in Terminal Buds of Plants

LARRY D. NOODÉN AND JAMES A. WEBER

DORMANCY AND DEVELOPMENTAL ARREST

I. INTRODUCTION

The development, maintenance, and release of dormancy in buds involves a complex interaction of a number of factors ranging from environmental to genetic. Even though gardeners in northern Europe had known for many years that cuttings from flowering trees brought into a warm environment in the winter would grow out and produce flowers after they had experienced a certain amount of cold (Knight, 1801), the importance of environmental factors in regulating dormancy was not appreciated until after the turn of the century. Since the late 1940's, a steadily increasing stream of literature has greatly expanded our knowledge of these factors and their impacts on with plant development. Two environmental factors, day length and temperature, appear to be of overriding importance in the control of dormancy. However, moisture, nutrients, and light intensity may also control at least the initial stages in the development of dormant buds. The response to some factors, especially photoperiod, has been found to vary even within a species, depending on the latitude and altitude. Environmental factors cue the plant to changes in the seasons, so that it can make the adaptations that will favor survival through periods of unfavorable weather. These cues must be predictive of the weather through the seasons and allow sufficient time before the onset of unfavorable conditions for the various morphological and physiological changes leading to dormancy. Once in the dormant state, the plant can withstand unfavorable environments better than in the growing state. For example, plants that would be killed by a frost when growing can withstand temperatures many degrees below freezing in the dormant state (see Weiser, 1970, for a review of cold hardiness).

The dormant condition (*sensu latu*) in buds can be characterized by both reduced respiration (Section II) and reduced or no growth (see Samish, 1954; Romberger, 1963). The decreased activity of buds either may be maintained by forces external to the bud (quiescence, summer dormancy, correlated dormancy) or may arise within the bud (innate dormancy, deep dormancy, rest, dormancy *sensu strictu*) (see Dorrenbos, 1953; Romberger, 1963; Wareing, 1969; Perry, 1971). Romberger (1963) differentiated between correlated dormancy, which he defined as being controlled within the plant but not in the bud, and quiescence, which is imposed by the environment. In species in which bud dormancy is induced by the leaves, the inability of defoliation to induce bud outgrowth has been taken as an indicator of deep dormancy (e.g., Doorenbos, 1953; Nesterov, 1962; Tinklin and Schwabe 1970; Witkowska-Žuk, 1970).

In this review, "dormancy" indicates that state of deep rest controlled within the bud, and "quiescence" indicates dormancy imposed from outside the bud. Thus, axillary buds that are repressed through apical dominance are not considered dormant any more than buds that are inactive as a result of an unfavorable environment. Plants that develop dormancy generally go through a period of quiescence before and after the period of dormancy. Dormancy generally develops in organs with special structural adaptations, and the development of the "dormant" morphology may be considered an important phase (probably even obligatory) in the sequence leading to dormancy.

Trees have a wide range of bud types that may become dormant, including apical buds, adventitious buds, and in some cases axillary buds after the first year of existence (see Romberger, 1963; Kozlowski, 1971). Furthermore, in some trees the apical bud dies and an axillary bud forms the new leader, producing a sympodial growth pattern, e.g., citrus (Schroeder, 1951) and black locust (*Robinia pseudoacacia*) (Kozlowski, 1971). Dormant buds are also found in bulbs, tubers, rhizomes, and turions (or hibernaculae of many aquatic vascular plants). Because of (1) the diversity of structures associated with dormant buds and (2) the repeated evolution of dormancy in buds, it should be expected that the mechanisms controlling dormancy may likewise be diverse. In terms of general morphology, a dormant buu can be summarized as a shortened axis with appressed foliar structures. There are also changes in the cellular structure associated with the reduced growth (Gentel' and Barskaya, 1960; Dereuddre, 1971; Shilova, 1974; also see Kozlowski, 1971). Finally, looking at the progression within a tree, the development of dormant buds on European ash (*Fraxinus excelsior*) (Witkowska-Żuk and Kozlowska, 1973, 1974) and *Populus* × *berolinensis* (Witokowska-Żuk and Kapuscinski, 1969) proceeds from the base of the tree toward the apex.

II. METABOLISM IN RELATION TO BUD DORMANCY

Buds undergo some very marked metabolic changes both as they develop toward the dormant state and as they emerge from it. Even though these are of fundamental importance, they have received very little attention in recent times, certainly nowhere near the amount of effort directed toward analogous processes in seeds. Furthermore much of what is known was learned at a time when basic biochemistry and its technology were much less advanced. Although these results sometimes seem less rigorous than more modern methods could produce, they did establish many important points. These older studies have been discussed in earlier reviews (Samish, 1954;

Romberger, 1963; Vegis, 1964; Hemberg, 1965) and will only be summarized here to provide a basis for comparison with the other systems
reviewed in this volume. Recently, some effort has been made to reexamine
the older findings using newer methods and in some cases correlating
metabolism more closely with degree of dormancy, but the basic picture
remains the same (Bachelard and Wightman, 1973; Klenovska *et al.*, 1974;
Sizov, 1975; Szalai *et al.*, 1975).

Like other dormant systems, buds are generally relatively dehydrated;
there are, of course, some exceptions, notably turions (the dormant buds of
aquatic vascular plants). Rehydration and a very large gain in fresh weight
accompany the outgrowth that occurs when dormancy is broken and
followed by favorable conditions. Reserve foods, particularly starch, accumulate during dormant bud development and are depleted as outgrowth
proceeds. Termination of dormancy by natural or artificial means may
cause an increase in free sugars, amino acids, and organic acids. The
activity of many enzymes or groups of enzymes (most notably hydrolytic
and oxidative enzymes) decreases as dormancy develops and increases again
as (or after) dormancy is broken. Very extensive studies show that O_2
consumption follows a similar pattern. In poplar (*Populus balsamifera*)
vegetative buds, the rate of O_2 absorption increases slowly as dormancy is
broken and then surges upward rapidly when outgrowth starts (Bachelard
and Wightman, 1973). The respiratory quotient (ratio of volume of CO_2
produced to volume of O_2 consumed per unit of time) may increase or
decrease as dormancy develops or is broken (see Samish, 1954; Hemberg,
1965; Bachelard and Wightman, 1973). The exact values and timing seem to
differ depending on the species, but this does indicate some changes in
the major respiratory substrates.

More recently, it was shown that dormancy-breaking treatments—
ethylene chlorohydrin and giberellic acid (GA_3)—increase RNA and
DNA precursor incorporation in potato tuber buds (Tuan and Bonner,
1964; Shih and Rappaport, 1970) and the RNA template activity of chromatin isolated from buds of treated (ethylene chlorohydrin) tubers (Tuan
and Bonner, 1964). The increased nucleic acid precursor incorporation in
potato buds, analyzed by autoradiography, may be largely related to a reactivation of cell division (Shih and Rappaport, 1970). In addition, inhibitors
of RNA and protein synthesis (actinomycin D and puromycin) inhibit
potato bud sprouting (Taun and Bonner, 1964; Madison and Rappaport,
1968), but this could be due to an inhibition of cell division or enlargement
rather than the primary dormancy-breaking processes.

At various times, one or another of these metabolic changes, particularly
respiration or release of stored food, has been invoked as the key reaction

that maintains dormancy or allows outgrowth to proceed (see Hemberg, 1965, for a summary), but too many discrepancies exist. For example, treatments that inhibit respiration may terminate dormancy of tree buds (Weber, 1916a,b; Boresch, 1926; Miller, 1934). Similarly, large amounts of the starch in woody plants may be converted to sugar without breaking the dormancy of these tissues (Johannsen, 1906). Much consideration has been given to the possibility that the bud coverings cause dormancy by restricting the O_2 supply (see Vegis, 1964; Wareing, 1965), but this idea is now less widely accepted for a variety of reasons. First, dormancy may develop before the bud coverings are completely formed (Wareing, 1965). In addition. O_2 deficiencies may break dormancy (see Section III.A). Furthermore, replacement of bud scales with lanolin as a barrier to O_2 diffusion does not maintain dormancy, at least in black currant (Tinklin and Schwabe, 1970).

At present, cause cannot be distinguished from effect among these changes in metabolic activity in relation to dormancy, yet it seems important to work out the sequence of changes in the metabolic step-down leading into dormancy and in the step-up going out as well as determining which reactions arrest bud outgrowth. Unfortunately, the experiments with radioactively labeled compounds that would be very helpful in analyzing the metabolic control of dormancy may in many cases be complicated by the epiphytes generally occurring on or in these structures; however, these epiphytes might be controlled by antibiotics (see, for example, Watts and King, 1973).

III. ENVIRONMENTAL CONTROLS

A. Temperature

The possible importance of temperature was first observed in studies on cuttings or entire plants that were transferred into a warm environment after exposure to different periods of cold. One of the earliest reports (Knight, 1801) noted that grapevines maintained in a greenhouse through the winter do not sprout, whereas those brought in after some exposure to cold weather grow. Both Krašan (1873a,b) and Askenasy (1877a,b,c) found that cuttings of dark-leaved willow (*Salix nigricans*) and sweet cherry (*Prunus avium*), respectively, would sprout more readily when brought indoors later during the winter, but it was not until much later that the effect of temperature was studied carefully.

Observations that dormant buds develop at about the same time as the temperature begins decreasing in the fall led to the hypothesis that tempera-

ture controls the initiation of dormancy. The studies of Coville (1920) and
Weber (1921) indicated, however, that induction of dormant buds in many
plants is not controlled primarily by cold temperatures. At about the same
time, the importance of cold treatment in the release of dormant tree buds
was being studied. Howard (1910) presented information that he interpreted
as showing that low temperatures maintain dormancy, but these data can
better be explained as showing that low temperatures break dormancy.
Simon (1906) and Molisch (1908) found that plants of a number of species
placed in a greenhouse during the fall sprout later than those left outside
during part of the winter; they concluded that chilling is required for breaking dormancy. Weber (1916c) found that, if woody plants (*Tilia* spp. and
Fraxinus spp.) are protected from the cold, bud dormancy can be maintained for more than 18 months. Coville (1920) and Weber (1921) found
that chilling can release dormancy in terminal buds of woody plants and
that the effect is very localized. Chilling the roots of woody plants has no
effect (Weber, 1921).

Since gardeners had used immersion of plants in warm-water baths
(30°–40°C) to break dormancy, some investigations were carried out using
this technique. Molisch (1909b) found that the effect of a warm-water bath
(8–16 hr at 30°–40°C) varies depending on the season, species, and bud
type. Furthermore, he (Molisch, 1908, 1909a) discovered that neither high
air temperature nor submergence alone could break dormancy. Although he
was aware of the reduced concentration of oxygen in water at high temperatures, Molisch (1909b) did not consider this to be a contributing factor.
However, Boresch (1924, 1926) found that high air temperature (30°C)
combined with reduced air pressure (50 mm Hg) could break dormancy in
plants not submersed and that high water temperature combined with
increased air pressure over the water to increase the oxygen concentration
could not break dormancy in submersed plants.

The effect of temperature on the induction of dormancy has received
comparatively little study, especially since the studies of Coville (1920) and
Weber (1921), which indicated that temperature was not important in the
development of dormancy. However, more recent studies have shown that
there can be an interaction between temperature and photoperiod on the
induction of dormancy of Norway spruce (*Picea abies*), (Dormling *et al.*,
1968), grasses (Laude, 1953), azalea (Pettersen, 1972), Douglas fir
(Schaedle, 1959), *Populus* spp. (van der Veen, 1951), frogbit (*Hydrocharis
morsus-ranae*) (Vegis, 1953, 1955), and water milfoil (*Myriophyllum verticillatum*) (Weber and Noodén, 1976b). Van der Veen (1951) found that at
5°C normally inductive 9-hr photoperiods would not induce dormancy in
Populus spp. In Douglas fir, high or low temperatures during the matura-

tion of buds reduce later growth (Dormling *et al.*, 1968). It is clear, then, that temperature is important in the induction of dormancy at least in the sense that some temperatures are permissive, while others are not.

Several studies indicated that chilling is required to release bud dormancy in several species, including several orchard trees (Weldon, 1935; Chandler *et al.*, 1937), peach (*Prunus persica*) (Chandler and Tufts, 1934), rose (Yerkes and Gardner, 1934), and frogbit (*Hydrocharis morsus-ranae*) (Simon, 1928; Matsubara, 1931). From Table I, it is apparent that the response to temperature varies widely from species to species. Generally, the temperature requirements reflect the environment to which the species is adapted and serve to coordinate development with climate. There is a pronounced bias in the data toward controls found in the North Temperate Zone, where the season to be avoided is winter, because most of the work on dormancy has been done on plants from this zone. Plants native to the regions where hot, dry summers prevail, e.g., the grass, *Poa scabrella* (Laude, 1953), react to high temperature by becoming dormant and require a cool, but not freezing, temperature to be released. There are also plants that have no deep dormancy but that become quiescent in response to cooling or warming (see Laude, 1953). For these plants to resume growth, it is only necessary that favorable temperatures for growth return.

The breaking of dormancy often requires particular temperature treatments. Freezing and nearly freezing temperatures often release the dormancy of buds (Table I). Thomas and Wilkinson (1964) and Tinklin and Schwabe (1970) found that 5°C for several weeks breaks dormancy of the buds of black currant (*Ribes nigra*); however −15°C for 10 min also breaks dormancy of the buds. Tinklin and Schwabe (1970) proposed that at 5°C there is a gradual metabolic breakdown of the growth inhibitors in the bud scales, but the −15°C treatment, which kills the bud scales, accelerates this breakdown. Temperatures of the order of −15°C are not uncommon in cold temperate and boreal areas. Similarly, complete removal of the scales also breaks dormancy; merely trimming the scales back does not.

Fluctuating day/night temperatures can also affect the dormancy of buds. Blake (1972) found that wide fluctuation in day/night temperatures (30°/15°C) stimulates sprouting in the lignotubers of *Eucalyptus obliqua*. In contrast, fluctuating day/night temperatures slow the flushing of Douglas fir (Campbell and Sugano, 1975). However, many cool temperate species reenter the dormant state in response to warm day temperatures, even though the night temperatures are favorable for breaking dormancy (Chandler *et al.*, 1937; Bennett, 1950; Erez and Lavee, 1971; Nienstaedt, 1966; Weinberger, 1954; Vegis, 1963). In the case of peach buds, Erez and Lavee (1971) found that 21°C day temperatures nullify the effect of low temperature at night,

TABLE I

Effect of Temperature on Dormancy of Various Species[a]

Species	Initiate	Maintain	Release	Interact with photoperiod	References
Acer saccharum (sugar maple)			F	x	Kriebel and Wang (1962), Taylor and Dumbroff (1975)
Allium cepa (onion)	H		C	x	Heath and Holdsworth (1948), Sinnandurai and Amuti (1971)
Begonia evansiana	C		F		Esashi and Nagao (1973)
Citrus sp.	C	C	W		Cooper *et al.* (1969)
Citrus grandis (grapefruit)	C				Young and Peyando (1965)
Cyperus esculentus (yellow nutsedge)			F		Tumbleson and Kommedahl (1962)
Eucalyptus obliqua			W/C-A		Blake (1972)
Fraxinus sp. (ash)			F		Weber (1961c)
Fraxinus excelsior (European ash)			F		Witkowska-Zuk and Koslowska (1974)
Gladiolus sp.			H, C		Loomis and Evans (1928), Tsukamoto and Konoshima (1972)
Hordeum bulbosum	W			x	Koller (1970), Koller and Highkin (1960), Ofir *et al.* (1967)
Hydrocharis morsus-ranae (frogbit)			F, C	x	Kummerow (1958), Matsubara (1931), Simon (1928), Vegis (1953, 1955)
Iris spp.			W-C, W		Halevy *et al.* (1964), Tsukmoto and Ando (1973a)
Lilium longiflorum			F		Wang and Roberts (1970)
Liquidambar styraciflua (sweetgum)			F	x	Farmer (1968)
Myriophyllum verticillatum (green water milfoil)	C	C	F	x	Weber and Noodén (1976a)
Oxalis spp.			F-C		Chawdhry and Sagar (1974)

228

Species					References
Picea abies (Norway spruce)			x	F	Dormling et al. (1968), Worrall and Meyer (1967)
Picea canadensis (white spruce)	W-RI			F	Nienstaedt (1966)
Picea spp. (spruce)			x	F	Nienstaedt (1967)
Pinguicula grandiflora (greater butterwort)	C		x	F	Heslop-Harrison (1962)
Pinus densiflora (Japanese red pine)			x	F	Nagata (1967)
Pinus sylvestris (scots pine)			x	F	Hoffman and Lyr (1967), Jensen and Gatherum (1965), Wareing (1951)
Poa scabrella	W		x	C	Laude (1953)
Populus spp.	F		x	F	van der Veen (1951)
Prunus avium (sweet cherry)				F	Askenasy (1877a,b)
Prunus persica (peach)				F, C	Chandler and Tufts (1934), Erez and Lavee (1971), Flemion (1959), Weinberger (1969)
Pseudotsuga menziesii (Douglas fir)				F	Lavender and Hermann (1970), Sugano (1971), van den Driesche (1975), Womack (1964)
Pyrus spp.	W-RI			F	Bennett (1950)
Pyrus malus (apple)				F	Il'in (1971)
Rheum rhaponticum (rhubarb)				F	Tompkins (1965)
Rhododendron simsii (azalea)	W		x	F	Pettersen (1972)
Rhododendron spp.				F	Schneider, E. (1970)
Ribes nigrum (black currant)			x	F	Tinklin and Schwabe (1970), Thomas and Wilkinson (1964)
Salix nigricans (dark-leaved willow)				F	Krašan (1873a,b)
Solanum tuberosum (potato)	C		x	C	Bialek (1974), Engelbrecht and Bielinska-Czarnecka (1972), Werner (1934)
Sorghum halepense	C	C		W	Hull (1970)
Spirodela polyrhiza (greater duckweed)				F	Perry (1968)
Theobroma cacao (chocolate)				W	Humphries (1944)
Tilia platyphyllos (large-leaved lime)				W	Lyr et al. (1970)
Tilia sp. (line, linden, basswood)				F	Weber (1916c)
Tsuga canadensis (eastern hemlock)				F	Olson and Nienstaedt (1956)

229

[a] Key to abbreviations: F, Temperatures less than 5°C; C, temperatures 5°–15°C; W, temperatures 15°–30°C; H, temperatures greater than 30°C; A, alternating, e.g., W/C-A denotes alternating warm day and cool night temperatures; RI, reinduce dormancy; W-C, warm followed by cool temperatures.

whereas 18°C temperatures do not. Once dormancy is broken and growth is underway, warm day temperatures no longer inhibit outgrowth.

The effect of high temperatures on the outgrowth of buds has been used by Vegis (1964) along with other data as the basis for a hypothesis of changing temperature requirements for growth as buds develop into and out of dormancy. According to this postulate, as dormancy develops, the temperature range for growth narrows until either it does not coincide with the prevailing temperature or the plant enters deep dormancy, in which case it will not grow at all. After a period of rest and/or specific dormancy-releasing requirements are met, the temperature range for growth widens until this restriction is essentially nonexistent in the plant's normal environment. Ofir *et al.* (1967) found that the bulbs of *Hordeum bulbosum*, a perennial barley grass of the Mediterranean region, behave this way with the rate of sprouting at high temperature (25°–30°C) increasing as the duration of rest lengthens.

The most effective temperature for release of dormancy in those plants requiring a cold treatment is near 5°C (Erez and Lavee, 1971; Campbell and Sugano, 1975). Mathematical formulations for quantifying the effect of chilling in satisfying the chilling requirement are presented by Erez and Lavee (1971) and Campbell and Sugano (1975). Erez and Lavee (1971) defined weighted chilling hours (*WCH*) as the summation of the number of hours at a given temperature (t_x) times the relative efficiency of that temperature (RE_x):

$$WCH = \sum_{x=i}^{j} t_x RE_x$$

Where i and j are the lower and upper limits of the effective temperature range for breaking dormancy (3° and 10°C in peach), and RE_x is the percentage bud break after treatment at temperature, t_x, divided by the percentage bud break after treatment at the optimal temperature times 100. On the basis of a study of several ecotypes of Douglas fir, Campbell and Sugano (1975) proposed daily average rate of development (DARD) as a measurement of dormancy release by chilling. DARD is obtained by taking the reciprocal of the number of days to sprouting under various conditions (*W*) and multiplying by 100 to give the percentage per day. When the chilling requirement is satisfied, the DARD should reach a maximum and remain constant. This formulation is very useful for comparing the effects of various temperature and/or photoperiod requirements during bud break. Van den Driesche (1975) also found that increasing the number of hours of chilling produces more rapid flushing. Once the chilling requirement has been met, the sprouting of the buds may be controlled entirely by temperature (Lavender and Hermann, 1970). However, the buds of some plants may begin to grow even at 0°–5°C, e.g., water milfoil (*Myriophyllum verticillatum*) (Weber and Noodén, 1976b). Chilling may be needed for more than breaking dormancy; Flemion (1959) found that peach seedlings will grow but remain dwarf unless they have had a chilling treatment.

More complex temperature treatments are sometimes required for optimal sprouting in bulbs. Dutch iris (Tsukamoto and Ando, 1973a) and wedgewood iris bulbs (Halevy *et al.*, 1964) require a period of warm temperatures (greater than 15°C) before cool temperatures (5°–10°C) for sprouting. Chawdhry and Sagar (1974) found that bulbs of *Oxalis latifolia* and *O. pes-caprae* require a complex treatment of low temperatures (1°–10°C) and then warm temperatures (20°C) for sprouting.

B. Photoperiod

Before the studies of Garner and Allard (1923), little attention was paid to the importance of light in controlling bud dormancy in plants. The turions of *Hydrocharis morsus-ranae* were found to respond differently to light of different colors produced by passing light through variously colored solutions (Terras, 1900), and yellow to orange light being most effective. Jost (1894) found that exclusion of light inhibits bud break in beech (*Fagus sylvatica*). Further studies on this species by Klebs (1914) showed that some plants could be maintained in the growing state during the winter with continuous light in greenhouses. He also found that moving plants from continuous light to normal winter photoperiods induces dormancy. In one respect, these researchers were lucky in their choice of material, for beech does not have a chilling requirement for the breaking of dormancy (Wareing, 1953, 1954). Klebs (1914) did find that for plants grown outdoors the length of time to bud break in continuous light varies with the time of year the plants are brought indoors (i.e., 10 days in September and February, 36–38 days in November). He proposed that the total length of illumination and the intensity are the important factors controlling dormancy.

It was not until the epochal studies of Garner and Allard (1923) that the length of the photoperiod itself was implicated in the control of dormancy. These authors found that prolonged days prevent dormancy in tulip tree and *Aster linariifolius* and that short days promote tuberization in potato. Since then, many investigators have found that short days promote the development of dormancy in trees (Bogdanov, 1931; Kramer, 1936; Jester and Kramer, 1939; Sylvén, 1940a,b,c) and that long days can in some cases release it (Kramer, 1936; Gustafson, 1938). In addition, Moshkov (1935) found that short days are required for the development of cold hardiness in black locust (*Robinia pseudoacacia*). Table II lists the response of a number of species to photoperiod. A similar table has been constructed by Nitsch (1957b) using a modification of Chouard's (1946) classification of response to day length. A glance at Table II reveals that most plants studied to date become dormant in response to short days. However, three other categories can also be distinguished: those species that become dormant in response to long days, those that eventually become dormant irrespective of day length, and those that do not become dormant. Nitsch (1957b) noted that plants which eventually become dormant irrespective of photoperiod become dormant faster in short days than in long days. Wareing and Black (1958) found that axillary buds of birch (*Betula pubescens*) initially repressed due to apical dominance develop dormancy in response to short days. The axillary buds of black currant behave in a similar way (Nasr and Wareing, 1961). It must be noted, however, that most of the long-day effects have not been rigorously shown to be responses to photoperiod rather than to more light. For this purpose, the use of short-days with dark-period interruptions seems preferable to the use of long days or extensions of photoperiod using low light intensity.

The receptor of the photoperiodic response has been shown to occur in either leaves or buds. The studies of Wareing (1953, 1954) showed that the leaves are the receptors of the photoperiodic stimulus for the development of dormant buds in seedlings of beech (*Fagus sylvatica*), sycamore maple (*Acer pseudoplatanus*), and birch (*B. pubescens*). During the period before the development of dormancy, the buds are often maintained in a state of quiescence by an inhibitory influence from the leaves as long as they experience dormancy-inducing photoperiods. When the leaves are removed before dormancy has set in, the buds may grow out. The effect of defoliation has been used as a measure of the state of dormancy (Section I). A minimum number of leaves has been found to be necessary in some species for photoperiodic induction, indicating the importance of leaves in the response to day length; e.g., Douglas fir (Wareing, 1948) and beech and English oak (Dostál, 1927).

Although many of the studies of the importance of photoperiod in development of dormancy have been conducted on north temperate tree species that have a short-day response, photoperiodism as a control of dormancy has been found in many other plants. Plants growing in regions with hot, dry summers often become dormant in response to long days, e.g., onion (*Allium cepa*) bulbs (Heath and Holdsworth, 1948; Mann and Lewis, 1956) and the liverwort, *Lunularia cruciata* (Valio and Schwabe, 1969; Schwabe and Valio, 1970b). Whereas some plants of these regions appear to be held in the nongrowing state by lack of water, a number are truly dormant

TABLE II

Effect of Photoperiod on Bud Dormancy in Various Species

Species	Induces	Maintains	Releases	Night interruption tested	References
Abelia grandiflora	SD			x	Waxman (1957)
Acer rubrum (red maple)	SD				Downs and Borthwick (1956a), Jester and Kramer (1939)
Acer palmatum (Japanese maple)	SD				Nitsch (1957a)
Acer pseudoplatanus (sycamore maple)	SD		—		Wareing (1954)
Acer saccharum (sugar maple)			LD		Olmsteadt (1951)
Aesculus hippocastanum (horse chestnut)	NE, D	NE, D	NE		Downs and Borthwick (1956)
Albizzia julibrissium	SD				Nitsch (1957a)
Allium cepa (onion)	LD				Heath and Holdsworth (1948)
Aster linarifolium	SD				Garner and Allard (1923)
Betula spp.	SD				Johnsson (1951)
	SD				Downs and Borthwick (1956a)
Betula mandshurica	SD	SD	LD		Wareing (1954)
Betula pubescens (brown birch)	SD	SD	LD		Warden (1970)
Bryophyllum crenatum	SD			x	Downs and Borthwick (1956a)
Catalpa bignonioides (southern catalpa)	SD				Downs and Borthwick (1956a)
Catalpa speciosa (northern catalpa)	SD				Nitsch (1957a)
Cercis canadensis (red bud)	SD				
Cornus florida (flowering dogwood)	SD			x	Downs and Borthwick (1956a), Waxman (1957)
Cornus kousa	SD				Waxman (1957)
Euonymus alata	NE, C	NE, C	NE, C		Nitsch (1957a)
Fagus sylvatica (beech)	SD	SD	LD		Klebs (1914), Kramer (1936), Lona and Borghi (1957), Wareing (1953, 1954)
Fagus grandifolia (American beech)	SD				Jester and Kramer (1939)

(Continued)

TABLE II (Continued)

Species	Induces	Maintains	Releases	Night interruption tested	References
Fraxinus spp. (ash)	SD				Jester and Kramer (1939)
Hibiscus syriacus	SD				Nitsch (1957a)
Hordeum bulbosum (bulbous barley)	LD				Koller and Highkin (1960), Ofir and Koller (1974)
Hydrocharis morsus-ranae (frogbit)	SD				Vegis (1953, 1955)
Ilex opaca (American holly)	SD				Nitsch (1957a)
Juniperus horizontalis (creeping juniper)	SD			x	Waxman (1957)
Kleinia articulata	LD	LD	SD		Schwabe (1970)
Kolkwitzia amabilis	SD			x	Mahlstede (1956)
Larix decidua (European larch)	SD	SD	LD		Wareing (1954)
Liquidambar styraciflua (sweetgum)	NE, D	NE, D	NE		Downs and Borthwick (1956a)
Liriodendron tulipifera (tulip tree)	SD			x	Garner and Allard (1923), Zahner (1955), Downs and Borthwick (1956a)
Lunularia cruciata	LD	LD	SD		Schwabe and Valio (1970b), Valio and Schwabe (1969)
Myriophyllum verticillatum (green water milfoil)	SD	SD	LD		Weber and Noodén (1976a,b)
Paulownia tomentosa (royal paulownia)	NE, D	NE, D	NE		Downs and Borthwick (1956a)
Picea abies (Norway spruce)	SD		LD		Nitsch (1957a), Worrall and Meyer (1967)
Picea spp.			LD		Nienstaedt (1967)
Pinguicula grandiflora (greater butterwort)	SD		LD		Heslop-Harrison (1962)
Pinus banksiana (jack pine)	SD				Jester and Kramer (1939)
Pinus caribaea	SD				Jester and Kramer (1939)
Pinus densiflora			LD		Nagata (1967)
Pinus echinata (shortleaf pine)	SD				Jester and Kramer (1939)
Pinus resinosa (red pine)	SD		LD		Jester and Kramer (1939), Gustafson (1938)

234

Species				References
Pinus sylvestris (scots pine)	SD			Hoffman and Lyr (1967)
Pinus taeda (loblolly pine)	SD			Phillips (1941)
Platanus occidentalis (American sycamore)				Nitsch (1957a)
Poa scabrella	LD	LD		Laude (1953)
Populus canadensis (Carolina poplar)	SD			van der Veen (1951)
Populus lasiocarpa	SD			van der Veen (1951)
Populus nigra (black poplar)	SD			van der Veen (1951)
Populus robusta	SD			van der Veen (1951)
Populus tremula (aspen)	SD			Sylvén (1940a,b,c), van der Veen (1951)
Populus trichocarpa (black cottonwood)	SD			van der Veen (1951)
Populus tomentosa	SD	SD		Jester and Kramer (1939), Nitsch (1957a)
Prunus spp.	NE, C	NE, C		Nitsch (1957b)
Pseudotsuga menziesii (Douglas fir)	SD			Lavender and Hermann (1970)
Pyracantha coccinae (fire thorn)	NE, C	NE, C		Nitsch (1957a)
Quercus borealis	SD			Nitsch (1957a)
Quercus robur (pedunculate oak)	SD	—		Wareing (1954)
Quercus rubra (northern red oak)	NE, D	NE		Jester and Kramer (1939)
Rhododendron mucronulatum	SD			Nitsch (1957a)
Rhus typhina (staghorn sumac)	SD			Nitsch (1957a)
Robina pseudoacacia (black locust)	SD	—	x	Bodganov (1931), Jester and Kramer (1939), Klebs (1914), Kramer (1936), Moshkov (1935), Phillips (1941), Schneider (1973), Wareing (1954)
Solanum tuberosum (potato)	SD		x	Garner and Allard (1923), Kumar and Wareing (1973), Milthorpe (1963), Werner (1934)
Spiraea froebelii	SD			Mahlstede (1956)
Spirodela polyrhiza	SD	LD		Perry (1968)
Syringa vulgaris (lilac)	NE, C	NE, C		Nitsch (1957a)
Thuja occidentalis (northern white cedar)	SD		x	Phillips (1941), Waxman (1957)
Tsuga canadensis (eastern hemlock)	SD	LD		Nitsch (1957a), Olson *et al.* (1959)

(Continued)

TABLE II *(Continued)*

Species	Induces	Maintains	Releases	Night interruption tested	References
Ulmus americanus (American elm)	SD				Downs and Borthwick (1956a)
Viburum opulus (European cranberry bush)	SD			x	Waxman (1957)
Weigela florida	SD	SD	LD	x	Downs and Borthwick (1956b)

[a] Key to abbreviations: SD, Short-day response; LD, long-day response; NE, no effect; C, continuous growth; D, dormancy develops irrespective of photoperiod.

(Laude, 1953). Koller and Highkin (1960) found that long days induced dormant bulb formation and flowering in bulbous barley (*Hordeum bulbosum*). Induction of both flowering and bulb development required about 24 long days (16 hr), but bulb formation alone required only 10–18 long days (Ofir and Koller, 1974).

When noninductive conditions are maintained for very long periods, i.e., significantly longer than normally experienced by the plants, abnormal development will often take place, as evidenced by the incidence of abnormal leaves. If maintained long enough, this treatment can be fatal [e.g., silver fir and beech (Balut, 1956)]. It is also well known among plant growers that many species which normally experience a rest or dormant period do poorly unless allowed to rest.

In a few cases, pretreatment with noninductive photoperiods has been reported to enhance dormant bud formation. The butterwort, *Pinguicula grandiflora*, requires a period of vegetative growth before special buds ("brood buds") are produced. Either long days or warm nights can provide the needed stimulus for a vegetative period (Heslop-Harrison, 1962). A long-day pretreatment (16 hr) promotes short-day induction (8 hr) of dormant buds in water milfoil (*Myriophyllum verticillatum*) (Weber and Noodén, 1976a). Finally, increasing the number of short days (8 hr) decreases the number of long days required for induction of bulbs in bulbous barley (Ofir and Koller, 1974). After 60 short days, however, this species forms bulbs under short-day conditions without the long-day treatment.

There has been a lengthy search for the chemical photoreceptor. Initially, some process associated with photosynthesis was thought to be involved. Jost (1894), who found that continuous light breaks dormancy of beech buds in an atmosphere free of carbon dioxide, concluded that carbon dioxide fixation per se is not required. However, these data do not preclude the possibility of other reactions associated with photosynthesis being the important reaction(s) in controlling dormancy. Much later, Phillips (1941) showed that a light break during the dark period counteracts the inductive effect of short days in seedlings of several tree species. Furthermore, he found that red light is more effective than blue. Wareing (1950), Zahner (1955), Mahlstede (1956), and Waxman (1957) reported that night interruptions during short-day treatment prevent dormant bud development in several woody species. According to Nitsch (1957b), the most effective portions of the spectrum for controlling dormancy are blue (about 400 nm) and red (about 650 nm). In a personal communication to Wareing (1969), van der Veen stated that he had found red–far-red reversibility of dormant bud formation in European larch (*Larix europea*). Schwabe and Valio (1970b) demonstrated red–far-red reversibility in the control of dormancy in the desert liverwort (*Lunularia cruciata*). Schneider (1973) reported data that show red–far-red reversibility in the effect of light breaks on dormancy in black locust. Thus, phytochrome would appear to be the photoreceptor involved in the control of dormancy through photoperiod, but further studies are needed to more firmly establish the role of phytochrome in controlling dormancy.

The role of photoperiod in the release of dormancy has been studied less intensively than for induction. However, a number of species are maintained in a quiescent state by photoperiod, e.g., birch (*B. pubescens*) (Wareing, 1954) and *Kleinia articulata* (Schwabe, 1970). Furthermore, some species that have chilling requirements may be induced to sprout under long-day conditions without chilling, e.g., sweetgum (Farmer, 1968). To our knowledge, there have been no studies that attempted to break dormancy by using night interruptions. This type of data would be a stronger indication of a true photoperiodic effect and not simply an increase in photosynthesis, especially for evergreens like Norway spruce (Dormling *et al.*, 1968; Worrall and Meyer, 1967).

C. Interaction of Temperature and Photoperiod

It is not surprising that temperature can modify the effect of photoperiod and vice versa. In a number of species, long days substitute for the chilling requirement (Worrall and Meyer, 1967; Olsen *et al.*, 1959; Nienstaedt, 1966, 1967; Farmer, 1968). Furthermore, the effect of inductive photoperiods can be counteracted by low temperature (Moshkoy, 1935; van der Veen, 1951). In water milfoil (*M. verticillatum*) a temperature below 15°C is required for induction of dormant buds (Weber and Noodén, 1976a). *Poa scabrella*, a grass native to central California, where hot, dry summers prevail, requires both rising temperatures and long days for induction of dormancy (Laude, 1953). Vernalized seedlings of bulbous barley produce more bulbs in response to long days than do nonvernalized seedlings (Ofir and Koller, 1974). Both cool nights and short days are required for dormancy in a butterwort (*Pinguicula grandiflora*) (Heslop-Harrison, 1962).

Hydrocharis morsus-ranae possesses one of the most complex responses to temperature and photoperiod. Raising the temperature from 15°C to 25°C increases the critical day length for dormant bud formation from 15 hr to 21 hr; above 25°C even continuous light does not block dormant bud development (Vegis, 1953, 1955).

Perry (1962) found that red maple seedlings from Vermont become dormant with night temperatures greater than 23°C or less than 10°C, but he also found that low light intensities (less than 1000 fc) found produce dormant buds at temperatures between 10° and 23°C. In an extensive study of Norway spruce, Dormling *et al.* (1968) found that reduced growth occurs with high or low temperature or short photoperiods during the final stages of bud development or with short photoperiods during flushing. Heide (1974), using Norway spruce from several areas, found that the dormant bud formation is a short-day response, with the critical day length varying with temperatures in the range of 18°–24°C.

D. Other Environmental Factors

There are reports of several other factors that induce dormant-type buds or that break dormancy. These factors include water stress, nutrient stress, low light intensity, and leaching. The first three factors cause the formation of dormant-type buds and can be characterized as stress conditions. The fourth factor is involved in the release of dormancy.

It has been known for some time (Wareing, 1948; see Romberger, 1963 and Kozlowski, 1971) that mature trees in the North Temperate Zone cease elongation and develop dormant-type buds long before unfavorable weather occurs or the critical day length for the induction of dormancy is reached. Studies by Kozlowski (1962, 1964) and Priestley (1962) indicate that there may be a depletion of internal reserves of both inorganic and organic materials before and during induction of dormant type buds. More direct evidence is presented by Czopek (1963) and Perry (1968), who found that nutrient deficiency, especially of nitrogen, could induce dormant bud formation in the duckweed, *Spirodela polyrhiza.* Conversely, applications of mineral fertilizers, particularly nitrogen, are well known among gardeners to inhibit dormant bud development.

Since moisture stress is a common occurrence in some environments, the ability to develop a dormant structure that can withstand dry conditions in response to drought is of obvious adaptive significance. Many desert species become dormant in response to water stress (Laude, 1953), and moisture stress is apparently required to break dormancy in coffee bud flowers (see Browning *et al.*, 1970). Some of the plants can begin growth at any time that there is sufficient moisture, e.g., the grass, *Bromus carinatus;* others require a specific rest period, e.g., *Poa scabrella.* Drought also affects the development of dormancy in plants from more mesic regions. Water stress can induce or at least hasten dormant bud formation in *Citrus* sp. (Cooper *et al.*, 1969), Douglas fir (Lavender *et al.*, 1968), potato (Milthorpe, 1963), and loblolly pine (Zahner, 1962).

Low light intensities (less than 1,000 fc) induce dormant buds in northern ecotypes of red maple but not southern ecotypes (Perry, 1962). Reduced light intensity is frequently used by floriculturists to induce dormancy or rest in plants that normally require it. This phenomenon is probably common in plants growing on the floor of deciduous forests, where the light intensity is much lower during the summer than in the spring.

Finally, loss of an inhibitor though leaching has been reported to allow germination of the gemmae of *Lunularia cruciata* (Valio and Schwabe, 1969) and the tubers of yellow nutsedge (Tumbleson and Kommedahl, 1961, 1962). By requiring leaching for germination, these plants assure themselves of an adequate water source at the start of their growth.

E. Genetic Factors

A variety of woody plants—scots pine (Langlet, 1944; Vaartaja, 1956; Wassink and Wiersma, 1955), Douglas fir (Irgens-Moller, 1957; Womack, 1964; Campbell, 1974; Campbell and Sugano, 1975), *Populus* spp. (Pauley and Perry, 1954), Norway spruce (Vaartaja, 1959), red maple (Perry and Wu, 1960; Perry, 1962), sugar maple (Kriebel and Wang, 1962), and tulip tree and bur oak (*Quercus macrocarpa*) (Vaartaja, 1961)—have differing requirements depending on the latitude and altitude of the various ecotypes. The response of an aquatic vascular plant, *Spirodela polyrhiza*, to photoperiod also varies according to the latitude of the collection site (Perry, 1968). The trees from nearer the poles generally become dormant at longer photoperiods than do those from nearer the equator (Wassink and Wiersma, 1955; Vaartaja, 1954; Irgens-Moller, 1957; Pauley and Perry, 1954). A longer cold treatment is also required for the breaking of dormancy in trees from colder climates (Womack, 1964; Perry and Wu, 1960; Kriebel and Wang, 1962). A very interesting response to temperature by red maple was reported by Perry (1962). Northern types became dormant with temperatures above 23°C and below 10°C, and those from southern latitudes did not. Thus, appropriate temperature and photoperiod dormancy responses have been selected for ecotypes as well as for species.

This variation has particular importance for foresters and horticulturists who are involved in developing hybrids and introducing new varieties, for example, fruit tree varieties from colder climates may not acclimate well to warmer climates (Chandler and Tufts, 1934).

F. Summary

From the research on the enviromental control of dormancy to date, it appears that temperature and photoperiod are the primary factors controlling dormancy. Photoperiod appears to be most important during the induction of dormancy, whereas temperature is most important in the breaking of dormancy. However, it is incorrect to think that the temperature and photoperiod act independently. As we have shown, there is abundant evidence to support the hypothesis that during all phases of dormancy each factor can exert a modifying effect on the other. Stress, e.g., mineral nutrient deficiency and lack of moisture, may also induce the initial stages of dormancy, such as the development of the dormant bud morphology.

The adaptive significance of dormancy is indicated by the studies of ecotypes of various species. The fact that there is both latitudinal and altitudinal variation in response to temperature and photoperiod strongly suggests that there is strong pressure for the development of dormancy responses that are adaptive to the local climate.

IV. HORMONAL CONTROLS

A. Introduction

Studies on correlative effects in dormant systems have provided much of the impetus in the search for hormones, and even today there remains considerable opportunity to learn more about the controls of dormancy through careful analysis of correlative interactions. It has long been known that in certain woody species defoliation may cause outgrowth of terminal buds that have stopped or are slowing their elongation growth (Goebel, 1880; Gulisashvili, 1948; Olmstead, 1951; van der Veen, 1951). These observations fostered the idea that the subtending leaves may govern the development of a dormant apical bud and its entry into the dormant state. It remained for Wareing (1954) to show that mature leaves and not the terminal buds of sycamore maple and black locust perceive the short-day photoperiod that ultimately induces the terminal bud to become dormant. Dormancy is induced in these plants when the leaves are exposed to short days even when the apical buds receive continuous illumination. In *Weigela florida*, the younger leaves are most effective, whereas older leaves are much less effective (Downs and Borthwick, 1956a; Waxman, 1957). For some species such as birch, *Betula pubescens*, the apical buds themselves and not the expanded leaves perceive the inductive photoperiod (Wareing, 1954).

Inasmuch as tubers are often dormant structures with dormant buds, it is of interest that tuber development (and possible dormancy) is induced by a readily transmitted stimulus generated by short photoperiods perceived in the aerial portions of domesticated and wild potatoes (Gregory, 1956; Chapman, 1958; Kumar and Wareing, 1973).

Correlative interaction between different parts within the bud also maintains dormancy; removal of the scales from the buds of maple (Pollock,

1953), beech (Wareing, 1953), pondweed (*Potamogeton*), (Frank, 1966), rhododendron (Schneider, 1968), and sycamore maple (DeMaggio and Freeberg, 1969) and black currant (Tinklin and Schwabe, 1970) caused elongation. Furthermore, scale removal from one of two coterminal rhododendron buds affects only the treated bud (Schneider, 1968). Thus, it appears that the scales maintain the arrest of development in these buds, and this influence is also confined within the same bud. The dormant buds of potato tubers may also be under correlative control of the surrounding tissues, for peeling breaks dormancy (Schmid, 1901; Koltermann, 1927; see Hemberg, 1965); however, peeling could initiate changes other than just lifting correlative inhibition.

Regarding the distribution of bud-break promoters, a number of investigators shortly after 1900 found the dormancy-breaking effects of warm baths and several chemicals to be quite localized (see Vegis, 1965b). Subsequently, Denny and Stanton (1928) found that when the dormancy of a single lilac bud is broken with ethylene chlorohydrin, even nearby buds are not affected. The dormancy-breaking effect of cold treatment is also confined to the chilled tissues of blueberry (*Vaccinium corymbosum*) bushes (Coville, 1920). Similar experiments with warm bath treatments also show a localized effect on poplar buds (Witkowska-Žuk, 1970). More recent experiments with forked Douglas fir seedlings also show that the effect of cold treatment on dormancy is confined to the treated branch (Timmis and Worrall, 1974). These results appear to be in conflict with the report of Krasnosselskaya and Richter (1942), who found that long-term (17- to 24-hr), warm-water (30°–35°C) treatments produced a bud-growth stimulus that was transmitted to nearby untreated buds. As a greater amount of stem was treated, the stimulus spread further. They also reported that a forced bud grafted onto a dormant branch could break dormancy in neighboring buds. Local application of aqueous extracts from heat-treated buds produced even more widespread effects (Richter and Krasnosselskaya, 1945). Still, a local stimulus of shorter duration (9–12 hr) did give a highly localized effect similar to that reported by others. In general, it seems that dormancy-breaking treatments can exert localized effects (possibly except cases in which large amounts of tissue are involved), and therefore whatever growth promoters are involved do not spread readily.

Although the studies of correlative controls in bud dormancy are not yet complete, they do set the stage for the study of the hormonal controls of dormant bud development and give some indication of how the controlling hormones should behave.

In the following sections, the hormonal controls of dormancy are viewed in terms of groups of opposing factors, (1) growth inhibitors and (2) growth promoters, because a balance of these seems to be critical and because there is some similarity in the theoretical considerations for each group. The evidence bearing on hormonal control is further organized into (a) correlations with changes in the endogenous hormones and (b) effects of exogenous hormones. Although (a) may be more important, (b) may provide important corroborative evidence where development is induced or offset by exogenous hormones. The transport aspect of hormone function in dormancy is given little attention, because almost no data are available except some observations of changes in hormone levels in the vascular system correlated with dormancy. Likewise, the mode of action of hormones that seem to control dormant development is given limited coverage, because there is so little information on their effects in buds.

Unfortunately, it is often not possible to be sure of the dormancy status of the buds in which metabolic activity or hormone levels are analyzed. In addition, the tissue sample sometimes includes not only the bud or the parts with growth potential but surrounding tissue.

B. Growth Inhibitors

Because dormancy is a repressed or inhibited state, much effort has been expended on the study of growth inhibitors in relation to dormancy. However, the dormant organs often differ structurally from their normal, growing counterparts, and some have very specialized structural adaptations. Therefore, development of dormant structures is more than just a suppression of metabolism and growth; it is a whole new pathway of development, and this must be accounted for as well as the suppression of normal activities.

1. GROWTH INHIBITORS IN RELATION TO DEVELOPMENT OF DORMANT STRUCTURES

a. **Endogenous Inhibitors.** The correlative effects of leaves on dormant bud development (Section IV,A) and the early reports by Hemberg (1949a,b) on a possible role of inhibitors in dormancy led to the classic experiments of Wareing and his co-workers on inhibitors produced by leaves. Phillips and Wareing (1958) found that the inhibitor content of sycamore maple buds and leaves changed with the season. The inhibitor β content (80% methanol-soluble material chromatographing at about R_f 0.7 in isopropanol ammonia/water measured as inhibition of coleoptile section elongation) increased during the period when the dormant buds were developing. The levels of activity in leaves increased until mid-August and then decreased, whereas the activity in the apices also increased but remained at the same level through October. Since short days had previously been shown to induce dormant bud development in this species,

Phillips and Wareing (1959) also examined the effect of short days on this inhibitor. Only 2 short days (before any visible effect on growth) were sufficient to increase the level of inhibitor (apparently inhibitor β) in the mature leaves, but in apices 5 short days were needed. Subsequently, Eagles and Wareing (1963, 1964) found that short days produced a similar increase in the inhibitor content of birch (*Betula pubescens*) leaves and buds. More importantly, the partially purified inhibitor (applied continuously by immersing a young leaf) could inhibit extension growth and induce dormant morphology even under a noninductive photoperiod. After abscisic acid (ABA) was identified and continuous foliar applications of ABA were shown to induce the formation of dormant buds in birch, sycamore, and black currant under noninductive long days (El-Antably *et al.*, 1967), the picture seemed almost unassailable. However, Lenton *et al.* (1972), who measured the actual amounts of ABA by gas chromatography rather than bioassays, found no increase in the ABA contents of leaves or buds of red maple, sycamore, and birch when these plants were transferred from long days to short days (dormancy-inducing conditions). This raised questions about the role of ABA in the development of dormancy and about the possibility that a decrease in growth promoters in the inhibitor β fraction could cause the increase in its inhibitory activity. This conjecture was reinforced by the observations that gibberellins can offset the inhibitory effect of inhibitor β even though they do not directly promote the growth of coleoptiles in their final elongation phase (Thomas *et al.*, 1965) and some gibberellins may cochromatograph with inhibitor β (see Lenton *et al.*, 1972). Indeed, gibberellin activity may decrease as dormancy develops (Section IV,C).

Increased inhibitor β or ABA activity during development of dormancy has also been reported in sycamore maple (Dörffling, 1963a,b), peach flower buds (Corgan and Peyton, 1970), and the leaves of red maple (Perry and Hellmers, 1973). Dormancy-inducing short days increase inhibitor activity in buds of staghorn sumac (Nitsch, 1957a), flowering dogwood (Waxman, 1957), and two species of birch (*B. pubescens* and *B. lutea*) (Kawase, 1966). Both the methanol-extractable inhibitor and that obtained from excised birch apices by diffusion increase with the number of short days. The inhibitor content in the growing points did not increase until after elongation had stopped, but there was an early increase in the stems. Dörffling (1963b) described a gradient of inhibitor β activity that changes with dormancy, e.g., decreasing from tip to base in dormant shoots but the reverse in growing shoots. In the floral primordia (but not the scales) of sour cherry buds, the level of ABA rose sharply during or before the period of deepest dormancy (Mielke and Dennis, 1975b). The changes in inhibitor β activity are less clear in aspen (*Populus tremula*), in which short days do not seem to increase the inhibitor activity in shoot apices within 13 days (Eliasson, 1969). Similarly, short-day photoperiods, which inhibit shoot elongation in canoe birch and *Malus hupehensis*, do not increase the ABA content of these shoot tips (Powell, 1976).

The ABA-like inhibitor activity (bioassay determinations) in xylem and phloem of willows (*Salix fragilis*, *S. viminalis*) increases as the plants enter dormancy naturally (Davison, 1965; Bowen and Hoad, 1968). A marked increase in the levels of ABA in peach xylem sap was shown to coincide with leaf fall (Davison and Young, 1974), which is probably after the dormant buds have developed. Bowen and Hoad (1968) also observed that willow (*S. viminalis*) xylem sap contained a large amount of an inhibitor that chromatographed near the solvent front, well ahead of ABA, and this inhibitor decreased as the ABA-like activity increased. Short days, which induce dormancy in willow (*S. viminalis*), also increase the level of activity (apparently ABA) in phloem sap (Hoad, 1967). The metabolism of ^{14}C-ABA (administered to the base of cuttings) by birch (*Betula lutea*) shoots does not seem to be altered significantly by short days, and therefore increased degradation does account for the decrease in endogenous ABA levels (Loveys *et al.*, 1974). The formation of potato tubers also appears to be accompanied by an increase in endogenous inhibitors (Booth, 1963; Bielinska-Czarnecka and Bialek, 1972).

The apparent contradiction raised by Lenton *et al.* (1972) still has not been resolved; however, some data do suggest that ABA or ABA-like activity may increase during the development of dormancy and dormant structures. Changes in other endogenous inhibitors, e.g., xanthoxin, and also counteracting (or interfering) promoters should be investigated.

b. Effects of Applied Inhibitors.   Inhibitor β can inhibit shoot tip elongation and induce dormant bud formation in birch (Eagles and Wareing, 1963, 1964), and ABA (0.1 mM) has the same effect in birch, sycamore, and black currant (El-Antably *et al.*, 1967). Abscisic acid was ineffective when sprayed onto the foliage but was effective when supplied continuously by immersing the youngest fully expanded leaf in addition to spraying onto the shoot apex (El-Antably *et al.*, 1967). Probably, the sprayed ABA was ineffective due to lack of penetration, to break down, or both. Although ABA inhibits shoot elongation in tree of heaven (*Ailanthus glandulosa*), a species with the sympodial growth pattern, it does not cause the abscission of shoot apices that short days would. Hocking and Hillman (1975), however, reported that ABA (up to 0.2 mM) applied by foliar immersion (but without also spraying directly onto the apex) did not induce dormant bud development in birch (same species as studied by El-Antably *et al.*, 1967) or even inhibit elongation significantly. Similar results were obtained for alder

(*Alnus glutinosa*). At least under the conditions employed by Hocking and Hillman (1975), only a small proportion of the leaf-applied ABA reaches the apex. Perhaps the failure to induce dormant bud development in birch is due to omission of the supplementary direct applications to the apices, but it cannot be explained away as a difference in day lengths used; El-Antably *et al.* (1967) found the ABA effect under long (18-hr) days. In a northern ecotype of red maple, ABA (0.1 mM) inhibited elongation but induced a bud different from the normal dormant bud (Perry and Hellmers, 1973). Elongation was affected much less and no gross changes occurred in the buds of a southern race of red maple. Under conditions that do not induce dormancy, growth resumed as soon as the ABA treatment was stopped. Cathey (1968) also reported that ABA only partially duplicated the effects of short days in the growth of certain woody ornamentals and then only at relatively high concentrations (0.4 mM). Similarly, Junttila (1976) found that neither direct nor indirect (via a leaf) ABA treatments could induce cessation of stem growth in two species of *Salix* as short days do. At concentrations as low as 0.06 μM, ABA could induce turion formation in the duckweed (*Spirodella polyrhiza*) (Perry and Byrne, 1969; Stewart, 1969), but even 10μM ABA only enhanced turion formation in the water milfoil (*Myriophyllum verticillatum*) and then only in an inductive environment (Weber and Noodén, 1976a).

Foliar applications (spraying) of ABA reduced shoot growth and stimulated tuber formation in both wild and domesticated potatoes (El-Antably *et al.*, 1967). In dahlias, ABA also enhances tuber formation (Biran *et al.*, 1972). Since ABA does not induce tuberization of the isolated potato stolons (Palmer and Smith, 1969), it is possible that this effect of ABA is indirect. Perhaps, the ABA effect is a result of decreased shoot growth (or altered source–sink relations). In wild potato there is evidence for a specific tuberization stimulus, which evidently is not ABA (Kumar and Wareing, 1973).

Clearly, the role of inhibitors in the development of dormant buds is not fully known, but very likely ABA is involved at least in some species.

2. INHIBITORS IN RELATION TO TERMINATION OF DORMANCY IN BUDS

a. Endogenous Inhibitors.   The implication of endogenous inhibitors in the dormancy of buds has its origin in the pioneering work of Hemberg (1949a,b), who studied the levels of growth-inhibiting substances with respect to release from dormancy in potato tubers and terminal buds of ash (*Fraxinus excelsior*). Measuring inhibitors by their ability to inhibit bending in the oat coleoptile curvature test when applied in agar disks already containing auxin, Hemberg observed that ether extracts from potato peels from dormant potatoes treated with ethylene chlorohydrin to break dormancy contained less inhibitor. The decrease starts 1–2 days after the treatment and is maximal at 4 days (Hemberg, 1949a). Aqueous solutions of these inhibitors do not, however, reinstate dormancy when applied 5 months later to one-eyed pieces of potatoes whose dormancy had been broken by storage. Similarly, ethylene chlorohydrin causes a decreased inhibitor activity in both the peripheral and central parts of ash buds (Hemberg, 1949b). The decrease occurs between 4 and 7 days after the treatment. A similar drop of inhibitors also occurs as dormancy is broken naturally outdoors. From these observations, Hemberg postulated that the rest period of buds is regulated by growth-inhibiting substances. This theory later received further support from studies on the effect of dormancy-inducing photoperiods on the levels of extractable inhibitors.

Since these initial reports, many investigators have studied the levels of endogenous inhibitors before, during, and after bud emergence from dormancy. Most of these also depend on bioassays [mostly inhibition of elongation growth (cell enlargement) induced by auxin in coleoptile segments], which have been criticized because they may not represent dormancy well and interfering substances may alter the growth response. Nonetheless, this test has been extremely useful in studying endogenous inhibitors and their role in dormancy. Furthermore, dormancy does, to some extent, involve repression of cell enlargement. A number of refinements have been added to purify and characterize the inhibitors, e.g., partitioning of ether extracts and paper chromatography of the extracts. Endogenous inhibitor activity has been reported to decrease as dormancy is broken in potatoes (Hemberg, 1952, 1954, 1958a,b; Steward and Caplin, 1952; Blommaert, 1954; Varga and Ferenczy, 1956, 1957; Bielinska-Czarnecka and Bialek, 1972), peach buds (Blommaert, 1955; Hendershott and Bailey, 1955), sycamore maple (von Guttenberg and Leike, 1958; Phillips and Wareing, 1958), longleaf pine (Allen, 1960), birch (*Betula pubescens*) (Eagles and Wareing, 1964), red pine (Giertych and Forward, 1966), several woody plant buds (Kawase, 1966), onion bulbs (Thomas, 1969), apricot (Ramsay and Martin, 1970), black currant buds (Kuzina, 1970; Tinklin and Schwabe, 1970), gladiolus corms (Tsukamoto and Konoshima, 1972), willow (*Salix viminalis*) buds (Michniewicz and Galoch, 1972), Dutch iris bulbs (Tsukamoto and Ando, 1973b), grape buds (Düring and Alleweldt, 1973), purple nutsedge tubers (Teo *et al.*, 1974), poplar (*Populus balsamifera*) buds (Bachelard and Wightman, 1974), tulip bulbs (Syrtanòva, 1974), and water milfoil (*Myriophyllum verticillatum*) turions (Weber and Noodén, 1976b). Yet others report no change of inhibitor levels in lilac buds (von Guttenberg and Leike, 1958), potato peels (Burton, 1956; Buch and Smith, 1959), English oak (*Quercus pedunculata*) buds (Allary,

1960, 1961), apple buds (Pieniazek, 1964b; Pieniazek and Rudnicki, 1971), and sugar maple buds (Taylor and Dumbroff, 1975). Although these studies based on bioassays of crude or partially purified extracts have been useful in the initial analysis of the relation of inhibitors to dormancy, there are some problems. First, such extracts or even fractions may contain a large number of inhibitory substances, some not active in dormancy. Second, bioassays often do not have a linear dose–response relationship or may saturate, making accurate quantitative measurements difficult. Third, and perhaps most important, the extracts may contain promoters that offset the activity of the inhibitors and produce an underestimate of the inhibitor. The last problem was first encountered in connection with interference by auxins and led to the development of the now standard isopropyl alcohol/ammonia/water and other systems for paper chromatography to separate auxins from the predominant acidic inhibitors (see Hemberg, 1961, for discussion). This separation is not infallible, and failure to separate inhibitors from promotors may account for some of the above-mentioned reports of no decrease in inhibitor activity with breakage of dormancy. It must also be noted that increased promoter activity cochromatographing with the inhibitors could also account for some, or all, of the decreases in inhibitor activity.

The major acidic, ether-soluble inhibitor(s) in plant tissues chromatograph in the R_f region 0.4–0.9 and were termed inhibitor β (Bennet-Clark and Kefford, 1953). However, it has become evident that this fraction could contain gibberellin-like growth promoters that could also offset the inhibitor activity in some bioassays (Thomas et al., 1965). One partial solution might be to use more than one bioassay in the hope that the activity of the promoters is different, but the best is to identify the endogenous inhibitors and measure their levels by methods that are less subject to interference, e.g., highly sensitive physicochemical techniques or even bioassays of highly purified material.

The isolation and characterization of ABA have greatly facilitated the studies on endogenous inhibitors and have provided a focus (see Wareing and Saunders, 1971). At the same time, there has been a great deal of effort to identify the active constituents of inhibitor β.

Although the inhibitor β fraction contains many inhibitory compounds including phenolics, ABA appears to be an important component and may account for most of the inhibitory activity in some extracts (Robinson et al., 1963; Robinson and Wareing, 1964; Milborrow, 1967; Eliasson, 1969; Browning et al., 1970; Tsukamoto and Ando, 1973b; Mielke and Dennis, 1975a). Inhibitor β from potato peels contains several active substances (Housley and Taylor, 1958; Holst, 1971), one of which may be ABA (Holst, 1971). In any event, inhibitor β has greater activity in oat coleoptile straight growth assay than does ABA alone (Holst, 1971), which suggests that inhibitors in addition to ABA should also be considered.

The ABA content (determined by relatively specific methods) has been shown to decrease as dormancy is broken in a variety of dormant tissues, including peach flower buds (Corgan and Peyton, 1970; Corgan and Martin, 1971), coffee flower buds (Browning *et al.*, 1970), gladiolus cormels (Ginzberg, 1973), black currant and beech buds (Wright, 1975), birch (*Betula verrucosa*) (Harrison and Saunders, 1975), and sour cherry buds (Mielke and Dennis, 1975b).

Levels of ABA in the xylem sap of willow (*Salix fragilis*) (Davison, 1965), peach (Davison and Young, 1974), and birch (*B. verrucosa*) (Harrison and Saunders, 1975) also decrease as bud dormancy is broken naturally, and this probably results in less ABA being supplied to the buds. Harrison and Saunders (1975) found that an ABA derivative increases as free ABA decreases, raising the possibility that this conversion inactivates the ABA. In sour cherry flower buds, however, the level of bound ABA tends to parallel that of free ABA (Mielke and Dennis, 1975b). Still, there are some important discrepancies in the changes in free ABA relative to dormancy. For example, ABA levels drop only near the end of dormancy in peach flower buds (Corgan and Peyton, 1970). Furthermore, ABA increases when the buds start to grow out. Lesham *et al.* (1974) also found that the cis-trans form of ABA (thought to be the biologically active form) does not change as almond buds emerge from dormancy. The trans-trans form (which is also the predominant form in these buds) does decrease and, to some extent, this is reflected in an increase in bound (glycosylated) *trans-trans*-ABA. Harrison and Saunders (1975), however, did not find significant amounts of *trans-trans*-ABA in birch buds. Similarly, the ABA content (partially purified and assayed by coleoptile section growth inhibition) does not change in parallel with the dormancy of yam bulbils, but other inhibitors do (Hasegawa and Hashimoto, 1974, 1975). There is therefore ample reason to believe that inhibitors other than ABA may be involved in dormancy of some tissues or at least that more than a change in levels of ABA is required (e.g., a change in promoters).

Among the other inhibitors suspected to be involved in dormancy, the most extensively studied are the phenolic compounds naringenin and prunin (a glucoside of naringenin), both flavonones and neutral. Some investigators (Hendershott and Walker, 1959a,b; El-Mansy and Walker, 1969), using bioassays (oat coleoptile growth inhibition) and chemical methods for measuring naringenin, found a decrease correlated with termination of dormancy in peach flower buds. Others (Dennis and Edgerton, 1961; Corgan, 1965) did not find this relationship and offered several explanations for the discrepancy, but it is not entirely clear why their results differ. Kononenko *et al.* (1975) also found the flavonoid aglycone content of birch (*Betula verrucosa*) buds to be maximal during the dormant period. Further studies on peach flower buds using a relatively sensitive colorimetric method (Corgan and Peyton, 1970) indicate no detectable naringenin in the floral cup (an inhibited tissue), which corroborates the earlier findings of Dennis and Edgerton (1961). The acidic inhibitor activity, including ABA, from peach flower buds does decrease near the end of dormancy, but the total acidic inhibitor activity decreases more than the ABA fraction (Corgan and Peyton, 1970; Corgan and Martin, 1971).

In yam bulbils, where ABA levels do not correlate well with dormancy, three neutral inhibitors termed "batatasins" do (Hashimoto *et al.*, 1972; Hasegawa and Hashimoto, 1973, 1974). They decrease during dormancy-breaking treatments, and they, but not ABA, induce dormancy in bulbils. Two of the compounds appear to be phenols (Hashimoto *et al.*, 1972, 1974).

Dormancy of the liverwort, *Lunularia cruciata*, is known to be controlled by a phenolic acid, lunularic acid; the endogenous levels change in relation to dormancy, and reapplication induces dormancy (Valio *et al.*, 1969; Schwabe and Valio, 1970a; Valio and Schwabe, 1970). Abscisic acid does not appear to be involved here.

Because most of the ABA or other inhibitory activity is found in the scales of peach flower (Dennis and Edgerton, 1961), black currant (Tinklin and Schwabe, 1970), apricot (Ramsey and Martin, 1970), and tung buds (Spiers, 1973), it has been suggested that the inhibitors measured do not cause the dormancy of the central tissues and are not relevent. It must be recalled, however, that the scales may impose or maintain the dormancy of the central tissues (Section IV,A) and therefore could function as a supplier of inhibitor to maintain dormancy.

In reviewing the literature, it seems that many inhibitor peaks on chromatograms change in relation to dormancy and are still unidentified; therefore, it should not be surprising to see new inhibitors implicated in dormancy in the future. Similarly, it should be noted that the oat coleoptile curvature test originally used by Hemberg (1949a,b) in his pioneering studies on changes of inhibitor activity in relation to dormancy is not responsive to ABA (Hashimoto and Tamura, 1969), and therefore other inhibitors must have been involved.

b. Effects of Applied Inhibitors. Many attempts have been made to maintain or reinstate dormancy through application of naturally occurring inhibitors. Although early attempts (Hemberg, 1949a; Buch and Smith, 1959) to induce dormancy in potato buds with inhibitor β were unsuccessful, later attempts did succeed (Blumenthal-Goldschmidt and Rappaport, 1965).

In addition, ABA strongly suppresses the growth of excised potato buds over a 15-day period (El-Antably *et al.*, 1967). Abscisic acid suppresses sprouting in potatoes and many of the physiological changes associated with bud outgrowth (Madison and Rappaport, 1968; Vanes and Hartmans, 1969; Shih and Rappaport, 1970, 1971; Szalai *et al.*, 1975). It also inhibits outgrowth of gladiolus cormels (Ginzburg, 1973) and purple nutsedge tubers (Teo and Nishimoto, 1973; Teo *et al.*, 1974). Bud outgrowth is also retarded in a wide variety of woody species: birch (*Betula pubescens*), sycamore maple, black currant (El-Antably *et al.*, 1967), coffee (van der Veen, 1968), white ash (Little and Eidt, 1968), Norway maple (DeMaggio and Freeberg, 1969), and citrus (Cooper *et al.*, 1969; Young and Cooper, 1969; Altman and Goren, 1974). El-Antably *et al.* (1967) found that daily foliar applications of ABA (95 μM) for 15 days did not delay bud break in a variety of woody plants (probably due to difficulty in penetration plus breakdown), but application through the base of stem cuttings was very effective. Abscisic acid also inhibits sprouting when applied to excised tree buds in aseptic culture (DeMaggio and Freeberg, 1969; Altman and Goren, 1974). For example, ABA at 40 μM gives about 90% inhibition of cultured maple bud (without bud scales) growth, and the ABA appears to be able to duplicate the repressive effect of bud scales (DeMaggio and Freeberg, 1969).

In certain species, other naturally occurring inhibitors seem to function in dormancy and, when applied externally, can inhibit outgrowth. In yam bulbils, the batatasins suppress sprouting, whereas ABA cannot (Hashimoto *et al.*, 1972, 1974). Likewise, lunularic acid can block growth in *Lunularia* (Valio *et al.*, 1969; Schwabe and Valio, 1970a; Valio and Schwabe, 1970. Naringenin offsets the growth-stimulating effect of GA_3 on dormant peach buds (Phillips, 1962); however, its role in peach bud dormancy is somewhat doubtful.

In summary, there is now considerable and varied evidence to support the inhibitor theory of dormancy first articulated by Hemberg (1949a,b). Abscisic acid appears to be one such inhibitor, at least in some species, but other inhibitors may function in other species. Breakage of dormancy (and probably also development of dormancy) must, however, be considered not just in terms of inhibitors but also of promoters, which are discussed in Section IV,C. In fact, the occasionally observed failure of inhibitors to reinstate dormancy or prevent outgrowth may be due to increased levels of growth promoters, which counteract the inhibitors.

C. Growth Promoters

Even though the dormancy-breaking stimulus generally seems to be localized (Section II,A), it seems plausible that bud dormancy could be due, at least in part, to a decrease in the level of growth promoters or that out-growth results from an increase in these promoters, which do not move readily. By analogy, one would normally not expect a car to move simply by release of the brake; the accelerator must be pressed, too. Initially, support for the role of promoters came from the dormancy-breaking effects of hormone applications, but more recently endogenous growth-promoting hormones have been shown to change in parallel with dormancy.

1. PROMOTERS IN RELATION TO DEVELOPMENT OF DORMANT STRUCTURES AND DORMANCY

a. Auxin. Since auxin levels and auxin production appear to be low dur-ing the dormant phase, it can be inferred that decreased auxin levels or production might accompany development of dormancy. Dormancy-induc-ing short days cause a reduction of auxin activity in sumac apices (Nitsch, 1957a) and black currant leaves and buds (Kuzina, 1970). The auxin content and production drop during the senescence of leaves (see Noodén and Leopold, 1978), a time when the dormant buds are completing their development. During this period, the auxin activity in sycamore maple buds (Dörffling, 1963a) and walnut (*Juglans regia*) buds (Langrova and Sladky, 1971) also decreases.

Inasmuch as auxins, particularly synthetic auxins, are known to disrupt many phases of plant development, they could be expected to interfere with development of dormancy. The role of auxin in the development of dormant structures is not clear, but quite conceivably it functions in some facet because it seems to be involved in so many aspects of development.

b. Gibberellin. Numerous studies on the termination of dormancy have shown that gibberellin activity increases before or during outgrowth which implies that a decrease occurs during the transition to the dormant state. Dormancy-inducing short days cause a decrease in gibberellin activity in birch (*Betula pubescens*) stem tissue (Digby and Wareing, 1966) and in willow (*Salix viminilis*) phloem sap (Hoad and Bowen, 1968). During the growth phase of potato tubers, gibberellin activity is high, but it decreases as the tubers enter dormancy (Bialek and Bielinska-Czarnecka, 1975; Moorby and Milthorpe, 1975). The proportion of "neutral" and "polar" gibberellins also changes during this period. The gibberellin activity (lettuce seed germination assay) in terminal buds of walnut (*J. regia*) appears to decrease well before the development of dormant buds (Langrova and Sladky, 1971). On the other hand, the gibberellin activity in sitka spruce needles (Lorenzi *et al.*, 1975) increased during dormant bud development, but as the buds entered dormancy the amount of less polar gibberellin-like activity dropped, whereas the more polar activity remained unchanged at a high level.

After the initial reports that GA_3 blocked photoperiod-induced dormant bud development under controlled conditions (Lockhart and Bonner, 1957; Nitsch, 1957b), numerous publications showed that gibberellin could interfere with natural and induced development of dormant buds (for additional references, see Romberger, 1963; Vegis, 1964, 1965b). In citrus, GA_3 prevents the growth retardation and abortion of the terminal bud characteristic of the "sympodial" growth pattern (Cooper et al., 1969). Gibberellic acid (in lanolin) applied to the stem of wild potato stimulates stolon development (Kumar and Wareing, 1972), but GA_3 (applied differently) retards tuberization of domesticated potato (Tizio, 1964; Lovell and Booth, 1967); likewise, inhibitors of gibberellin synthesis, e.g., (2-chloroethyl) trimethylammonium chloride (CCC) promote tuberization (see Moorby and Milthorpe, 1975). Gibberellic acid (sprayed onto the aerial parts) also inhibits tuberization in dahlias (Biran et al., 1972). It does seem possible that a decrease in gibberellin activity is an important factor in development of dormancy, but the problem requires further investigation.

The study of the role of gibberellins is greatly complicated by the large number of gibberellins and uncertainty over which are functioning as hormones. In addition, studies on gibberellins in relation to dormancy have depended on bioassays which, as in the case of inhibitors (Section IV,B,1,a), are subject to interference.

c. Cytokinins. Since cytokinin activity increases as buds emerge from dormancy (Section IV,C,2,c), it can be inferred that cytokinin levels drop as dormant buds develop, but the data are scanty.

Before the period when dormancy probably develops in aspen (*Populus tremula*) and sycamore maple buds, the cytokinin activity in the leaves decreases, although in sycamore leaves there may be a late rise (September 23) in cytokinin activity (Engelbrecht, 1971). The cytokinin that increases is not zeatin or its riboside but could be the nucleotide. Cytokinin activity changes during development of water milfoil turions but in a complex manner, increasing and then decreasing (Weber and Noodén, 1976a).

Cytokinin treatments interfere with dormant bud development in water milfoil turions (Weber and Noodén, 1976a), but cytokinin does promote tuberization in potato stolons (Palmer and Smith, 1969, 1970), and this effect is inhibited by direct applications of ABA (Palmer and Smith, 1969).

d. Other Promoters. Ethylene could have some role in the development of dormancy, at least in potatoes (Catchpole and Hillman, 1969) and dahlias (Biran *et al.*, 1972), where ethylene (or Ethephon) promotes tuberization when applied to whole plants. In contrast to its effect on whole potato plants, ethylene inhibits both the 8% CO_2- and kinetin-induced tuberization of isolated stolons. Curiously, the ethylene antagonist, CO_2, stimulates tuberization both in whole potato plants (Paterson, 1970) and in isolated stolons (Mingo-Castel *et al.*, 1974). Leaves (and possibly adjacent stem and bud tissues) may produce more ethylene during the period when dormant buds are completing their development (see Noodén and Leopold, 1978). Similarly, tuber-inducing short days cause dahlia plants to produce more ethylene (Biran *et al.*, 1972). The interesting possibility that ethylene plays a role in the growth of dormant structures and developmental arrest seems worth pursuing.

2. PROMOTERS IN RELATION TO TERMINATION OF DORMANCY IN BUDS

a. Auxin.   Both the discovery of the role of auxin in apical dominance and its predominance in all explanations of hormonal control until the discovery of other plant hormones led to an early emphasis on its role in bud dormancy (for more extensive discussions, see Samish, 1954; Hemberg, 1961, 1965; Romberger, 1963). Initially, it was considered that terminal bud dormancy was imposed by superoptimal auxin concentrations; however, resting pear buds yielded no detectable diffusible auxin (Bennett and Skoog, 1938). Similarly, other dormant tree buds, corms, and potato tubers all have a relatively low diffusible and extractable auxin content, which refuted the superoptimal auxin theory (see Samish, 1954; Hemberg, 1965).

Instead, an increase in diffusible auxin is correlated with breaking of dormancy in pear buds (Bennett and Skoog, 1938). Subsequently, numerous reports have shown that emergence from dormancy is accompanied (or preceded) by increased diffusible and extractable auxin in a variety of structures including tree buds (see Samish, 1954; Hemberg, 1965; and several recent references—Dörffling, 1963a; Thomas, 1969; Kuzina, 1970; Tsukamoto and Konoshima, 1972; Syrtanòva, 1974). Unfortunately, these determinations generally depend on bioassays or colorimetric determinations of crude extracts and are subject to many of the same problems as the measurement of inhibitors. Thus, it might be valuable to measure indoleacetic acid (IAA) in relation to dormancy by some of the more direct methods developed recently.

Auxins are, however, generally not able to break dormancy, and frequently they actually retard outgrowth. The data on this point are very numerous but tend to be scattered as minor points in larger, related reports on potato buds (Guthrie, 1939), peach buds (Marth *et al.*, 1947), grape buds

(Nigond, 1957; Weaver *et al.*, 1961), *Hydrocharis turions* (Kummerow, 1958), buds of several trees (Larson, 1960), sycamore buds (Wareing, 1965), apple buds (Pieniazek and Jankiewicz, 1967), onion bulbs (Thomas, 1969), rhododendron flower buds (Schneider, 1970), and nutsedge tubers (Teo *et al.*, 1973). Still, some exceptions exist (see Samish, 1954). Very high doses of IAA (2.9 mM) break dormancy of *Potamogeton* (pondweed) winter buds (Frank, 1966), and auxin enhances the effect of cytokinin on apple buds (Pieniazek and Jankiewicz, 1967). Thus, auxin probably plays an important role in the outgrowth of buds, but dormancy is not due to an auxin deficiency.

b. Gibberellin. Considerable interest in the role of gibberellin has developed as a result of early findings that gibberellins can break bud and seed dormancy in some species.

Gibberellin-like activity has been shown (by bioassay) to increase before or as buds emerge from dormancy in potatoes (Smith and Rappaport, 1960; Bialek, 1974; Bialek and Bielinska-Czarnecka, 1975), sycamore (Eagles and Wareing, 1964), birch (*Betula pubescens*), and black currant (Tumanov *et al.*, 1970), cranberry, terminal buds and leaves (Eady and Eaton, 1972), and Douglas fir (Lavender *et al.*, 1973). In addition, the activity of gibberellin-like substances in the xylem of Douglas fir increases with bud activity (Lavender *et al.*, 1973).

On the other hand, gibberellin activity has been reported to decrease before sprouting in onion bulbs with an increase accompanying sprouting (Thomas, 1969) and after dormancy is broken in peach flower buds (Ramsey and Martin, 1970). In popular buds (*Populus balsamifera*), the gibberellin activity does not increase until after dormancy is broken and then only on a per bud basis, not per gram fresh weight (Bachelard and Wightman, 1974).

Not surprisingly, the different gibberellins may show different patterns of change as dormancy is broken. Preceding sprouting potato tubers, both the acidic and "neutral" fractions of gibberellin activity increase, with the acidic fraction starting first (Bialek and Bielinska-Czarnecka, 1975). In sitka spruce needles, both the more polar and less polar gibberellin-like activities increase after dormancy is broken, the less polar increasing later than the more polar (Lorenzi *et al.*, 1975). In the buds (including young needles), however, the more polar activity shows no change or a slight decrease, whereas the less polar activity increases. The gibberellin activity in the less polar fraction may be due to GA_9.

Since the early reports that gibberellic acid could break dormancy in potatoes (Brian *et al.*, 1955; Rappaport, 1956) and certain trees (Marth *et al.*, 1956), there have been many similar observations (for summaries of the older literature see Romberger, 1963; Vegis, 1965b). Gibberellins seem to be able to break dormancy induced by short days as well as that broken by a cold treatment. It is of interest that GA_1 and GA_3 are able to break photoperiod-induced dormancy in *Weigela*, but GA_2 and GA_4 could not (Bukovac and Wittwer, 1961), which supports the idea that not all gibberellins function equally in dormancy.

Even though gibberellins do break dormancy in a very wide range of higher plants, they do not in all species. For example, gibberellic acid does not break dormancy in certain conifer buds (Lockhart and Bonner, 1957), onion bulbs (Rappaport, 1956; Thomas, 1969), apple buds (Hull and Lewis, 1959), large-leaved lime buds (*Tilia platyphyllos*) (Lyr *et al.*, 1970), excised rhododendron flower buds (Schneider, 1970), and purple nutsedge tubers (Teo *et al.*, 1973). It is interesting that in onion bulbs, where sprouting is accompanied by increased gibberellin activity, several gibberellins tested do not break dormancy (Thomas, 1969). In a number of cases, gibberellic acid breaks dormancy only in conjunction with some other factor or when dormancy is already partially broken. In several woody species GA_3 is able to stimulate bud outgrowth only before or after but not during the phase of deep dormancy (Leike, 1967; Leike and Lau, 1967). Similarly, GA_3 (10 μM) promotes outgrowth of water milfoil turions only during the quiescent phases before and after dormancy is broken by cold treatment (Weber and Noodén, 1976b). Turions in the dormant phase may be simply less sensitive or less permeable to GA_3, for a very high concentration (1 mM) is able to induce outgrowth of the dormant turions without a cold treatment. A number of woody species are also made more responsive to gibberellin by a cold treatment (Marth *et al.*, 1956; Donoho and Walker, 1957; Larson, 1960; Lyr *et al.*, 1970). In some cases, gibberellin treatments actually prolong dormancy, for example, aerial tubers of begonia (Esashi and Nagao, 1959; Nagao and Mitsui, 1959; Nagao and Okagami, 1966), grape buds (Weaver, 1959), yam bulbils (Okagami and Nagao, 1971), and gladiolus cormels (Ginzburg, 1973). (2-Chloroethyl)trimethylammonium chloride (CCC), an inhibitor of gibberellin synthesis, releases the begonia tubers from dormancy (Nagao and Okagami, 1966). In addition, application of GA_3 to a number of woody plants in the late summer or fall before leaf drop delays bud break in the subsequent spring (Brian *et al.*, 1959; Corgan and Widmoyer, 1971).

Several lines of evidence suggest that an increase in gibberellins may be important in terminating dormancy in some species but not in others. It does seem essential to analyze further which gibberellins are active internally.

How the gibberellins act to break or reinforce dormancy is still unknown; however, activation of metabolism commonly occurs where gibberellins break dormancy (see, for example, Shih and Rappaport, 1970; Szalai *et al.*, 1975), but these changes could be indirect effects. Gibberellic acid overcomes the inhibitory effect of ABA on bud growth at least partly in black currant (El-Antably *et al.*, 1967), potato buds (El-Antably *et al.*, 1967; Vanes and Hartmans, 1969), and excised Norway maple buds (DeMaggio and Freeberg, 1969). It also offsets (apparently competitively) the inhibitory effect of naringenin on peach bud growth (Phillips, 1962) and inhibitor β on birch (*B. pubescens*) buds (Eagles and Wareing, 1964) or potato buds (Blumenthal-Goldschmidt and Rappaport, 1965). Inasmuch as growth inhibitors tend to decrease as dormancy is broken, GA may cause a decrease in inhibitor content as it breaks bud dormancy. Breaking dormancy of potato tubers with GA_3 does decrease inhibitor β (Boo, 1961). In yam bulbils, where GA_3 increases dormancy, it also increases the inhibitors. Batatasins (Hashimoto *et al.*, 1972) and inhibitors are likewise increased in aerial begonia tubers (Okagami and Nagao, 1973). Whether gibberellin changes the inhibitor content of other buds and whether these changes are primary effects of gibberellin remain to be determined.

   c. Cytokinins.   Because of the ability of cytokinins to overcome correlative inhibition of auxillary buds (Sachs and Thimann, 1967), considerable interest has developed in the possible role of cytokinins in bud dormancy.

   The cytokinin activity (chlorophyll-retention assay) in dormant birch (*Betula papyrifera*) and poplar (*Populus balsamifera*) buds is increased by warm-water bath or 2-chloroethanol treatments, which break dormancy (Domanski and Kozlowski, 1968). When water extracts of birch and poplar buds were chromatographed on paper in *n*-butanol/88% formic acid/H_2O (1:1:1, v/v), the increased activity of the birch bud extracts was mainly in one peak (R_f 0.3–0.5) and that of the poplar bud extract was in two peaks (R_f 0.5–0.6 and 0.9–1.0). During natural termination of dormancy and preceding outgrowth, the cytokinin activity (tobacco callus bioassay) increased in buds of aspen (*Populus tremula*), and this activity chromatographed like zeatin (Engelbrecht, 1971). Similarly, the cytokinin activity (including zeatin-like activity) increased in the apical regions of potato tubers as a result of dormancy-breaking treatments (Engelbrecht and Bielinska-Czarnecka, 1972). Hewett and Wareing (1973) found that dormant poplar (*Populus* × *robusta*) buds contained no detectable cytokinin activity (soybean callus bioassay), but just before bud swelling (outdoors) a sharp rise occurred. After bud swelling was well underway, the cytokinin content dropped sharply. The cytokinin activity seemed to be distributed into five distinct peaks (chromatography on Sephadex LH-20), but the predominant peaks seemed to be zeatin and its riboside. In buds of sugar maple seedlings exposed to dormancy-breaking low temperatures, the activity of one cytokinin (R_f 0.95 in isopropanol/ammonia/water, 10:1:1, v/v) decreased until the chilling requirement was fulfilled and then increased sharply (Taylor and Dumbroff, 1975). Another cytokinin (R_f 0.6) increased only slightly at this time. The authors noted that the increase coincided with the renewal in the growth of roots, which may have been the source of the cytokinin.

The cytokinin activity in the xylem sap of cut grape canes increased as dormancy was broken by cold storage (Skene, 1972). The activity increased in an n-butanol-soluble fraction corresponding chromatographically to zeatin riboside and also a single chromatographic peak in the n-butanol-insoluble (water-soluble) fraction. It should be noted that these canes had been separated from the roots before the increase occurred; so the roots could not produce these cytokinins (at least not directly). Hewett and Wareing (1973) also observed increased cytokinin activity in sap (apparently xylem sap) of field-grown poplar (*Populus* × *robusta*) before bud break and an increase in bud cytokinin activity. Both the main cytokinin component, apparently zeatin riboside, and a minor component, zeatin, increase. The spring sap from sycamore maple does not contain significantly greater quantities of cytokinins just before or during bud break (Purse *et al.*, 1976). As was found for grape canes, excised poplar shoots showed increased cytokinin activity in both their buds and sap after about 3 months of storage at 2°C; however, bud break did not occur. The cytokinin activity in sap of the chilled cuttings did not ultimately reach the high levels of intact, naturally chilled plants. In addition, plants that were not chilled but maintained under 10-hr days with 15°C day and 10°C night temperatures showed an increase in sap cytokinin activity (perhaps even sooner than the chilled plants). This correlates with root growth, but bud break was erratic without the cold treatment. These data suggest that the increased cytokinins in sap and buds of cold-treated plants and increased cytokinin levels do not necessarily lead to bud outgrowth.

Considerable evidence now indicates that cytokinins, particularly the synthetic cytokinins benzyladenine and kinetin, can break bud dormancy in many species. Exogenous cytokinins break dormancy of *Hydrocharis morsus-ranae* turions (Kurz and Kummerow, 1957; Kummerow, 1958), *Utricularia* turions (Kurz, 1959), apple buds (probably lateral buds) (Chvojka *et al.*, 1962, 1963; Pieniazek, 1964a; Pieniazek and Jankiewicz, 1967; Kender and Carpenter, 1972), excised terminal buds from apple (Jones, 1967), Monterey pine (*Pinus radiata*) buds (Kummerow and de Hoffmann, 1963), grape buds (Weaver, 1963), Dutch iris bulbs

(Tsukamoto and Ando, 1973a), citrus buds (Cooper et al., 1969), potato tubers (Hemberg, 1970), gladiolus corms (Tsukamoto, 1972; Ginzberg, 1973), and purple nutsedge tubers (Teo and Nishimoto, 1973; Teo et al., 1973, 1974).

In citrus, benzyladenine produced bud break and some growth of lateral buds but not elongation (Cooper et al., 1969). At least in some circumstances, benzyladenine induces bud break in apple shoots but not much elongation unless gibberellic acid is added (Williams and Billingsley, 1970). Benzyladenine and kinetin produce a similar break of dormancy with limited elongation of water milfoil turions (Weber and Noodén, 1976b).

In comparing the effects of cytokinins and gibberellins (Section IV,C,2,b) on a particular species, it can be seen that in many cases both cytokinins and gibberellins can break dormancy, but there are many cases in which only the cytokinin seems to break dormancy, e.g., apple buds and nutsedge tubers. In addition, cytokinins can break dormancy of several buds that become more dormant in response to gibberellin.

Cytokinins do not act with equal effect at all stages. Kinetin, for example, does not break dormancy of coffee flower buds (van der Veen, 1968), unless the trees have been given a partial dormancy-breaking dry spell (see Browning et al., 1970). However, GA_3 acts on these buds after less drought (van der Veen, 1968; see Browning et al., 1970). Kinetin, like GA_3, appears to induce outgrowth of a number of woody plant buds only during the pre- and postdormant phases and not during full dormancy (Leike, 1967; Leike and Lau, 1967). In contrast, GA_3 (but not cytokinin) breaks dormancy of unchilled peach buds, whereas a synthetic cytokinin (SD 8339) works only after the chilling requirement has been partially fulfilled (Weinberger, 1969). Partial cold treatment also enhances the effect of kinetin in dormant apple buds (Pieniazek, 1964a). In cultured citrus buds, benzyladenine promotes or slightly retards outgrowth, depending on when during the annual cycle it is given (Altman and Goren, 1974).

Thus, cytokinin deficiency may be a factor in the dormancy of some but apparently not all buds. Moreover, even if it is necessary for bud break, it may not be sufficient; other factors may be required. In addition, the increased cytokinin activity correlated with bud break probably does not come from the roots, a point that needs further investigation, as does the role of the different cytokinins.

As is the case for gibberellins, the molecular mechanism by which the cytokinins work is uncertain. Inasmuch as metabolic activation goes along with emergence from dormancy, cytokinins no doubt produce some metabolic activation. For dormant potatoes at least, cytokinins decrease the inhibitor β content (Hemberg, 1970) and, in nutsedge tubers, inhibitor β reduces the growth-promoting effect of benzyladenine (Teo et al., 1974).

Abscisic acid offsets the growth-promoting effect of cytokinins in purple nutsedge tubers (Teo *et al.*, 1973, 1974) and gladiolus cormels (Ginzburg, 1973), but its effect on growth of citrus buds is not clear (Cooper *et al.*, 1969). The possibility that cytokinins act by counteracting or causing the breakdown of inhibitors deserves further investigation.

d. Other Promoters. A very large number of different chemicals have been shown to break dormancy in buds (see Vegis, 1965b, for a compilation). These range from inorganic salts to chlorocarbons, including ethylene and the well-known ethylene chlorohydrin, and it is difficult to see any pattern in these except possibly that they may produce a general trauma.

Interest in the effects of ethylene in dormant structures has increased, because ethylene may function as a hormone. Ethylene breaks dormancy of potato tubers (Rosa, 1925), gladiolus cormels (Harvey, 1927; Denny, 1930; Ginzburg, 1974), birch and beech buds (Börgström, in Abeles, 1973), and black currant buds (Tinklin and Schwabe, 1970). Ethephon, which is thought to act through conversion to ethylene (see Abeles, 1973), breaks dormancy of Dutch iris bulbs, but only after they have been heat-cured (Tsukamoto and Ando, 1973a), but not purple nutsedge tubers (Teo *et al.*, 1974). Curiously, CO_2, which is an antagonist of ethylene (Abeles, 1973), also promotes outgrowth of potato tubers, the optimal effect occurring at 40–60% (Thornton, 1933). Ethylene could play some role in termination of dormancy, but this remains to be determined.

Many early studies showed changes in growth-promoting substances (based on combinations of paper chromatography and bioassays) during termination of dormancy. Romberger (1963) has discussed some other possible promoters in detail. Many of these have not yet been identified and could turn out to be promoters different from those discussed above.

D. Summary

The environmental controls of bud dormancy appear to be mediated by hormones. A balance between growth inhibitors (ABA and probably others) and growth promoters (gibberellin, cytokinin, and possibly others) seems to be important in regulating both entry into and release from the dormant phase in buds. Without a doubt, the hormonal controls of dormancy differ from one species to another; e.g., gibberellins break bud dormancy in some species but reinforce it in others. Likewise, the correlative controls may differ, for nearby organs (e.g., leaves) may induce or impose dormancy in the terminal buds of some species but not others. Whereas the dormancy-inducing hormones may come from outside the terminal buds, the hormones that break dormancy seem to be produced within the bud.

ACKNOWLEDGMENT

This study was supported in part by research grant 416-15-79 from the USDA Cooperative State Research Service under P.L. 89-106 to L.D.N.

REFERENCES

Abeles, F. B. (1973). "Ethylene in Plant Biology." Academic Press, New York.
Allary, S. (1960). C. R. Hebd. Seances Acad. Sci. 250, 911.
Allary, S. (1961). C. R. Hebd. Seances Acad. Sci. 252, 930.
Allen, R. M. (1960). Physiol. Plant. 13, 555.
Altman, A., and Goren, R. (1974). Physiol. Plant. 30, 240.
Askenasy, E. (1877a). Bot. Ztg. 35, 793.
Askenasy, E. (1877b). Bot. Ztg. 35, 817.
Askenasy, E. (1877c). Bot. Ztg. 35, 833.
Bachelard, E. P., and Wightman, F. (1973). Can. J. Bot. 51, 2315.
Bachelard, E. P., and Wightman, F. (1974). Can. J. Bot. 52, 1483.
Balut, S. (1956). Ekol. Pol., Ser. A 4, 225.
Bennett, J. P. (1950). Calif. Agric. 4, 11.
Bennett, J. P., and Skoog, F. (1938). Plant Physiol. 13, 219.
Bennet-Clark, T. A., and Kefford, N. P. (1953). Nature (London) 171, 645.
Bialek, K. (1974). Z. Pflanzenphysiol. 71, 370.
Bialek, K., and Bielinska-Czarnecka, M. (1975). Bull. Acad. Pol. Sci., Ser. Sci. Biol. 23, 213.
Bielinska-Czarnecka, M., and Bialek, K. (1972). Bull. Acad. Pol. Sci., Ser. Sci. Biol. 20, 809.
Biran, I., Gur, I., and Halevy, A. H. (1972). Physiol. Plant. 27, 226.
Blake, T. J. (1972). New Phytol. 71, 327.
Blommaert, K. L. J. (1954). Nature (London) 174, 970.
Blommaert, K. L. J. (1955). S. Afr., Dep. Agric. Tech. Serv., Sci. Bull. 368, 1.
Blumenthal-Goldschmidt, S., and Rappaport, L. (1965). Plant Cell Physiol. 6, 601.
Bogdanov, P. L. (1931). Khozyaystvu i Lesnoy Promyshlennosti (Leningr.) 10, 21.
Boo, L. (1961). Physiol. Plant. 14, 676.
Booth, A. (1963). In "The Growth of the Potato" (J. D. Ivins and F. L. Milthorpe, eds.), p. 99 Butterworth, London.
Boresch, K. (1924). Biochem. Z. 153, 313.
Boresch, K. (1926). Biochem. Z. 170, 466.
Bowen, M. R., and Hoad, G. V. (1968). Planta 81, 64.
Brian, P. W., Hemming, H. G., and Radley, M. (1955). Physiol. Plant. 8, 899.
Brian, P. W., Petty, J. H. P., and Richmond, P. T. (1959). Nature (London) 184, 69.
Browning, G., Hoad, G. V., and Gaskin, P. (1970). Planta 94, 213.
Buch, M. L., and Smith, O. (1959). Physiol. Plant. 12, 706.
Bukovac, M. J., and Wittwer, S. H. (1961). In "Plant Growth Regulation" (4th Internat. Conf. on Plant Growth), p. 505. Iowa State University Press, Ames.
Burton, W. G. (1956). Physiol. Plant. 9, 567.
Campbell, R. K. (1974). J. Appl. Ecol. 11, 1069.
Campbell, R. K., and Sugano, A. I. (1975). Bot. Gaz. (Chicago) 136, 290.
Catchpole, A. H., and Hillman, J. (1969). Nature (London) 273, 1387.
Cathey, H. M. (1968). Proc. Am. Soc. Hortic. Sci. 93, 693.
Chandler, W. H., and Tufts, W. P. (1934). Proc. Am. Soc. Hortic. Sci. 30, 180.

Chandler, W. H., Kimball, M. H., Philp, G. L., Tufts, W. P., and Weldon, G. P. (1937). *Calif., Agric. Exp. Stn., Bull.* **611**, 1.
Chapman, H. W. (1958). *Physiol. Plant.* **11**, 215.
Chawdhry, M. A., and Sagar, G. R. (1974). *Weed Res.* **14**, 349.
Chouard, P. (1946). *C. R. Hebd. Seances Acad. Sci.* **223**, 1174.
Chvojka, L., Trávníček, M., and Zakouřilová, M. (1962). *Biol. Plant.* **4**, 203.
Chvojka, L., Kračmar, P., Beneš, J., and Belea, A. (1963). *Biologia (Bratislava)* **18**, 579.
Cooper, W. C., Young, R. H., and Henry, W. H. (1969). *Proc. 1st Int. Citrus Symp.*, Vol. 1, p. 301.
Corgan, J. N. (1965). *Proc. Am. Soc. Hortic. Sci.* **86**, 129.
Corgan, J. N., and Martin, G. C. (1971). *Hort. Science* **6**, 405.
Corgan, J. N., and Peyton, C. (1970). *J. Am. Soc. Hortic. Sci.* **95**, 770.
Corgan, J. N., and Widmoyer, F. B. (1971). *J. Am. Soc. Hortic. Sci.* **96**, 54.
Coville, F. V. (1920). *J. Agric. Res. (Washington, D.C.)* **20**, 151.
Czopek, M. (1963). *Acta Soc. Bot. Pol.* **32**, 199.
Davison, R. M. (1965). *Aust. J. Biol. Sci.* **18**, 475.
Davison, R. M., and Young, H. (1974). *Plant Sci. Lett.* **2**, 79.
DeMaggio, A. E., and Freeberg, J. A. (1969). *Can. J. Bot.* **47**, 1165.
Dennis, F. G., and Edgerton, L. J. (1961). *Proc. Am. Soc. Hortic. Sci.* **77**, 107.
Denny, F. E. (1930). *Am. J. Bot.* **17**, 602.
Denny, F. E., and Stanton, E. N. (1928). *Am. J. Bot.* **15**, 337.
Dereuddre, J. (1971). *C. R. Hebd. Seances Acad. Sci.*, Ser. D **273**, 2239.
Digby, J., and Wareing, P. F. (1966). *Ann. Bot. (London)* **30**, 607.
Domanski, R., and Kozlowski, T. T. (1968). *Can. J. Bot.* **46**, 397.
Donoho, C. W., and Walker, D. R. (1957). *Science* **126**, 1178.
Doorenbos, J. (1953). *Meded. Landbouwhogesch. Wageningen* **53**, 1.
Dörffling, K. (1963a). *Planta* **60**, 390.
Dörffling, K. (1963b). *Planta* **60**, 413.
Dormling, I., Gustafsson, A., and von Wettstein, D. (1968). *Silvae Genet.* **17**, 44.
Dostál, R. (1927). *Ber. Dtsch. Bot. Ges.* **45**, 436.
Downs, R. J., and Borthwick, H. A. (1956a). *Proc. Am. Soc. Hortic. Sci.* **68**, 518.
Downs, R. J., and Borthwick, H. A. (1956b). *Bot. Gaz. (Chicago)* **117**, 310.
Düring, H., and Alleweldt, G. (1973). *Vitis* **12**, 26.
Eady, F. C., and Eaton, G. W. (1972). *Can. J. Plant Sci.* **52**, 263.
Eagles, C. F., and Wareing, P. F. (1963). *Nature (London)* **199**, 874.
Eagles, C. F., and Wareing, P. F. (1964). *Physiol. Plant.* **17**, 697.
El-Antably, H. M. M., Wareing, P. F., and Hillman, J. (1967). *Planta* **73**, 74.
Eliasson, L. (1969). *Physiol. Plant.* **22**, 1288.
El-Mansy, H. I., and Walker, D. R. (1969). *J. Am. Soc. Hortic. Sci.* **94**, 658.
Engelbrecht, L. (1971). *Biochem. Physiol. Pflanz.* **162**, 547.
Engelbrecht, L., and Bielinska-Czarnecka, M. (1972). *Biochem. Physiol. Pflanz.* **163**, 499.
Erez, A., and Lavee, S. (1971). *J. Am. Soc. Hortic. Sci.* **96**, 711.
Esashi, Y., and Nagao, M. (1959). *Sci. Rep. Tohoku Univ.*, Ser. 4 **25**, 191.
Esashi, Y., and Nagao, M. (1973). *Plant Physiol.* **51**, 504.
Farmer, R. E. (1968). *Physiol. Plant.* **21**, 1241.
Flemion, F. (1959). *Boyce Thompson Inst. Plant Res., Contrib.* **20**, 57.
Frank, P. A. (1966). *J. Exp. Bot.* **17**, 546.
Garner, W. W., and Allard, H. A. (1923). *J. Agric. Res. (Washington, D.C.)* **23**, 871.
Gentel', P. A., and Barskaya, E. I. (1960). *Fiziol. Rast.* **7**, 645.

Giertych, M. M., and Forward, D. F. (1966). *Can. J. Bot.* **44,** 717.

Ginzburg, C. (1973). *J. Exp. Bot.* **24,** 558.

Ginzburg, C. (1974). *Plant Sci. Lett.* **2,** 133.

Goebel, K. (1880). *Bot. Ztg.* **38,** 801.

Gregory, L. E. (1956). *Am. J. Bot.* **43,** 281.

Gulisashvili, V. Z. (1948). *Priroda (Moscow)* **3,** 63.

Gustafson, R. G. (1938). *Plant Physiol.* **13,** 655.

Guthrie, J. D. (1939). *Boyce Thompson Inst. Plant Res., Contrib.* **11,** 29.

Halevy, A. H., Shoub, J., and Rakati-Aaylon, D. (1964). *Isr. J. Agric. Res.* **14,** 11.

Harrison, M. A., and Saunders, P. F. (1975). *Planta* **123,** 291.

Harvey, R. B. (1927). *Off. Bull. Am. Gladiolus Soc.* **4,** 10.

Hasegawa, K., and Hashimoto, T. (1973). *Plant Cell Physiol.* **14,** 369.

Hasegawa, K., and Hashimoto, T. (1974). *Plant Cell Physiol.* **15,** 1.

Hasegawa, K., and Hashimoto, T. (1975). *J. Exp. Bot.* **26,** 757.

Hashimoto, T., and Tamura, S. (1969). *Bot. Mag.* **82,** 327.

Hashimoto, T., Hasegawa, K., and Kawarada, A. (1972). *Planta* **108,** 369.

Hashimoto, T., Hasegawa, K., Yamaguchi, H., Saito, M., and Ishimoto, S. (1974). *Phytochemistry* **13,** 2849.

Heath, O. V. S., and Holdsworth, M. (1948). *Symp. Soc. Exp. Biol.* **2,** 326.

Heide, O. M. (1974). *Physiol. Plant.* **30,** 1.

Hemberg, T. (1949a). *Physiol. Plant.* **2,** 24.

Hemberg, T. (1949b). *Physiol. Plant.* **2,** 37.

Hemberg, T. (1952). *Physiol. Plant.* **5,** 115.

Hemberg, T. (1954). *Physiol. Plant.* **7,** 312.

Hemberg, T. (1958a). *Physiol. Plant.* **11,** 610.

Hemberg, T. (1958b). *Physiol. Plant.* **11,** 615.

Hemberg, T. (1961). *In* "Handbuch der Pflanzenphysiologie" (W. Ruhland, ed.), Vol. 14, pp. 1162–1184. Springer-Verlag, Berlin and New York.

Hemberg, T. (1965). *In* "Handbuch der Pflanzenphysiologie" (W. Ruhland, ed.), Vol. 15, Part 2, pp. 669–698. Springer-Verlag, Berlin and New York.

Hemberg, T. (1970). *Physiol. Plant.* **23,** 850.

Hendershott, C. H., and Bailey, L. F. (1955). *Proc. Am. Soc. Hortic. Sci.* **65,** 85.

Hendershott, C. H., and Walker, D. R. (1959a). *Science* **130,** 798.

Hendershott, C. H., and Walker, D. R. (1959b). *Proc. Am. Soc. Hortic. Sci.* **74,** 121.

Heslop-Harrison, Y. (1962). *Proc. R. Ir. Acad., Sect. B* **62,** 23.

Hewett, E. W., and Wareing, P. F. (1973). *Physiol. Plant.* **28,** 393.

Hoad, G. V. (1967). *Life Sci.* **6,** 1112.

Hoad, G. V., and Bowen, M. R. (1968). *Planta* **82,** 22.

Hocking, J. J., and Hillman, J. R. (1975). *Planta* **125,** 235.

Hoffman, A., and Lyr, H. (1967). *Flora (Jena), Abt. A* **158,** 373.

Holst, U. (1971). *Physiol. Plant.* **24,** 392.

Housley, S., and Taylor, W. C. (1958). *J. Exp. Bot.* **9,** 458.

Howard, W. L. (1910). *Mo., Agric. Exp. Stn., Res. Bull.* **1,** 1.

Hull, J., Jr., and Lewis, L. N. (1959). *Proc. Am. Soc. Hortic. Sci.* **74,** 93.

Hull, R. J. (1970). *Weed Sci.* **18,** 118.

Humphries, E. C. (1944). *Ann. Bot. (London)* **8,** 259.

Il'in, V. S. (1971). *Fiziol. Rast.* **18,** 369.

Irgens-Moller, H. (1957). *For. Sci.* **3,** 79.

Jensen, K. F., and Gatherum, G. E. (1965). *For. Sci.* **11,** 189.

Jester, J. R., and Kramer, P. J. (1939). *J. For.* **37**, 796.
Johannsen, W. (1906). "Das Ather-Verfahren beim Frühtreiben mit besonderer Berücksichtigung der Fliedertreiberei," 2nd ed. Fischer, Jena.
Johnsson, H. (1951). *Sven. Papperstidn.* **54**, 379.
Jones, O. P. (1967). *Nature (London)* **215**, 1514.
Jost, L. (1894). *Ber. Dtsch. Bot. Ges.* **12**, 188.
Junttila, O. (1976). *Physiol. Plant.* **38**, 278.
Kawase, M. (1966). *Proc. Am. Soc. Hortic. Sci.* **89**, 752.
Kender, W. J., and Carpenter, S. (1972). *J. Am. Soc. Hortic. Sci.* **97**, 377.
Klebs, G. (1914). *Abh. Heidelberger Akad. Wiss., Math.-Naturwiss. Kl.* **3**, 1.
Klenovska, S., Pastyrik, L., and Peterkova, J. (1974). *Physiol. Plant.* **9**, 55.
Knight, T. A. (1801). *Philos. Trans. R. Soc. London, Ser. B* **91**, 333.
Koller, D. (1970). *Symp. Soc. Exp. Biol.* **23**, 449.
Koller, D., and Highkin, H. R. (1960). *Am. J. Bot.* **47**, 843.
Koltermann, A. (1927). *Angew. Bot.* **9**, 289.
Kononenko, G. P., Popravka, S. A., and Wulfson, N. S. (1975). *Bioorg. Khim.* **1**, 506.
Kozlowski, T. T. (1962). "Tree Growth." Ronald Press, New York.
Kozlowski, T. T. (1964). *Bot. Rev.* **30**, 335.
Kozlowski, T. T. (1971). *In* "Growth and Development of Trees" (T. T. Kozlowski, ed.), Vol. 1, pp. 179–185 and 362–386. Academic Press, New York.
Kramer, P. J. (1936). *Plant Physiol.* **11**, 127.
Krašan, F. (1873a). *Sitzungsber. Akad. Wiss. Wien, Math.-Naturwiss. Kl., Abt. 1* **67**, 143.
Krašan, F. (1873b). *Sitzungsber. Akad. Wiss. Wien, Math.-Naturwiss. Kl., Abt. 1* **67**, 252.
Krasnosselskaya, T. A., and Richter, A. A. (1942). *Dokl. Ross. Akad. Nauk, Ser. A* **35**, 184.
Kriebel, H. B., and Wang, C. (1962). *Silvae Genet.* **11**, 125.
Kumar, D., and Wareing, P. F. (1972). *New Phytol.* **71**, 639.
Kumar, D., and Wareing, P. F. (1973). *New Phytol.* **72**, 283.
Kummerow, J. (1958). *Beitr. Biol. Pflanz.* **34**, 293.
Kummerow, J., and de Hoffmann, C. A. (1963). *Ber. Dtsch. Bot. Ges.* **76**, 189.
Kurz, L. (1959). *Beitr. Biol. Pflanz.* **35**, 111.
Kurz, L., and Kummerow, J. (1957). *Naturwissenschaften* **44**, 121.
Kuzina, G. V. (1970). *Fiziol. Rast.* **17**, 76.
Langlet, O. (1944). *Medd. Statens Skogsforskningsinst. (Swed.)* **33**, 295.
Langrova, V., and Sladky, Z. (1971). *Biol. Plant.* **13**, 361.
Larson, P. R. (1960). *Forest Sci.* **6**, 232.
Laude, H. M. (1953). *Bot. Gaz. (Chicago)* **114**, 284.
Lavender, D. P., and Hermann, R. K. (1970). *New Phytol.* **69**, 675.
Lavender, D. P., Ching, K. K., and Hermann, R. K. (1968). *Bot. Gaz. (Chicago)* **129**, 70.
Lavender, D. P., Sweet, G. B., Zaern, J. B., and Hermann, R. K. (1973). *Science* **182**, 838.
Leike, H. (1967). *Flora (Jena), Abt. A* **158**, 351.
Leike, H., and Lau, R. (1967). *Flora (Jena), Abt. A* **157**, 467.
Lenton, J. R., Perry, V. M., and Saunders, P. F. (1972). *Planta* **106**, 13.
Lesham, Y., Philosoph, S., and Wurzberger, J. (1974). *Biochem. Biophys. Res. Commun.* **57**, 526.
Little, C. H., and Eidt, D. C. (1968). *Nature (London)* **220**, 498.
Lockhart, J. A., and Bonner, J. (1957). *Plant Physiol.* **32**, 492.
Lona, F., and Borghi, R. (1957). *Ateneo Parmense* **28**, 116.
Loomis, W. E., and Evans, M. M. (1928). *Proc. Am. Soc. Hortic. Sci.* **25**, 73.
Lorenzi, R., Horgan, R., and Heald, J. K. (1975). *Planta* **126**, 75.
Lovell, P. H., and Booth, A. (1967). *New Phytol.* **66**, 525.

Loveys, B. R., Leopold, A. C., and Kriedeman, P. E. (1974). *Ann. Bot. (London)* **38**, 85.
Lyr, H., Hoffmann, G., and Richter, R. (1970). *Biochem. Physiol. Pflanz.* **161**, 133.
Madison, M., and Rappaport, L. (1968). *Plant Cell Physiol.* **9**, 147.
Mahlstede, J. P. (1956). *Proc. Plant Propagation Soc.* **6**, 130.
Mann, L. K., and Lewis, D. A. (1956). *Hilgardia* **26**, 161.
Marth, P. C., Harris, L., and Batjer, L. P. (1947). *Proc. Am. Soc. Hortic. Sci.* **49**, 49.
Marth, P. C., Audia, W. V., and Mitchell, J. W. (1956). *Bot. Gaz. (Chicago)* **118**, 106.
Matsubara, M. (1931). *Planta* **13**, 695.
Michniewicz, M., and Galoch, E. (1972). *Bull. Acad. Pol. Sci., Ser. Sci. Biol.* **20**, 333.
Mielke, E. A., and Dennis, F. G., Jr. (1975a). *J. Am. Soc. Hortic. Sci.* **100**, 285.
Mielke, E. A., and Dennis, F. G., Jr. (1975b). *J. Am. Soc. Hortic. Sci.* **100**, 287.
Milborrow, B. V. (1967). *Planta* **76**, 93.
Miller, L. P. (1934). *Boyce Thompson Inst. Plant Res., Contrib.* **6**, 279.
Milthorpe, F. L. (1963). *In* "The Growth of the Potato" (J. D. Ivins and F. L. Milthorpe, eds.), p. 121. Butterworth, London.
Mingo-Castel, A. M., Negm, F. B., and Smith, O. E. (1974). *Plant Physiol.* **53**, 798.
Molisch, H. (1908). *Sitzungsber. Kais. Akad. Wiss. Wien, Math.-Naturwiss. Kl., Abt. 1* **117**, 87.
Molisch, H. (1909a). *Sitzungsber. Kais. Akad. Wiss. Wien, Math.-Naturwiss. Kl., Abt. 1* **118**, 637.
Molisch, H. (1909b). "Das Warmbad als Mittel zum Treiben der Pflanzen." Fischer, Jena.
Moorby, J., and Milthorpe, F. L. (1975). *In* "Crop Physiology: Some Case Studies" (L. T. Evans, ed.), pp. 225–257. Cambridge Univ. Press, London and New York.
Moshkov, B. S. (1935). *Planta* **23**, 774.
Nagao, M., and Mitsui, E. (1959). *Sci. Rep. Tohoku Univ., Ser. 4* **25**, 199.
Nagao, M., and Okagami, N. (1966). *Bot. Mag.* **79**, 687.
Nagata, H. (1967). *J. Jpn. For. Soc.* **49**, 415.
Nasr, T. A. A., and Wareing, P. F. (1961). *J. Hortic. Sci.* **36**, 1.
Nesterov, Ya. A. (1962). *Izd. Sel'skokhoz. Lit. Zh. Plak.*
Nienstaedt, H. (1966). *For. Sci.* **12**, 374.
Nienstaedt, H. (1967). *Silvae Genet.* **16**, 65.
Nigond, J. (1957). *C. R. Seances Acad. Agric. Fr.* **46**, 452.
Nitsch, J. P. (1957a). *Proc. Am. Soc. Hortic. Sci.* **70**, 512.
Nitsch, J. P. (1957b). *Proc. Am. Soc. Hortic. Sci.* **70**, 526.
Noodén, L. D., and Leopold, A. C. (1978). *In* "Plant Hormones and Related Compounds" (D. S. Letham, J. J. Higgins, and P. B. Goodwin, eds.). Elsevier, Amsterdam (in press).
Ofir, M., and Koller, D. (1974). *Aust. J. Plant Physiol.* **1**, 259.
Ofir, M., Koller, D., and Negbi, M. (1967). *Bot. Gaz. (Chicago)* **128**, 25.
Okagami, N., and Nagao, M. (1971). *Planta* **101**, 91.
Okagami, N., and Nagao, M. (1973). *Plant Cell Physiol.* **14**, 1064.
Olmstead, C. E. (1951). *Bot. Gaz. (Chicago)* **112**, 365.
Olson, J. S., and Nienstaedt, N. (1956). *Science* **125**, 492.
Olson, J. S., Streans, F. W., and Neinstaedt, N. (1959). *Conn., Agric. Exp. Stn., New Haven, Bull.* **620**.
Palmer, C. E., and Smith, O. E. (1969). *Plant Cell Physiol.* **10**, 657.
Palmer, C. E., and Smith, O. E. (1970). *Plant Cell Physiol.* **11**, 303.
Paterson, D. R. (1970). *HortScience* **5**, 333.
Pauley, S. S., and Perry, T. O. (1954). *J. Arnold Arbor., Harv. Univ.* **35**, 167.
Perry, T. O. (1962). *For. Sci.* **8**, 336.
Perry, T. O. (1968). *Plant Physiol.* **43**, 1866.

Perry, T. O. (1971). *Science* **171**, 29.

Perry, T. O., and Byrne, O. R. (1969). *Plant Physiol.* **44**, 784.

Perry, T. O., and Hellmers, H. (1973). *Bot. Gaz. (Chicago)* **134**, 283.

Perry, T. O., and Wu, W. C. (1960). *Ecology* **41**, 790.

Pettersen, H. (1972). *J. Am. Soc. Hortic. Sci.* **97**, 17.

Phillips, I. D. J. (1962). *J. Exp. Bot.* **13**, 213.

Phillips, I. D. J., and Wareing, P. F. (1958). *J. Exp. Bot.* **9**, 350.

Phillips, I. D. J., and Wareing, P. F. (1959). *J. Exp. Bot.* **10**, 504.

Phillips, J. E. (1941). *J. For.* **39**, 55.

Pieniazek, J. (1964a). *Acta Agrobot.* **16**, 157.

Pieniazek, J. (1964b). *Bull. Acad. Pol. Sci., Ser. Sci. Biol.* **12**, 227.

Pieniazek, J., and Jankiewicz, L. S. (1967). *Wiss. Z. Univ. Rostock, Math.-Naturwiss. Reihe* **4/5**, 651.

Pieniazek, J., and Rudnicki, R. (1971). *Bull. Acad. Pol. Sci., Ser. Sci. Biol.* **19**, 201.

Pollock, B. M. (1953). *Physiol. Plant.* **6**, 47.

Powell, L. E. (1976). *HortScience* **11**, 498.

Priestley, C. A. (1962). *Comm. Bur. Hortic. Plantation Crops, Tech. Commun.* p. 27.

Purse, J. G., Horgan, R., Horgan, J. M., and Wareing, P. F. (1976). *Planta* **132**, 1.

Ramsay, J., and Martin, G. C. (1970). *J. Am. Soc. Hortic. Sci.* **95**, 569.

Rappaport, L. (1956). *Calif. Agric.* **10**, 4.

Richter, A. A., and Krasnosselskaya, T. A. (1945). *Dokl. Ross. Akad. Nauk, Ser. A* **47**, 218.

Robinson, P. M., and Wareing, P. F. (1964). *Physiol. Plant.* **17**, 315.

Robinson, P. M., Wareing, P. F., and Thomas, T. H. (1963). *Nature (London)* **199**, 875.

Romberger, J. A. (1963). *U.S., Dep. Agric., Tech. Bull.* **1293**.

Rosa, J. T. (1925). *Potato News Bull.* **2**, 363.

Sachs, T., and Thimann, K. V. (1967). *Am. J. Bot.* **54**, 136.

Samish, R. M. (1954). *Annu. Rev. Plant Physiol.* **5**, 183.

Schaedle, M. (1959). M.S.A. Thesis, University of British Columbia, Vancouver.

Schmid, B. (1901). *Ber. Dtsch. Bot. Ges.* **19**, 76.

Schneider, E. F. (1968). *J. Exp. Bot.* **19**, 817.

Schneider, E. F. (1970). *J. Exp. Bot.* **21**, 799.

Schneider, M. J. (1973). *Yearb. Am. Philo. Soc.* pp. 350–352.

Schroeder, C. A. (1951). *Calif. Citrogr.* **37**, 16.

Schwabe, W. W. (1970). *Ann. Bot. (London)* **34**, 29.

Schwabe, W. W., and Valio, I. F. M. (1970a). *J. Exp. Bot.* **21**, 112.

Schwabe, W. W., and Valio, I. F. M. (1970b). *J. Exp. Bot.* **21**, 122.

Shih, C. Y., and Rappaport, L. (1970). *Plant Physiol.* **45**, 33.

Shih, C. Y., and Rappaport, L. (1971). *Plant Physiol.* **48**, 31.

Shilova, N. V. (1974). *Bot. Zh. (Leningrad)* **59**, 206.

Simon, S. (1906). *Jahrb. Wiss. Bot.* **43**, 1.

Simon, S. V. (1928). *Jahrb. Wiss. Bot.* **68**, 149.

Sinnadurai, S., and Amuti, S. K. (1971). *Exp. Agric.* **7**, 17.

Sizov, S. S. (1975). *Biol. Nauki (Moscow)* **18**, 83.

Skene, K. G. (1972). *Planta* **104**, 89.

Smith, O. E., and Rappaport, L. (1960). *Adv. Chem. Ser.* **28**, 42.

Spiers, J. M. (1973). *J. Am. Soc. Hortic. Sci.* **98**, 237.

Steward, F. C., and Caplin, S. M. (1952). *Ann. Bot. (London)* **16**, 477.

Stewart, G. R. (1969). *Nature (London)* **221**, 61.

Sugano, A. I. (1971). M.Sc. Thesis, Oregon State University, Corvallis.

Sylvén, N. (1940a). *Sven. Papperstidn.* **43**, 317.

Sylvén, N. (1940b). *Sven. Papperstidn.* **43**, 332.
Sylvén, N. (1940c). *Sven. Papperstidn.* **43**, 350.
Syrtanòva, G. A. (1974). *Izv. Akad. Nauk SSR*, *Ser. Biol.* **4**, 15.
Szalai, I., Nagy, M., and Helfrich, M. (1975). *Acta Agron. Acad. Sci. Hung.* **24**, 35.
Taylor, J. S., and Dumbroff, E. B. (1975). *Can. J. Bot.* **53**, 321.
Teo, C. K. H., and Nishimoto, R. K. (1973). *Weed Res.* **13**, 118.
Teo, C. K. H., Bendixen, L. E., and Nishimoto, R. K. (1973). *Weed Sci.* **21**, 19.
Teo, C. K. H., Nishimoto, R. K., and Tang, C. S. (1974). *Weed Res.* **14**, 173.
Terras, J. A. (1900). *Trans. Proc. Bot. Soc. Edinburgh* **21**, 318.
Thomas, G. G., and Wilkinson, E. H. (1964). *Hortic. Res.* **4**, 78.
Thomas, T. H. (1969). *J. Exp. Bot.* **20**, 124.
Thomas, T. H., Wareing, P. F., and Robinson, P. M. (1965). *Nature (London)* **205**, 1270.
Thornton, N. C. (1933). *Boyce Thompson Inst. Plant Res., Contrib.* **5**, 408.
Timmis, R., and Worrall, J. (1974). *Can. J. For. Res.* **4**, 229.
Tinklin, I. G., and Schwabe, W. W. (1970). *Ann. Bot. (London)* **34**, 691.
Tizio, R. (1964). *C. R. Hebd. Seances Acad. Sci.* **259**, 1187.
Tompkins, D. R. (1965). *Proc. Am. Soc. Hortic. Sci.* **87**, 371.
Tsukamoto, Y. (1972). *Proc. Jpn. Acad.* **48**, 34.
Tsukamoto, Y., and Ando, T. (1973a). *Environ. Control Biol.* **11**, 69.
Tsukamoto, Y., and Ando, T. (1973b). *Proc. Jpn. Acad.* **49**, 627.
Tsukamoto, Y., and Konoshima, H. (1972). *Physiol. Plant* **26**, 244.
Tuan, D. Y. H., and Bonner, J. (1964). *Plant Physiol.* **39**, 768.
Tumanov, I. I., Kuzina, G. V., and Karnikova, L. D. (1970). *Fiziol. Rast.* **17**, 885.
Tumbleson, M. E., and Kommedahl, T. (1961). *Weeds* **9**, 646.
Tumbleson, M. E., and Kommedahl, T. (1962). *Bot. Gaz. (Chicago)* **123**, 186.
Vaartaja, O. (1954). *Can. J. Bot.* **32**, 392.
Vaartaja, O. (1956). *Can. J. Bot.* **34**, 377.
Vaartaja, O. (1959). *Ecol. Monogr.* **29**, 91.
Vaartaja, O. (1961). *Can. J. Bot.* **39**, 649.
Valio, I. F. M., and Schwabe, W. W. (1969). *J. Exp. Bot.* **20**, 615.
Valio, I. F. M., and Schwabe, W. W. (1970). *J. Exp. Bot.* **21**, 138.
Valio, I. F. M., Burdon, R. S., and Schwabe, W. W. (1969). *Nature (London)* **223**, 1176.
van den Driessche, R. (1975). *B. C. Forest Serv., Res. Note* **71**, 1.
van der Veen, R. (1951). *Physiol. Plant.* **4**, 35.
van der Veen, R. (1968). *Acta Bot. Neerl.* **17**, 373.
Vanes, A., and Hartmans, K. J. (1969). *Eur. Potato J.* **12**, 59.
Varga, M. B., and Ferenczy, L. (1956). *Nature (London)* **178**, 1075.
Varga, M. B., and Ferenczy, L. (1957). *Acta Bot. Acad. Sci. Hung.* **3**, 111.
Vegis, A. (1953). *Experientia* **9**, 462.
Vegis, A. (1955). *Symb. Bot. Ups.* **14**, 1.
Vegis, A. (1963). *In* "Environmental Control of Plant Growth" (L. T. Evans, ed.), pp. 265–287. Academic Press, New York.
Vegis, A. (1964). *Annu. Rev. Plant Physiol.* **15**, 185.
Vegis, A. (1965a). *In* "Handbuch der Pflanzenphysiologie" (W. Ruhland, ed.), Vol. 15, Part 2, pp. 499–533. Springer-Verlag, Berlin and New York.
Vegis, A. (1965b). *In* "Handbuch der Pflanzenphysiologie" (W. Ruhland, ed.), Vol. 15, Part 2, pp. 534–668. Springer-Verlag, Berlin and New York.
von Guttenberg, H., and Leike, H. (1958). *Planta* **52**, 96.
Wang, S. Y., and Roberts, A. N. (1970). *J. Am. Soc. Hortic. Sci.* **95**, 554.
Warden, J. (1970). *Port. Acta Biol., Ser. A* **11**, 319.

Wareing, P. F. (1948). *Forestry* **22**, 211.
Wareing, P. F. (1950). *Physiol. Plant.* **3**, 258.
Wareing, P. F. (1951). *Physiol. Plant.* **4**, 41.
Wareing, P. F. (1953). *Physiol. Plant.* **6**, 692.
Wareing, P. F. (1954). *Physiol. Plant.* **7**, 261.
Wareing, P. F. (1965). *Sci. Prog. (Oxford)* **53**, 529.
Wareing, P. F. (1969). *Symp. Soc. Exp. Biol.* **23**, 241.
Wareing, P. F., and Black M. (1958). *In* "The Physiology of Forest Trees" (K. V. Thimann, ed.), pp. 539–556. Ronald Press, New York.
Wareing, P. F., and Saunders, P. F. (1971). *Annu. Rev. Plant Physiol.* **22**, 261.
Wassink, E. C., and Wiersma, J. H. (1955). *Acta Bot. Neerl.* **4**, 657.
Watts, J. W., and King, J. M. (1973). *Planta* **113**, 271.
Waxman, S. (1957). Ph.D. Thesis, Cornell University, Ithaca, New York.
Weaver, R. J. (1959). *Nature (London)* **183**, 1198.
Weaver, R. J. (1963). Nature (London) **198**, 207.
Weaver, R. J., McCune, S. B., and Coombe, B. G. (1961). *Am. J. Enol. Vitic.* **12**, 131.
Weber, F. (1916a). *Sitzungsber. Kais. Akad. Wiss. Wien, Math.-Naturwiss. Kl., Abt. 1* **125**, 189.
Weber, F. (1916b). *Sitzungsber. Kais. Akad. Wiss. Wien, Math.-Naturwiss. Kl., Abt. 1* **125**, 311.
Weber, F. (1916c). *Sitzungsber. Kais. Akad. Wiss. Wien, Math.-Naturwiss. Kl., Abt. 1* **127**, 57.
Weber, F. (1921). *Ber. Dtsch. Bot. Ges.* **39**, 152.
Weber, J. A., and Noodén, L. D. (1976a). *Am. J. Bot.* **63**, 936.
Weber, J. A., and Noodén, L. D. (1976b). *Plant Cell Physiol.* **17**, 727.
Weinberger, J. H. (1954). *Proc. Am. Soc. Hortic. Sci.* **63**, 191.
Weinberger, J. H. (1969). *HortScience* **4**, 125.
Weiser, C. J. (1970). *Science* **169**, 1269.
Weldon, G. P. (1935). *Calif., Dep. Agric., Mon. Bull.* **23**, 160.
Werner, H. O. (1934). *Nebr., Agric. Exp. Stn., Res. Bull.* **75**, 1.
Williams, M. W., and Billingsley, H. D. (1970). *J. Am. Soc. Hortic. Sci.* **95**, 649.
Witkowska-Žuk, L. (1970). *Acta Soc. Bot. Pol.* **39**, 285.
Witkowska-Žuk, L., and Kapuscinski, L. (1969). *Acta Soc. Bot. Pol.* **38**, 615.
Witkowska-Žuk, L., and Kozlowska, W. (1973). *Acta Soc. Bot. Pol.* **42**, 627.
Witkowska-Žuk, L., and Kozlowska, W. (1974). *Acta Soc. Bot. Pol.* **43**, 421.
Womack, D. E. (1964). Ph.D. Thesis, Oregon State University, Corvallis.
Worrall, J., and Meyer, F. (1967). *Physiol. Plant.* **20**, 733.
Wright, S. T. C. (1975). *J. Exp. Bot.* **26**, 161.
Yerkes, G. E., and Gardner, F. C. (1934). *Proc. Am. Soc. Hortic. Sci.* **32**, 347.
Young, R., and Cooper, W. C. (1969). *J. Am. Soc. Hortic. Sci.* **94**, 8.
Young, R., and Peynado, A. (1965). *Proc. Am. Soc. Hortic. Sci.* **86**, 244.
Zahner, R. (1955). *For. Sci.* **1**, 193.
Zahner, R. (1962). *For. Sci.* **8**, 345.

6

Sleep and Torpor—Homologous Adaptations for Energy Conservation

H. C. HELLER, G. L. FLORANT, S. F. GLOTZBACH

J. M. WALKER AND R. J. BERGER

I. INTRODUCTION

Three forms of dormancy or hypometabolic conditions are generally recognized in mammals and birds: sleep, shallow torpor, and deep torpor

DORMANCY AND DEVELOPMENTAL ARREST

Copyright © 1978 by Academic Press, Inc.
All rights of reproduction in any form reserved
ISBN 0-12-177050-8

(hibernation). In this chapter we discuss these states in terms of their adaptive significance, their thermoregulatory physiology, and their probable evolutionary relationships. We present the hypothesis that these different forms of dormancy are homologous adaptations that evolved in response to selective pressures favoring energy conservation. Our attention is focused on comparative electrophysiological and thermoregulatory studies.

A. The Three Forms of Dormancy

1. SLEEP

Electrophysiological studies in birds and mammals reveal that sleep is not a unitary phenomenon, but consists of two relatively discrete states, which are categorized as rapid eye movement (REM) and nonrapid eye movement (NREM) sleep. In some animals, notably higher primates, NREM sleep can be subdivided into four stages (1–4). The cortical electroencephalogram (EEG), eye movements as measured by the electrooculogram (EOG), and skeletal muscle electromyogram (EMG) are the variables most commonly used for identifying arousal states in animals. Other parameters include hippocampal EEG, theta activity, and the occurrence of distinctive monophasic spiking potentials in the pons, lateral geniculate and occipital cortex (PGO spikes). Since the EEG in both REM sleep and wakefulness are similar, consisting of a high-frequency (>10 Hz), low-amplitude (<100 μV) pattern, other criteria such as atonia of the skeletal muscles must be used to distinguish REM sleep from wakefulness. Rapid eye movement sleep has also been called paradoxical sleep (PS) because of the conjunction of some physiological signs of activation with others indicating inhibition. The EEG during NREM sleep is of a lower frequency (<5 Hz) and higher amplitude (>100 μV) than that of either REM sleep or wakefulness, and the EMG is usually intermediate in amplitude. Nonrapid eye movement sleep is often referred to as slow-wave sleep (SWS) because of the characteristic slow-wave, high-amplitude EEG pattern. In human beings, SWS is essentially synonymous with stages 3 and 4 (also called delta sleep) of NREM sleep. Detailed discussions of the phenomenology of sleep can be found in many places (e.g., Berger, 1969; Dement, 1958; Rechtschaffen and Kales, 1968).

It is now well established that sleep is also characterized by a decline in body temperature (T_b). Before this conclusion could be drawn, it was necessary to distinguish between circadian influences on T_b independent of sleep and changes in T_b specifically related to sleep (Heller and Glotzbach, 1977). It is clear that there are thermoregulatory adjustments specific to sleep that are independent of and normally superimposed on a circadian rhythm of T_b.

1. SHALLOW TORPOR

Shallow torpor is a phenomenon primarily seen in small mammals and birds. It is manifested by a profound fall in body temperature—as much as 15°–20°C below normal. Since shallow torpor usually occurs on a daily basis during the normal period of inactivity and sleep, it is frequently referred to as daily torpor. It may, however, extend to multiday torpor, in which case the most probable time of arousal is the time that arousal would normally occur from a daily bout plus some multiple of 24 hr, thus revealing an underlying circadian pattern. Bouts of shallow torpor usually occurring seasonally in the hotter, drier portions of the year are frequently called estivation.

1. DEEP TORPOR, OR HIBERNATION

Deep torpor, or hibernation, is a seasonal phenomenon manifested as multiday bouts of torpor and even greater declines in body temperature than are seen in shallow torpor. The body temperature of hibernators may approximate ambient temperatures as low as 0°C. Phylogenetically, hibernation is not a widely occurring phenomenon. It is found in less than one-third of the orders of mammals and only in relatively few species in each of those orders. Hibernators are small mammals, the largest being the marmot, which weighs 3–5 kg. A wealth of information on mammalian torpor has been assembled in various reviews and in the published proceedings of the four international symposia on hibernation (Lyman and Chatfield, 1955; Lyman and Dawe, 1960; Kayser, 1961; Soumalainen, 1964; Fisher et al., 1967; South et al., 1972; Hudson, 1973).

B. Dormancy and Energy Conservation

Two characteristics that these three forms of dormancy have in common are immediately obvious. They all involve a reduction in body temperature, and they all occur during periods when environmental conditions are not conducive to activity. In the case of sleep, birds and mammals are generally specialized physiologically, morphologically, and behaviorally for either the light or the dark portion of the day and are poorly equipped for activity during the other portion. This is also true of those that enter shallow or deep torpor, but in addition harsh environmental conditions such as excessive cold, drought, or lack of food usually prevail at the time that these animals undergo torpor.

Inactivity precludes the search for and ingestion of food. The seriousness of this hiatus in the acquisition of energy depends on the animal's rate of metabolism relative to its energy reserves and therefore can be quite critical

for small mammals. Mammals and birds are endothermic; they use metabolically produced heat to maintain a fairly constant body temperature,
usually in the range of 36°–40°C. Possible adaptive advantages derived
from maintenance of a constant, high T_b have been discussed by Heinrich
(1977) and Hochachka and Somero (1973). Nevertheless, this strategy is
energetically costly. The basal metabolic rate of a mammal is 8–10 times
higher than that of a reptile of similar size and at the same T_b as the mammal. The maximum metabolic rate in an active reptile barely reaches the
basal metabolic rate of a mammal of the same size. As ambient temperature (T_a) falls, the metabolic rate of the mammal increases to offset the
increasing rate of heat loss and its T_b is maintained at approximately the
same level, whereas the T_b and the metabolic rate of the reptile declines.
Energy expenditures of mammals during periods of inactivity become a
more serious problem as body size decreases. Thermal conductance, rate of
weight-specific heat loss, and therefore rate of weight-specific heat production increase as body size decreases. Basal metabolic rate is a power function of body weight (W), with a factor of approximately 2/3. Moreover, the
potential energy reserves that an animal can carry in the form of fat are
directly proportional to W but limited to about 50% of total body weight.
Given these two relations, it is clear that maximal fasting time decreases as
body size decreases ($\propto W^{1/3}$) (Morrison, 1960). These calculations underestimate the actual relationship between body size and fasting time because
they are based on basal metabolic rates. In addition, the smaller the animal
the higher will be the lower limit of its thermoneutral zone. Therefore, as T_a
falls a small mammal incurs a thermoregulatory increase in metabolic rate
before a larger mammal does.

The influences of body size, T_a, and T_b on the maximal period of time
that mammals can go without food are illustrated in Fig. 1. The assump-
tions on which these calculations were based are that each animals begins
the fast with a fat reserve equal to 20% of its lean body weight, and it
engages in no activity during the fast. Furthermore, it is assumed that the
Q_{10} of the animals' metabolic rates is 2.5 and that the animals' thermal
conductances equal $3.57W^{0.49}$ (Bartholomew, in Gordon, 1977). It can be
seen from the calculations that if a 100-g mammal fasts within its thermal
neutral zone it can be expected to survive almost 9 days, but if the animal
experiences a T_a of 0°C during its fast, it will survive for only slightly over 4
days. If, however, the animal allows its T_b to fall 2°C during the fast, a fall
commonly seen during sleep in a small mammal, the fast can be lengthened
to 5 days. If the animal undergoes shallow torpor and allows T_b to drop
10°C, it can fast for 6.8 days. Finally, if the animal hibernates and regulates
T_b at 7°C it can survive a 65-day fast. These calculations are, of course,

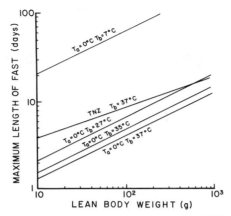

Fig. 1.   Estimations of the maximal time that mammals of different body sizes can survive without food at thermoneutrality (TNZ) and at 0°C with body temperature regulated at different levels. It is assumed that stored fat at the beginning of the fast is equal to 20% of lean body weight.

gross approximations, but nevertheless they point out the adaptive advantage of lowering T_b during periods of food unavailability. This adaptive advantage may be especially crucial for small mammals because of their short potential fast times.

II. REGULATION OF BODY TEMPERATURE IN DORMANT STATES

The fact that body temperature decreases during sleep, shallow torpor, and hibernation does not in itself indicate that these three forms of dormancy are physiologically homologous, nor does it lend strong support to the hypothesis that energy conservation is a common adaptive function of dormant states. The fall in T_b during sleep could be a simple consequence of the decline in muscle metabolism associated with inactivity. The fall in T_b during hibernation and shallow torpor could be the consequence of the inability to maintain a high, normal mammalian T_b in the face of a large T_b–T_a gradient and limited energy reserves. In other words, sleep need not involve a regulated change in T_b, and the deeper forms of dormancy could involve an abandonment or a turning off of temperature regulation. If, on the other hand, it were established that all three forms of dormancy involved a regulated decline in T_b, the hypothesis of homologous adaptations for energy conservation would gain credence.

A. During Sleep

1. CIRCADIAN FLUCTUATIONS IN BODY TEMPERATURE

The earliest hypotheses purporting to explain the fall in T_b during sleep claimed that it was an indirect effect of the circadian sleep/activity rhythm. The higher T_b during the active phase was probably due to muscular activity and/or food consumption and digestion. However, evidence soon accumulated demonstrating the persistence of a daily fluctuation of T_b in people restricted to bed (Johansson, 1898), in paralyzed persons (Kleitman, 1923, 1963; Kreider, 1961; Timbal et al., 1972), and in food-deprived human beings and animals (Chossat, 1843; Jürgensen, 1873; Cumming and Morrison, 1960; Chew et al., 1965). Moreover, in studies on rats (Heusner, 1957, 1959) and hummingbirds (Morrison, 1962), T_b was measured as a function of level of circadian activity. A high correlation between T_b and activity was observed throughout the circadian cycle but, for any given level of activity, a higher T_b was observed when the activity occurred during the active phase than when it occurred during the inactive phase of the circadian cycle. When viewed together, these various studies reveal a circadian cycle of T_b independent of sleep, activity, or food intake and digestion. However, the most convincing evidence for a circadian T_b rhythm independent of the sleep/activity cycle is the observation that rhythms of sleep/activity and T_b of human beings can become desynchronized and free-run with different periodicities in the absence of environmental time cues (Aschoff et al., 1967; Lund, 1974; Wever, 1975).

2. THERMOREGULATORY ADJUSTMENTS ASSOCIATED WITH CHANGES IN AROUSAL STATE

The demonstration of a circadian rhythm of T_b that is independent of the sleep/activity cycle does not exclude the possibility of additional changes in T_b occurring as a function of changes in arousal state. In fact, there is abundant evidence for arousal-state-dependent changes in various T_b's. Many studies did not include measurements of electrophysiological variables together with T_b and thermoregulatory adjustments; so they shed light only on the differences between wakefulness and sleep as a whole. Nevertheless, they are quite interesting. In an early study, Day (1941) measured rectal and skin temperatures and sweat rate during afternoon naps in nine children between the ages of 5 months and 4 years. He consistently observed a decline in rectal temperature (mean of 0.55°C) beginning at the onset of sleep and coincident with increased sweating and a rise in skin temperature. Termination of the decline in rectal temperature was abrupt and was preceded by decreases in sweating and lowered skin temperatures. These results clearly indicated a decline in T_b during the time of day when the overall circadian rhythm of T_b was still in the rising phase.

Thermoregulatory adjustments associated with sleep have also been seen in adult human beings. In neutral or cool environments the onset of sleep is accompanied by a rise in skin temperature and a fall in rectal temperature (Kirk, 1931; Kreider et al., 1958; Kreider and Iampietro, 1959; Hammel et al., 1963). In neutral or warm environments sweating increases at the onset of sleep (Satoh et al., 1965; Geschickter et al., 1966; Takagi, 1970).

Animal studies also reveal the existence of thermoregulatory adjustments relateed to sleep that are independent of and normally superimposed on a separate circadian rhythm of body temperature. Euler and Söderberg (1957) induced episodes of sleep in cats and observed that the onset of EEG synchronization (i.e., SWS) coincided with vasodilation in the ear pinnae, cessation of shivering, and a fall in T_b. Adams (1963) also noted that the hypothalamic temperature (T_{hy}) of cats always fell during behavioral sleep. Abrams and Hammel (1964) continuously measured T_{hy} in unanesthetized albino rats and found that behavioral sleep was always accompanied by a fall in T_{hy}, with the magnitude of the fall being proportional to the duration of the sleep episode.

Interesting experiments demonstrating both circadian and sleep-related thermoregulatory changes in a rhesus monkey were reported by Hammel et al. (1963). The T_{hy} was recorded continuously over many days and showed a diurnal rhythm, the magnitude of which depended on T_a. The fact that T_{hy} fell at night even in the 35°C environment indicates that the nocturnal drop in T_b is an active, regulated decline. In a 7-hr experiment at a neutral T_a, ear pinna temperature and T_{hy} were recorded while activity and periods of eye closure were noted by an observer. It was assumed that eye closure signified episodes of sleep and that changes in ear pinna temperature reflected vasomotor adjustments. Fluctuations in T_{hy} and ear temperature did not correlate well with gross body movements, but they were highly correlated with eye closure. Whenever the animal closed its eyes, ear pinna temperature rose and T_{hy} fell. The reverse happened when the animal opened its eyes.

The studies discussed above did not distinguish between sleep states. However, many workers have described consistent changes in deep brain and other body temperatures as a function of sleep state. Most of these studies did not report the T_a at which experiments were conducted, nor did they monitor thermoregulatory adjustments. Therefore, an interpretation of the various results in terms of thermoregulatory changes is quite difficult. Here, we shall summarize a review and interpretation of this literature (Heller and Glotzbach, 1977). In general, a fall in brain temperature (T_{br})

has been observed in many species during the transition from wakefulness to SWS, and a rise in T_{br} has been observed during the transition from SWS to PS. Several hypotheses have been offered to explain these T_{br} changes, but the most convincing explanation is that changes in peripheral vasomotor activity occur as a function of sleep states (Baker and Hayward, 1967; Hayward and Baker, 1969). Vasomotor changes influence the rate of heat loss from the body and, consequently, T_b. Even if this is the correct explanation of T_{br} changes during sleep, it does not necessarily implicate the thermoregulatory system. Although peripheral vasomotor tone is under the influence of the thermoregulatory system, it is also influenced by various other control mechanisms.

More direct evidence of central nervous system (CNS) thermoregulatory adjustments during sleep is the cessation of shivering in a cool environment and an increase in sweating in a neutral or warm environment with the onset of sleep (Euler and Söderberg, 1958; Satoh et al., 1965; Geschickter et al., 1966; Ogawa et al., 1967; Takagi, 1970; Allison et al., 1972; Parmeggiani and Sabattini, 1972). These observations, together with the above-mentioned changes in peripheral vasomotor tone, indicate a decline in a regulated T_b during SWS in comparison to wakefulness.

Even though an increase in T_{br} is commonly seen during the transition from SWS to PS, thermoregulatory variables do not exhibit coordinated changes indicative of a shift in the regulated T_b during PS. Measurements of thermoregulatory variables during PS at different T_a's suggest an inhibition of thermoregulatory functions. In warm environments panting or sweating ceases during episodes of PS (Satoh et al., 1965; Ogawa et al., 1967; Parmeggiani and Rabini, 1967; Takagi, 1970; Parmeggiani and Sabattini, 1972; Shapiro et al., 1974; Van Twyver and Allison, 1974; Henane et al., 1977), and in cool environments shivering ceases during episodes of PS (Parmeggiani and Rabini, 1967; Affanni et al., 1972; Parmeggiani and Sabattini, 1972; Van Twyver and Allison, 1974) (Fig. 2). The fact that PS is associated with inhibition of heat loss at high T_a's and of heat production of low T_a's suggests that the changes in T_{br} associated with this sleep stage are not regulated changes, but instead reflect a *lack* of thermoregulation.

3. CHANGES IN HYPOTHALAMIC THERMOSENSITIVITY AS A FUNCTION OF AROUSAL STATE

Studies of the relationship between T_{hy} and thermoregulatory responses strongly support the hypothesis that T_b is regulated at a lower level during SWS than during wakefulness, but T_b is not regulated during PS. The T_{hy} is a major control variable in the regulation of T_b in mammals (Hammel, 1968; Hammel et al., 1973). If the T_{hy} of a mammal is gradually lowered in

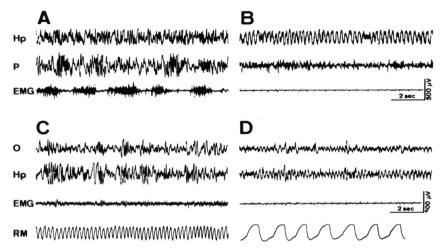

Fig. 2. Recordings of EEG, EMG, and respiratory movements in a cat during sleep. Traces labeled A and B were recorded at a T_a of 6°C during SWS and PS, respectively. Bursts of shivering are evident in the neck muscle EMG during SWS but not during PS. Traces labeled C and D were obtained at a T_a of 36.5°C during SWS and PS, respectively. Panting during SWS is evidenced by the high frequency of respiratory movements (255/min) in C. During the bout of PS shown in D there is no panting as evidenced by the slower, deeper respiratory movements. Abbreviations: HP, hippocampus dorsalis; P, parietal; EMG, neck electromyogram; O, occipital; RM, respiratory movements. (Reproduced from Parmeggiani and Rabini, 1967, with permission.)

a warm or thermoneutral environment, a threshold T_{hy} will be reached, below which the rate of metabolic heat production (MHP) is inversely proportional to T_{hy}. Similarly, if the T_{hy} of a mammal is gradually raised in a thermoneutral or cold environment, a threshold T_{hy} will be reached, above which the activity of heat loss effectors is directly proportional to T_{hy}. The term T_{set} can be used to designate the T_{hy} threshold for a thermoregulatory response, and the symbol α can be used to refer to the proportionality constant relating the rate of a response to T_{hy}. Results from a large number of species have established that the CNS control of a thermoregulatory response can be expressed by the following general equation:

$$R_1 - R_0 = \alpha(T_{hy} - T_{set})$$

where R_0 is the rate of the response when $T_{hy} = T_{set}$, and R_1 is the rate when T_{hy} is above the T_{set} for a heat loss response or below the T_{set} for a heat production response. Other factors such as skin temperature, spinal cord temperature, pyrogens, and even sleep, which all influence the thermoregulatory system, can be expressed as factors that modulate T_{set} and/or

α. Therefore, determination of the characteristics of hypothalamic thermosensitivity during sleep stages should provide an accurate quantitative description of changes in the thermoregulatory system associated with sleep.

Parmeggiani *et al.*, (1973) were the first to explore hypothalamic sensitivity during sleep states, using diathermic heating to elicit thermal polypnea and panting in cats. Results were not reported for wakefulness, but a comparison was made between SWS and PS. Marked thermal polypnea and even panting were elicited by hypothalamic heating during SWS, but the same range of T_{hy}'s had no effect on respiratory rate during PS. These results demonstrated that the central thermoregulatory system functions during SWS but is severely inhibited during PS. The next question to be answered was, How do the quantitative characteristics of hypothalamic thermosensitivity during wakefulness, SWS, and PS compare?

Glotzbach and Heller (1976) continuously recorded EEG, EMG, and metabolic rate in kangaroo rats while manipulating T_{hy}. It was therefore possible to plot metabolic rate as a function of T_{hy} for wakefulness, SWS, and PS. The results demonstrated the proportional regulation of the MHP response during SWS, but the T_{set} and the α for the response were lower than during wakefulness, constituting evidence for a decline in the regulated T_b during SWS. Thermoregulatory responses of MHP could not be elicited by lowering T_{hy} during PS, indicating once again that the thermoregulatory system is severely inhibited during PS (Fig. 3).

Changes in hypothalamic thermosensitivity associated with changes in arousal states have recently been obtained in a hibernator, the marmot, by Florant *et al.*, (1978). The results were very similar to those obtained on kangaroo rats. The plot of metabolic rate versus T_{hy} revealed a lower threshold of the MHP response and a lower slope during SWS than during wakefulness (Fig. 3). Hypothalamic thermosensitivity was absent during PS. In both the kangaroo rat and the marmot the proportionality constant for MHP, α_{MHP}, during SWS was about half the value during wakefulness. The decrease in the threshold T_{hy} for the metabolic response with transition from wakefulness to SWS was greater in the marmot than in the kangaroo rat. In three marmots this decline in threshold averaged 0.9°C, whereas in three kangaroo rats it averaged only 0.3°C. The significance of the fact that the arousal-state-related changes in hypothalamic thermosensitivity are common to both a hibernator and a nonhibernator will become evident in Section II, C.

4. SUMMARY

Changes in various body temperatures and in the activities of thermoregulatory effector mechanisms occur during sleep. The nature of these

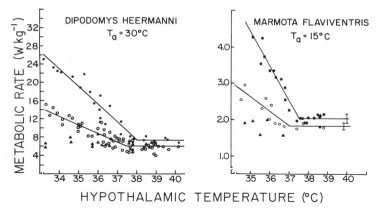

Fig. 3. Characteristics of hypothalamic thermosensitivity as a function of arousal states in a kangaroo rat (a nonhibernator) and in a marmot (a hibernator). Data points obtained during wakefulness are the solid circles, those obtained during SWS are the open circles, and those obtained during PS are the closed triangles. The α_{MHP} is -3.9 W kg^{-1} °C^{-1} for the wakeful kangaroo rat and -1.9 W kg^{-1} °C^{-1} for the kangaroo rat in SWS, values similar to those published by Glotzbach and Heller (1976) for other kangaroo rats. For the marmot the wakeful α_{MHP} is -0.9 W kg^{-1} °C^{-1}, and the SWS α_{MHP} is -0.4W kg^{-1} °C^{-1}.

changes is dependent on sleep state. Changes in thermoregulatory responses that occur during the transition from wakefulness to SWS indicate that T_b is regulated at a lower level during SWS than during wakefulness. During PS active thermoregulatory responses always decrease in comparison to SWS or wakefulness. Comparisons of thermoregulatory responses to manipulations of T_{hy} during different arousal states show that the characteristics of the CNS regulator of T_b are altered during SWS so as to cause a decline in the regulated T_b, and the thermoregulatory system appears to be severely inhibited during PS.

B. During Shallow Torpor

Shallow torpor, or daily torpor, has not been investigated as extensively as have sleep and hibernation, probably because most animals that undergo shallow torpor are small and therefore poor subjects for physiological studies. For example, the pygmy mice (*Baiomys taylori*) studied by Hudson (1965) weighed 6–9 g, the pocket mice studied by Bartholomew and Cade (1957) weighted 6.5–10 g, and the hummingbirds (*Panterpe insignis* and *Eugenes fulgens*) studied by Wolf and Hainsworth (1972) weighed between 4 and 10 g. Nevertheless, recordings of temperatures, metabolic rates, heart rates, and respiratory rates on these and other species allow us to conclude that the decline in body temperature during shallow torpor is a regulated decline and is not due to a failure or an inactivation of the thermoregulatory system.

The first line of evidence is the simple observation that animals in shallow torpor resist a decline in T_b below a certain level. Typically, T_b passively equilibrates with T_a down to a lower limit, but as T_a declines further there is an increase in rate of metabolic heat production that will maintain T_b at or above the lower limit. The pygmy mouse resisted a drop in T_b below 22°C (Hudson, 1965). The hispid pocket mouse (*Perognathus hispidus*) maintained its T_b around 14°C throughout the T_a range of 4°–14°C (Fig. 4) (Wang and Hudson, 1970). Five species of *Peromyscus* maintained T_b's above 13.4°C even at T_a's of 0°C (Morhardt, 1970a). In a study of 52 eastern chipmunks (*Tamias striatus*) during torpor, the average gradient between T_b and T_a increased from 2.8° to 13.4°C over a T_a range of 25.8°–2°C (Wang and Hudson, 1971). Similarly, the body temperatures of four species of hummingbirds were observed over a range of T_a's during nocturnal torpor and, in all four, T_b was regulated at some minimal level even though T_a sometimes fell considerably below that level. The minimal regulated T_b differed for different species, the high altitude species permitting a lower torpid T_b than the low-altitude species (Wolf and Hainsworth, 1972; Carpenter, 1974).

The second line of evidence that T_b is regulated during shallow torpor is actually a corollary of the first. If the torpid animal holds T_b at or above some minimal value as T_a falls below that value, the animal must be decreasing its rate of heat loss or increasing its rate of heat production. Wang and Hudson (1970, 1971) observed intermittent shivering during entrance into torpor in hispid pocket mice and in eastern chipmunks. They

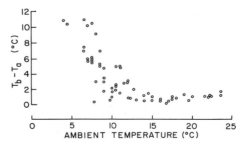

Fig. 4. The difference between the body temperature of torpid pocket mice (*Perognathus hispidus*) and ambient temperature plotted as a function of ambient temperature. Body temperatures were measured with chronically implanted thermocouples in 17 individuals. (Redrawn from Wang and Hudson, 1970, with permission.)

also noted an increase in shivering as T_b approached its minimum level. When T_a was decreased below the level of minimum T_b of torpid subjects, an intensification of shivering occurred. Similar heat production responses to low T_a's have been quantified in three species of hummingbirds during torpor (Hainsworth and Wolf, 1970; Wolf and Hainsworth, 1972). Metabolic rate declines as T_a is decreased down to the minimal tolerated T_b. However, as T_a falls below this level, metabolic rate increases and becomes inversely proportional to T_a. The metabolic rate versus T_a curve of the torpid animal is therefore qualitatively similar to that of the euthermic one except that the lower critical temperature during torpor is the minimal T_b tolerated.

A third line of evidence for shallow torpor being a regulated state comes from the studies of variables other than temperature and metabolism. Heart rate as a function of T_b shows hysteresis during entrance into and arousal from torpor in deer mice (Morhardt, 1970b), in hispid pocket mice (Wang and Hudson, 1970), and in eastern chipmunks (Wang and Hudson, 1971). During entrance heart rate is always lower than in an anesthetized, hypothermic animal at the same temperature, and during arousal the reverse is true. A similar hysteresis was shown for breathing rate during the torpor cycle in deer mice (Morhardt, 1970b). These results indicate dominant parasympathetic control during entrance and dominant sympathetic control during arousal. Morhardt (1970b) confirmed the existence of parasympathetic activity throughout torpor by demonstrating that the parasympathetic blocking agent, atropine (0.06 ml, 10 mg/kg), caused an increase in the rate and regularity of the heart beat when infused through indwelling catheters into the peritoneal cavity of torpid deer mice. Control injections of 0.06 ml of isotonic saline did not affect the heart rate of most torpid mice.

Although there have been no studies to date of the characteristics of the thermoregulatory system during shallow torpor, the available information is consistent with the hypothesis that the regulation of body temperature is qualitatively similar in shallow torpor and euthermia, but with the threshold T_{hy}'s for thermoregulatory responses lowered during the entrance into torpor.

C. During Hibernation

A controversy existed for many years over whether hibernation is an evolutionary primitive or advanced condition. Does hibernation constitute an abandonment of thermoregulation and a reversion to a poikilothermic mode of life, or is it an extension of the range over which most mammals can regulate T_b? Recent work that we shall review here demonstrates that in several well-studied species hibernation definitely constitutes a regulated lowering of T_b continuous throughout a bout of hibernation.

1. MEASUREMENTS OF BODY TEMPERATURE DURING HIBERNATION

Simple measurements of T_b and metabolic rate of hibernating animals subjected to a range of T_a's indicate that T_b is in some way protected against declines to dangerously low levels. As T_a approaches and falls below 0°C, many species elevate their metabolic rate but remain in hibernation and exhibit a larger gradient between T_b and T_a than would be expected if T_b were passively equilibrated with T_a and metabolic rate dependent only on tissue temperature (Wyss, 1932; Lyman, 1948; Chao and Yeh, 1950; Kayser, 1953; Pengelley, 1964; Kristoffersson and Soivio, 1964; Reite and Davis, 1966; Soivio et al., 1968; Wang and Hudson, 1971; Heller and Colliver, 1974; Pivorun, 1976). For example, hedgehogs maintained at a T_a of −5°C held their T_b's at least 5°C higher than T_a for 2–3 days without fully arousing from torpor (Soivio et al., 1968). Reite and Davis (1966) exposed two species of bats (*Myotis lucifugus* and *Lasiurus borealis*) to T_a's ranging from 10° to −5°C. Between T_a's of 10° and 7°C the T_b's of the dormant bats were only slightly above T_a, and the gradient between T_b and T_a remained relatively constant. At T_a's below 4°C, however, the gradient between T_b and T_a progressively increased, together with heart rate. A final example is that of hibernating eastern chipmunks (*Tamias striatus*), which maintained T_b between 10° and 24°C over a range of T_a's from 9.5 to 1.5°C, with the gradient between T_b and T_a increasing as T_a decreased (Wang and Hudson, 1971).

Strumwasser (1959a) showed that at the beginning of the hibernation season California ground squirrels exposed to T_a's of 5.5°–8.0°C entered a series of daily bouts of shallow torpor before undergoing a multiday bout of torpor. During each successive short bout of torpor, the animal's T_b decreased to a lower level than during the previous bout. Strumwasser termed each of these short bouts a "test drop," and the minimal T_b reached a "critical temperature." Although the term "test drop" has come into general usage in hibernation research, the function hypothesized for it has never been demonstrated. It is not clear what is being tested, much less how it is being tested. We do not favor this term as a general one to be applied to early, short, or shallow bouts of hibernation. Nevertheless, the fact that the critical temperatures emcompassed the range of T_b's from levels occurring during euthermia to those of deep hibernation suggested that this species retained thermosensitivity during hibernation, that this thermosensitivity could drive metabolic heat production, and that the threshold for this cold sensitivity changed during the hibernation season.

Another phenomenon also indicating thermoregulatory ability during hibernation and probably related to the changes in "critical temperatures" described by Strumwasser is "braking" of the decline in T_b during entrance into hibernation. Animals entering hibernation frequently do not exhibit as rapid a rate of cooling as would be excepted if their T_b's were passively equilibrating with T_a (Lyman, 1958; Strumwasser, 1959a; Heller et al., 1977). Strumwasser's continuous recordings of T_{br} during entrance revealed definite plateaus and small step increases and drops. Plateaus and rises in T_{br} were accompanied by low skin temperatures, indicating vasoconstriction, and were also sometimes preceded or accompanied by shivering. Declines in T_{br} were preceded by increases in skin temperature, reflecting vasodilation (Strumwasser, 1959a).

2. HYPOTHALAMIC THERMOSENSITIVITY DURING HIBERNATION

The above lines of evidence are only indications of thermoregulation during hibernation. They are all compatible with an alternate hypothesis, which is that the falls in T_b are due to inactivations of the thermoregulatory system but that these drops in temperature can stimulate periodic arousal responses with consequential reactivations of the thermoregulatory system. Such a mechanism would be analogous to an "on–off" temperature regulator and not the normal euthermic proportional thermoregulator. Therefore, the question that needed to be answered was, Are the characteristics of the thermoregulatory system qualitatively the same in hibernation as in euthermia, and, if so, are quantitative changes in the characteristics continuous between euthermia and deep hibernation?

A convenient and probably the most significant measure of the characteristics of the thermoregulatory system of small mammals is hypothalamic thermosensitivity. Experiments by South and colleagues involving manipulation of T_{hy} in euthermic and hibernating marmots revealed hypothalamic thermosensitivity during hibernation. Elevating T_{hy} resulted in decreased heart rate and EMG activity, whereas lowering T_{hy} resulted in increased heart rate and EMG activity (South and Hartner, 1971; Mills and South, 1972; South et al., 1975). Lyman and O'Brien (1972, 1974) also demonstrated a site of thermosensitivity located in the heads of hibernating ground squirrels and hamsters. They produced increases in heart and respiratory rates of hibernating animals by cooling their heads with a spoon-shaped external thermode. If T_a and therefore skin temperature over the rest of the body were cooled while the head was held at constant temperature, increases in heart and respiratory rates were absent. In the hibernating dormouse (*Glis glis*), however, cooling of the hind feet was very effective in evoking these responses. Although these studies of South and colleagues and of Lyman and O'Brien reveal thermosensitivity during hibernation, they do not enable one to make quantitative comparison of the thermoregulatory system during euthermia and hibernation.

In studies of golden-mantled ground squirrels, Heller *et al.* (1974) measured metabolic rate as a function of T_{hy} in both euthermic and hibernating animals. They found that the characteristics of the thermoregulatory system of the hibernating and euthermic animals were qualitatively the same (Fig. 5). During hibernation the hypothalamic threshold for the initiation of a thermogenic response was about 30°–35°C lower, and the proportionality constant for this response was about 16-fold lower than during euthermia. The T_{set} for the thermogenic response could be either above or below actual T_{hy}. When it was above actual T_{hy}, the animal exhibited periodic bursts of elevated metabolism, resulting in rises in T_{hy}. When T_{hy} fell again, another burst would occur. When T_{set} was below actual T_{hy}, metabolic rate remained stable at a minimal level, and T_b would closely approach T_a. These threshold hypothalamic temperatures are probably identical to the phenomenon of "critical temperatures" described by Strumwasser (1959a).

The quantitative comparison of the proportionality constants for the MHP response in euthermia and in hibernation was exactly that predicted

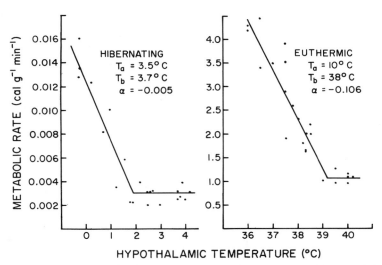

Fig. 5.   Characteristics of hypothalamic thermosensitivity for a golden-mantled ground squirrel (*Citellus lateralis*) during euthermia and hibernation.

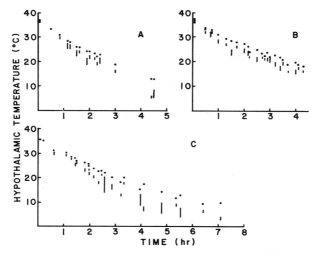

Fig. 6. Demonstration of the continuous decline in the threshold hypothalamic temperature for the metabolic heat production response during entrances into hibernation. Each graph consists of data obtained on one ground squirrel. Data were obtained during two entrances from A, during three entrances for B, and during four entrances for C. Each dot indicates the average T_a over a period of 15–30 min, and the vertical line under that dot connects the lowest T_{hy} that did not elicit an increase in MHP and the highest T_{hy} that did elicit an increase in MHP over that period of time. Therefore, the threshold T_{hy} for the MHP response at the time that each T_b was recorded falls on the vertical line beneath the dot.

from a neuronal model of the thermoregulatory system assumed to be continuous between these two states (Heller and Colliver, 1974). Therefore, an effort was made to demonstrate directly the continuity of the thermoregulatory system during entrance into hibernation (Heller *et al.*, 1977). In these experiments, T_{set} was continuously measured by manipulating T_{hy} as animals entered hibernation. Each series of manipulations was limited to a period of 10–20 min, with the longer series occurring late in the entrance due to slower response times. The T_{hy} was first lowered in steps until an increase in MHP occurred. Then T_{hy} was increased in steps to find the lowest T_{hy} that returned metabolic rate to the minimal level. If T_{set} increased during a series of manipulations, it was assumed that a partial arousal had occurred, and the results were discarded. In this fashion it was possible to demonstrate that a T_{set} for the MHP response exists throughout the entrance into hibernation (Fig. 6). This threshold progressively declines during entrance, and the rate of its decline partially limits the rate at which T_b can decline. If T_{set} falls more rapidly than T_b, the animal's cooling curve appears to be passive, but, if T_b tends to fall more rapidly than T_{set}, the entrance is slowed by episodes of elevated metabolic rate.

These observations on the continuity of the regulator of T_b during hibernation have been confirmed and extended in experiments on marmots (Florant and Heller, 1977). The marmot, like the ground squirrel, showed proportional regulation of T_b both in euthermia and in hibernation as evidenced by metabolic responses to manipulations of T_{hy}. The T_{set} also progressively dropped in the marmot during entrance into hibernation and, depending on the rate of decline of T_b, at times could be above actual T_{hy} and at other times below actual T_{hy}. Continued manipulations of T_{hy} throughout individual bouts of hibernation showed that a T_{set} existed at all times throughout the bout but underwent a consistent pattern of change, continuing to decline over the first 5 days of the bout. A large number of T_{hy} manipulations were not made on most subjects just before arousal but, on one animal for which sufficient data were obtained, there was a clear increase in T_{set} one day before spontaneous arousal occurred.

Changes in T_{set}'s during bouts of hibernation may be responsible for the changes in responsiveness to nonthermal stimuli described by Twente and Twente (1968) and by Beckman and Stanton (1976). In light of data reviewed in the next section, it is reasonable to suppose that T_{set} is partially influenced by the reticular activating system, and peripheral stimuli may influence T_{set} via this pathway. Any influence on T_{set} will not result in a metabolic response unless it is of sufficient magnitude to push that T_{set} above actual T_{hy}. Therefore, identical peripheral stimuli may result in different levels of metabolic responses at different times within a bout because of the constantly changing T_{set}. When T_{set} is just below actual T_{hy}, the animal will appear to be very responsive, and as threshold falls farther below T_{hy} the animal will appear to be less and less responsive to peripheral stimuli.

Studies of the marmot have yielded additional evidence for continuity of central thermoregulatory function during hibernation by quantification of the changes in the proportionality constant, α_{MHP}, during entrance into hibernation (Florant et al., 1978). The slow entrance of the marmot and occasional plateaus with regulation at intermediate T_b's allowed sufficient manipulations of T_{hy} to make possible the calculation of α_{MHP} for many levels of T_{set} between euthermia and deep hibernation. During such plateaus, multiple manipulations of T_{hy} provided data points for plotting usual response curves (Fig. 7). However, when T_{set} was continually changing, response curves could be plotted from only two or three responses. The calculations of α_{MHP} from both of these techniques for all three animals studied were in good agreement and illustrate a continual decline in α_{MHP} as T_{set} declines (Fig. 8). Of great interest is the fact that the regression of a α_{MHP} as a function of T_{set} projects to the SWS measurements for these animals in euthermia rather than to their wakeful measurements. This result is

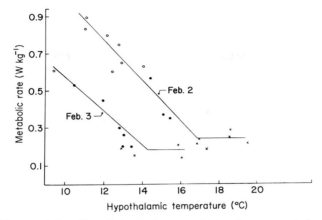

Fig. 7. Characteristics of hypothalamic thermosensitivity of a marmot during two plateaus in an entrance into hibernation ($T_a = 5°C$).

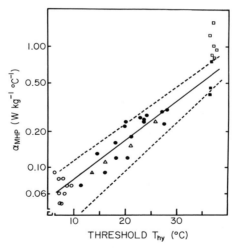

Fig. 8. Hypothalamic thermosensitivity of three marmots plotted as a function of the threshold T_{hy} for the metabolic heat production response. Data points obtained during deep hibernation are open circles, points obtained during plateaus in entrances are open triangles, points obtained by the two-point method during entrance are closed circles, euthermic SWS values are closed squares, and euthermic wakeful values are open squares. The regression line was fitted to the hibernating values and then extrapolated to the euthermic range. The dashed lines delineate the 99% confidence limits of the regression line.

consistent with the hypothesis presented in the next section, that the thermoregulatory adjustments that occur during entrance into hibernation are extensions of those that occur during SWS.

So far we have presented functional evidence that the same central regulatory mechanism is operative at all body temperatures experienced by the hibernator, but we have not provided direct neurophysiological data in support of this conclusion. Do hypothalamic neurons exist that exhibit continuous firing rates over this enormous range of temperatures? Wünnenberg *et al.* (1976) showed that they are present in a hibernator, the golden hamster, but are absent in a nonhibernator, the guinea pig (Fig. 9). In these studies T_{hy} of anesthetized, euthermic animals was manipulated by heating and cooling the heads of an implanted array of silver needles encased in a chamber above the skull of the animal, through which fluid at controlled temperatures was perfused. The authors recorded extracellularly from 6 warm-sensitive neurons in the guinea pig. These neurons had an average firing rate of 15.8 impulses/sec at 36.5°C and an average Q_{10} of 2.3. These

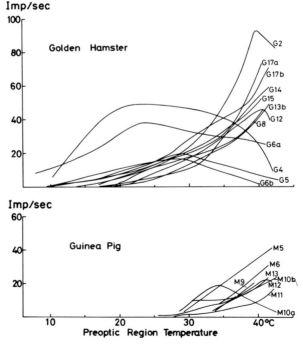

Fig. 9. Firing rate as a function of temperature recorded extracellularly from units in the preoptic region of a hibernator, the golden hamster, and a nonhibernator, the guinea pig. (Reproduced from Wünnenberg *et al.*, 1976, with permission.)

units, therefore, were totally silent when T_{hy} was cooled to 28°–30°C. In contrast, the 13 warm-sensitive units studied in the golden hamster had an average firing rate of 40 impulses/sec at 36.5°C. These units had an average Q_{10} of 3.7 between 40° and 30°C, 1.5 between 30° and 20°C, and 0.6 between 20° and 15°C. They displayed continuous firing rate/temperature curves down to a T_{hy} of 14°C. More recent experiments of Wünnenberg and Speulda (1977) combined skin cooling with manipulations of T_{hy}. A cool skin resembles actual hibernating conditions more closely. They were able to record from 7 warm-sensitive units at skin temperatures of 36° and 20°C. Cooling the skin reduced the thermosensitivity of these units and lowered the T_{hy} at which they reached minimal firing rates by 5°C. It is clear from these experiments that a neuronal basis does exist for a unitary mammalian thermoregulator that can regulate T_b over a range of 35°C.

3. SUMMARY

The studies of mammals reviewed in this section allow us to conclude that a regulated decrease in T_b occurs during SWS, shallow torpor, and deep hibernation. For SWS and deep hibernation this regulated decline in T_b is due to a downward resetting of the central nervous regulatory mechanism controlling effectors of heat loss and heat production. Furthermore, this resetting in the hibernator continuously extends over the entire range of T_b's experienced by the animal. We anticipate results in the near future that will document a similar resetting of the central nervous thermoregulator in animals underoing shallow torpor. The hypothesis that we draw from these conclusions is that these three forms of mammalian dormancy or hypometabolism are homologous. Electrophysiological evidence supporting this hypothesis is presented in the next section.

III. ELECTROPHYSIOLOGICAL CORRELATES OF SLEEP AND HIBERNATION

In spite of the behavioral and thermoregulatory similarities between sleep and hibernation, it is not clear how the neural systems subserving sleep operate in the hibernating animal, if indeed the two states are homologous. Comparisons of electrophysiological phenomena during hibernation and sleep have shed some light on this question. Electrophysiological events during various parts of hibernation have been described. Many of these studies emphasize the coordinated and dynamic changes in the CNS of hibernating animals compared to those produced by artificially induced hypothermia, without making any functional comparisons between hibernation and sleep (for review, see Lyman and Chatfield, 1953; Strumwasser, 1959b;

Strumwasser *et al.*, 1963; Massopust *et al.*, 1965; South *et al.*, 1969; Shtark, 1970; Allison and Van Twyver, 1972; Mihailovic, 1972). Although it has been stated or implied by many authors that hibernation is entered through a sleeping state or represents an extension of sleep, only a few studies have attempted to quantify both the amount and types of sleep patterns persisting through various phases of hibernation. South and his colleagues (1969) implanted marmots (*Marmota flaviventris*) with chronic EEG and EMG needle electrodes. On the basis of cortical and subcortical EEG and EMG/EKG records obtained, they concluded that, in the initial phase of the entry into hibernation down to a T_{br} of 25°C, the distribution of sleep states remained similar to that seen in the euthermic marmot, i.e., 80% SWS:20% PS. Throughout the remainder of the hibernation bout, normative sleep stages could not be identified due to the influence of temperature on the EEG. However, South *et al.* concluded that, during the entrance into hibernation at T_{br}'s of 20°C or below and during deep hibernation, the relative proportions of slow and fast EEG activity from the cortex were similar, indicating that perhaps what could be analogous to sleep during deep hibernation was equally divided between SWS and PS. Parameters used to establish a PS-like state included cardiac irregularity, increases in hypothalamic and midbrain reticular formation background activity, spikes or spindles from the occipital cortex, and bodily movement.

Satinoff (1970) recorded the cortical EEG, EMG, and EOG of ground squirrels (*Citellus lateralis*) entering hibernation. She concluded that hibernation may be entered through a state resembling the onset of SWS and on the basis of data showing few REM's, lack of neck muscle atonia, and absence of low-voltage, fast (LVF) EEG activity with 12–20 Hz spindles, is probably not an extension of PS. The range of T_b's to which these descriptions apply is not clear from her data.

Walker *et al.* (1977) extensively studied the electrophysiological correlates of sleep and wakefulness in ground squirrels (*Citellus lateralis* and *C. beldingi*) during various phases of the hibernation cycle and during the summer. In summer euthermic animals, daily total sleep time (TST) was 66% of which 19% was PS and 81% SWS. Recordings during the hibernating season showed that, although hibernation was entered through sleep, the distribution of sleep states differed from summer sleep. In the initial parts of the entry, between T_{br}'s of 35° and 25°C, TST was 88%, of which 10% was spent in PS (Fig. 10). Although sleep states could not be identified by classical criteria below a T_{br} of 25°C, computer-analyzed EEG revealed that activity in the 0–4 Hz band occupied a majority of the record even though the frequency of the EEG increased as T_{br} decreased. During deep hibernation (T_{br} of 7°–10°C), the EEG was isoelectric except for occasional bursts of spindles.

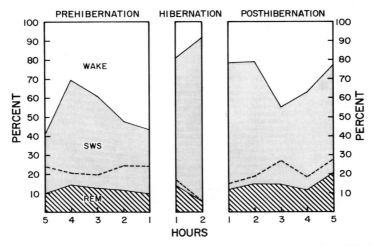

Fig. 10. Mean hourly percentages of wakefulness, SWS, and REM sleep (PS) for five ground squirrels for recording perioods just before entrance into hibernation, during early entrance, and immediately following arousal. Percent sleep time occupied by REM sleep is represented by a dashed line. (From Walker *et al.*, 1977.)

In three hibernating animals, Walker *et al.* were able to slowly raise T_a and allow T_{br} to come into equilibrium with T_a at about 26°C without initiating an arousal from hibernation. The EEG at this T_{br} was characterized by a mixed pattern of spindles (75–150 μV; 3–5 Hz), intermittent slow waves (75–125 μV; <3 Hz), and LVF activity (<25 μV; 5–8 Hz). In this state of "shallow hibernation" the EEG closely resembled that of SWS.

These electrophysiological studies of hibernation have recently been confirmed by similar experiments on a species that undergoes shallow torpor (J. M. Walker, A. Garber, and R. J. Berger, unpublished results). Estivation was induced by food deprivation in four round-tailed ground squirrels (*Citellus tereticaudus*) in the laboratory at a T_a of 25°C during June and July. One animal entered its first bout of torpor after 2 days of food deprivation, one after 3 days, one after 5 days, and one after 14 days. The first bout of torpor lasted from 8–10 hr. Subsequent entrances into torpor followed at 24- or 48-hr intervals. Entrance typically occurred during the first half of the night, and bouts lasted an average of 10 hr, timed from initiation to return to euthermic T_b. Electroencephalogram, EOG, EMG, and brain temperature recordings were obtained from 20 complete bouts of torpor (at least 4 on each subject) and also during euthermic periods between bouts of torpor.

The changes in the distributions of arousal states during entrance into shallow torpor in *C. tereticaudus* are similar to those described for *C. lateralis* and *C. beldingi* during entrance into hibernation. Although EEG amplitude declined as T_b declined, identification of arousal states according to conventional criteria was possible at all times. During entrance into torpor, TST increased to approximately 90%. The percentage of total sleep time spent in PS progressively decreased as T_{br} decreased, and for the entire bout PS comprised only 5–8% of TST. The SWS recordings consisted of slow-wave EEG activity (0–1 cps, > 100 μV). Increased spindling was also present during SWS and was especially prevalent during the transition from wakefulness to sleep. During the deepest portion of the bout, when T_{br} was stable at 26°–27°C, electrophysiological activity alternated between very brief periods of wakefulness and long episodes of SWS. The portion of the bout over which T_{br} was stable at the lowest level consisted of approximately 95% SWS and 5% wakefulness; PS was completely absent. Paradoxical sleep was also absent during the exit from torpor. We prefer not to use the common phrase "arousal from torpor," since approximately 70% of the return to euthermia consisted of SWS and only 30% consisted of wakefulness. This mode of termination of torpor contrasted with that of *C. lateralis* and *C. beldingi*, which typically returned to euthermia from hibernation through extended episodes of wakefulness accompanied by shivering (Walker *et al.*, 1977).

In summary, electrophysiological evidence from animals entering hibernation and animals in shallow torpor supports the hypothesis that these hypometabolic states are homologous with SWS. The decrease in PS during entrance into, and its virtual absence during, torpor are an unexpected and intriguing observation, especially since there does not appear to be any post-torpor PS rebound.

IV. CONCLUSIONS

The thermoregulatory and electrophysiological studies reviewed above point out the similarities between SWS, shallow torpor, and deep hibernation and lead us to the conclusion that they are homologous hypometabolic states. The question posed by this conclusion is, What are the evolutionary origins of these homologous states? Since SWS is ubiquitous in birds and mammals, it probably evolved early in the history of the endotherms and before the more specialized and phylogenetically sparsely distributed phenomena of daily torpor and hibernation. The endothermic mode of life is energetically costly, and perhaps one of the selective pressures leading to the evolution of SWS in early endotherms was the advantage of conserving energy during periods of obligate inactivity, i.e., the night for diurnal animals and the day for nocturnal animals (Berger, 1975). As climates cooled in the late Tertiary and early Quaternary periods (Flohn, 1969), selection would have greatly favored the energy-conserving function of SWS in some species of small mammals occupying harsh or highly seasonal habitats. The resulting extension of the SWS-related decline in the regulated T_b could have generated new selective pressures for additional adaptations ranging from behavioral to molecular levels. The sum total of these adaptations we view today as shallow torpor and hibernation.

In this chapter we have covered only a very small portion of the topics that would fit under the rubric of mechanisms of dormancy. It is hoped, however, that we have pointed out the value of a comparative approach and have shown the potential value of a confluence of the almost totally separate fields of sleep and hibernation research. Since neuroanatomical, neurophysiological, and neuropharmacological investigations of sleep have outstripped hibernation research, it would be quite valuable to cull these areas for more relationships and homologies between sleep and torpor. On the other hand, our understanding of functional brain mechanisms underlying sleep is far from complete or even clear. Perhaps investigations of the more extreme forms of torpor will provide insights into the more ephemeral phenomenon of euthermic sleep.

ACKNOWLEDGMENTS

The data on the kangaroo rat presented in Fig. 3 were obtained by Scott Sakaguchi. We are grateful to Dr. P. L. Parmeggiani, Dr. L. Wang, and Dr. W. Wünnenberg for permission to republish figures taken from their papers. Our own research which is presented in this chapter was supported by NIH Grants NS10367 and GM 26395 to H.C.H. and GM 26394 to R.J.B.

REFERENCES

Abrams, R., and Hammel, H. T. (1964). *Am. J. Physiol.* **206,** 641.
Adams, T. (1963). *Science* **139,** 609.
Affanni, J. M., Lisogorsky, E., and Scaravilli, A. M. (1972). *Experientia* **28,** 1046.
Allison, T., and Van Twyver, H. (1972). *Arch. Ital. Biol.* **110,** 185.
Allison, T., Van Twyver, H., and Goff, W. R. (1972). *Arch. Ital. Biol.* **110,** 145.
Aschoff, J., Gerecke, U., and Wever, R. (1967). *Pfluegers Arch.* **295,** 173.
Baker, M. A., and Hayward, J. N. (1967). *Science* **157,** 1586.
Bartholomew, G. A., and Cade, T. J. (1957). *J. Mammal.* **38,** 60.
Beckman, A. L., and Stanton, T. L. (1976). *Am. J. Physiol.* **231,** 810.

Berger, R. J. (1969). In "Sleep: Physiology and Pathology" (A. Kales, ed.), pp. 66–79. Lippincott, Philadelphia, Pennsylvania.

Berger, R. J. (1975). Fed. Proc., Fed. Am. Soc. Exp. Biol. 34, 97.

Carpenter, F. L. (1974). Science 183, 545.

Chao, I., and Yeh, C. J. (1950). Chin. J. Physiol. 17, 343.

Chew, R. M., Lindberg, R. G., and Hayden, P. (1965). J. Mammal. 46, 477.

Chossat, C. (1843). Ann. Sci. Nat. [2] 20, 293.

Cumming, M. C., and Morrison, S. D. (1960). J. Physiol. (London) 154, 219.

Day, R. (1941). Am. J. Dis. Child. 61, 734.

Dement, W. (1958). Electroencephalogr. Clin. Neurophysiol. 10, 291.

Euler, C. V., and Söderberg, U. (1957). Electroencephalogr. Clin. Neurophysiol. 9, 391.

Euler, C. V., and Söderberg, U. (1958). Acta Physiol. Scand. 42. 112.

Fisher, K. C., Dawe, A. R., Lyman, C. P., Schönbaum, E., and South, F. E., eds. (1967). "Mammalian Hibernation III." Am. Elsevier, New York.

Flohn, H. (1969). Eiszeitalter Ggw. 20, 204.

Florant, G. L., and Heller, H. C. (1977). Am. J. Physiol. 232, R203.

Florant, G. L., Turner, B. M., and H. C. Heller (1978) Am. J. Physiol. In Press.

Geschickter, E. H., Andrews, P. A., and Bullard, R. W. (1966). J. Appl. Physiol. 21, 623.

Glotzbach, S. F., and Heller, H. C. (1976). Science 194, 537.

Gordon, M. (1977). "Animal Physiology: Principles and Adaptation." Macmillan, New York.

Hainsworth, F. R., and Wolf, L. L. (1970). Science 168, 368.

Hammel, H. T. (1968). Annu. Rev. Physiol. 30, 641.

Hammel, H. T., Jackson, D. C., Stolwijk, J. A. J., Hardy, J. D., and Strømme, S. B. (1963). J. Appl. Physiol. 18, 1146.

Hammel, H. T., Heller, H. C., and Sharp, F. R. (1973). Fed. Proc. Fed. Am. Soc. Exp. Biol. 32, 1588.

Hayward, J. N., and Baker, M. A. (1969). Brain Res. 16, 417.

Heinrich, B. (1977). Am. Nat. 111, 623.

Heller, H. C., and Colliver, G. W. (1974). Am. J. Physiol. 227, 583.

Heller, H. C., and Glotzbach, S. F. (1977). Int. Rev. Physiol. 15, 147–188.

Heller, H. C., Colliver, G. W., and Anand, P. (1974). Am. J. Physiol. 227, 576.

Heller, H. C., Colliver, G. W., and Beard, J. (1977). Pfluegers Arch. 369, 55.

Henane, R., Buguet, A., Roussel, B., and Bittel, J. (1977). J. Appl. Physiol. 42, 50.

Heusner, A. (1957). C.R. Seances Soc. Biol. Ses. Fil. 150, 1246.

Heusner, A. (1959). C.R. Seances Soc. Biol. Ses Fil. 153, 1258.

Hochachka, P., and Somero, G. (1973). "Strategies of Biochemical Adaptation." Saunders, Philadelphia, Pennsylvania.

Hudson, J. W. (1965). Physiol. Zool. 38, 243.

Hudson, J. W. (1973). In "Comparative Physiology of Thermoregulation" (G. C. Whittow, ed.), pp. 98–165. Academic Press, New York.

Johansson, D. E. (1898). Skand. Arch. Physiol. 8, 85.

Jürgensen, T. (1873). "Die Körperwärme des gesunden Menschen." Leipzig.

Kayser, C. (1953). Annee Biol. [3] 29, 109.

Kayser, C. (1961). "The Physiology of Natural Hibernation." Pergamon, Oxford.

Kirk, E. (1931). Skand. Arch. Physiol. 61, 71.

Kleitman, N. (1923). Am. J. Physiol. 66, 67.

Kleitman, N. (1963). "Sleep and Wakefulness." Univ. of Chicago Press, Chicago, Illinois.

Kreider, M. B. (1961). Fed. Proc. Fed. Am. Soc. Exp. Biol. 20, 214.

Kreider, M. B., and Iampietro, P. F. (1959). J. Appl. Physiol. 14, 765.

Kreider, M. B., Buskirk, E. R., and Bass, D. E. (1958). J. Appl. Physiol. 12, 361.

Kristoffersson, R., and Soivio, A. (1964). *Ann. Acad. Sci. Fenn. Ser. A4* **82,** 3.
Lund, R. (1974). Ph.D. Thesis, Technische Universität München.
Lyman, C. P. (1948). *J. Exp. Zool.* **109,** 55.
Lyman, C. P., and Chatfield, P. O. (1953). *Science* **117,** 533.
Lyman, C. P., and Chatfield, P. O. (1955). *Physiol. Rev.* **35,** 403–425.
Lyman, C. P., and Dawe, A. R. (1960). *Bull. Mus. Comp. Zool.* **124,** 1.
Lyman, C. P., and O'Brien, R. C. (1972). *Am. J. Physiol.* **222,** 864.
Lyman, C. P., and O'Brien, R. C. (1974). *Am. J. Physiol.* **227,** 218.
Massopust, L. G., Wolin, L. R., and Neder, J. (1965). *J. Exp. Neurol.* **12,** 25.
Mihailovic, L. T. (1972). *In* "Hibernation-Hypothermia: Perspectives and Challenges" (F. E. Smith *et al.,* eds.), pp. 487–534. Elsevier, Amsterdam.
Mills, S. H., and South, F. E. (1972). *Cryobiology* **9,** 393.
Morhardt, J. E. (1970a). *Comp. Biochem. Physiol.* **33,** 423.
Morhardt, J. E. (1970b). *Comp. Biochem. Physiol.* **33,** 441.
Morrison, P. (1960). *Bull. Mus. Comp. Zool.* **124,** 75.
Morrison, P. (1962). *Condor* **64,** 315.
Ogawa, T., Satoh, T., and Takagi, K. (1967). *Jpn. J. Physiol.* **17,** 135.
Parmeggiani, P. L., and Rabini, C. (1967). *Brain Res.* **6,** 789.
Parmeggiani, P. L., and Sabattini, C. (1972). *Electroencephalogr. Clin. Neurophysiol.* **33,** 1.
Parmeggiani, P. L., Franzini, C., Lenzi, P., and Zamboni, G. (1973). *Brain Res.* **52,** 189.
Parmeggiani, P. L., Franzini, C., and Lenzi, P. (1976). *Brain Res.* **111,** 253.
Pengelley, E. T. (1964). *Nature* (*London*) **203,** 892.
Pivorun, E. B. (1976). *Comp. Biochem. Physiol. A* **53,** 265.
Rechtschaffen, A., and Kales, A., eds. (1968). "A Manual of Standardized Terminology, Techniques, and Scoring System for Sleep Stages of Human Subjects." Public Health Serv., U.S. Govt. Printing Office, Washington, D.C.
Reite, O. B., and Davis, W. H. (1966). *Proc. Soc. Exp. Biol. Med.* **121,** 1212.
Satinoff, E. (1970). *Prog. Physiol. Psychol.* **3,** 201–236.
Satoh, T., Ogawa, T., and Takagi, K. (1965). *Jpn. J. Physiol.* **15,** 523.
Shapiro, C. M., Moore, A. T., Mitchell, D., and Yodaiken, M. L. (1974). *Experientia* **30,** 1279.
Shtark, M. B. (1970). "The Brain of Hibernating Animals." Nauka Press, Novosibirsk (Engl. Transl. NASA TT F-619).
Soivio, A., Tähti, H., and Kristoffersson, R. (1968). *Ann. Zool. Fenn.* **5,** 224.
Soumalainen, P. (1964). *Ann. Acad. Sci. Fenn Ser. A4* **71,** 1.
South, F. E., and Hartner, W. C. (1971). *Cryobiology* **8,** 389.
South, F. E., Breazile, J. E., Dellman, H. D., and Epperly, A. D. (1969). *In* "Depressed Metabolism" (X. J. Mussachia and J. F. Saunders, eds.), pp. 277–312. Am. Elsevier, New York.
South, F. E., Hannon, J. P., Willis, J. R., Pengelley, E. T., and Alpert, N. R., eds. (1972). "Hibernation and Hypothermia, Perspectives and Challenges." Elsevier, Amsterdam.
South, F. E., Hartner, W. C., and Luecke, R. H. (1975). *Am. J. Physiol.* **229,** 150.
Strumwasser, F. (1959a). *Am. J. Physiol.* **196,** 15.
Strumwasser, F. (1959b). *Am. J. Physiol.* **196,** 23.
Strumwasser, F., Smith, J., Gilliam, J., and Schlechte, F. R. (1963). *Proc. Int. Congr. Zool., 16th, 1963 Vol.* **2,** p. 53.
Takagi, K. (1970). *In* "Physiological and Behavioral Temperature Regulation" (J. D. Hardy, A. P. Gagge, and J. A. J. Stolwijk, eds.), pp. 669–675. Thomas, Springfield, Illinois.
Timbal, J., Colin, J., Boutelier, C., and Guieu, J. D. (1972). *Pfluegers Arch.* **335,** 97.
Twente, J. W., and Twente, J. A. (1968). *Comp. Biochem. Physiol.* **25,** 467.

Van Twyver, H., and Allison, T. (1974). *Brain, Behav. Evol.* **9**, 107.
Walker, J. M., Glotzbach, S. F., Berger, R. J., and Heller, H. C. (1977). *Am. J. Physiol.* **233** (5): R213–R221.
Wang, L. C. -H., and Hudson, J. W. (1970). *Comp. Biochem. Physiol.* **32**, 275.
Wang, L. C. -H., and Hudson, J. W. (1971). *Comp. Biochem. Physiol. A* **38**, 59.
Wever, R. (1975). *Int. J. Chronobiol.* **3**, 19.
Wolf, L. L., and Hainsworth, F. R. (1972). *Comp. Biochem. Physiol. A* **41**, 167.
Wünnenberg, W., and Speulda, E. (1977) *Proc. Int. Union Physiol. Sci.* **13**, 822.
Wünnenberg, W., Merker, G., and Speulda, E. (1976). *Pfluegers. Arch.* **363**, 119.
Wyss, O. A. M. (1932). *Pfluegers Arch. Gesarnet Physiol. Menschen Tiere* **229**, 559.

7

Dormancy and Development

IAN M. SUSSEX

In the life cycle of most organisms there is a phase in which growth is arrested and the organism is dormant. Beyond this simple statement what generalizations can be made about dormancy as a developmental event?

It should be recognized that there is an almost bewildering array of changes in structure and cellular biochemistry associated with dormancy. In plants, shoot buds which become seasonally dormant may undergo a long period of developmental preparation during which vegetative growth is suppressed and specialized bud scales that cover the dormant bud are differentiated (Foster, 1936). The formation of dormant spores by bacteria, and cysts by amoebae and acellular slime molds involves the differentiation of specialized walls that are chemically different from those of the vegetative cells (Neff and Neff, 1969; Sauer, 1973). And hibernation of arctic mammals is preceded by elaborate physiological and behavioral changes (Irving, 1969). However, for each of these examples there are others that appear to be contradictory in that the dormant state is achieved with a minimum of preparatory alteration of structure and metabolism. Thus it does not seem probable that further study of structural and functional specializations will lead to a better understanding of dormancy as a developmental event.

DORMANCY AND DEVELOPMENTAL ARREST

A different approach to the study of dormancy is to consider it as a specific developmental pathway that can be subdivided into a sequence of stages. This has been done for seed dormancy by Amen (1968) who recognized four stages: induction, maintenance, triggering, and germination, and for the encystment cycle in *Acanthamoeba* by Neff and Neff (1969) who recognized the following stages: induction, wall synthesis, dormition, dormancy, encystment, and vegetative growth. Consideration of schemes such as these suggest that each stage will be subject to its own regulatory controls, and that these will be ultimately gene controlled. Thus dormancy can now be seen not simply as a phase of arrested growth, but as a genetically determined sequence of developmental events. The morphological, physiological, and behavioral specializations that occur are, on this view, manifestations of the underlying genetic events, and are not the causal events leading to dormancy.

There is now good evidence for the genetic control of parts, although not yet for all of the dormancy process in a variety of organisms. This evidence includes the transcription of messenger ribonucleic acid (mRNA) specific for the enzymes that synthesize the cyst wall in *Acanthamoeba* only as the organism enters dormancy (Neff and Neff, 1969), the modification of RNA polymerase permitting it to transcribe different genes in bacteria entering dormancy (Losick and Sonenshein, 1969), the existence of nonsporulating bacterial mutants which can only grow vegetatively (Szulmajster, 1973), and the viviparous mutants of corn in which the embryos germinate in the ear (Robertson, 1955). An interesting experimental demonstration of the nuclear control of dormancy was carried out by Neff and Neff (1969) who cut amoebae of *Acanthamoeba* into halves and found that only the nucleated half formed dormant cysts when subjected to an inductive stimulus. The nonnucleated half failed to respond in any detectable way.

Next to be considered is how the developmental pathway leading to dormancy relates to other developmental pathways of the organism. Is dormancy obligate or optional development for an organism? This question can be addressed by considering the dormancy of seeds in relation to embryogeny and germination. The primitive land plants produced embryos that were not dormant, and mosses and ferns retain this character today (Delevoryas, 1966). The developmental arrest of the seed could have evolved as an intercalated piece of development separating embryogeny from germination, thus being obligate (Fig. 1a). However, studies on plant embryos indicate that their dormancy does not seem to fit this concept. Immature embryos of many species have been removed from the seed and cultured aseptically on nutrient media where they germinate without passing through dormancy (Raghavan, 1976). In addition, the embryo of mangroves germinates naturally while still attached to the maternal plant and without

EMBRYOGENY ⟶ DORMANCY ⟶ GERMINATION

EMBRYOGENY ⟶ GERMINATION

DORMANCY

Fig. 1. Two hypothetical relationships between embryogeny, dormancy, and germination. In Fig. 1a (upper) dormancy intervenes as an obligate developmental stage between the other two. In Fig. 1b (lower) dormancy and the continuation of embryogeny directly into germination are shown as alternative developmental pathways, and dormancy is an optional stage of development.

any diminution of its rate of growth which is exponential at this time (Sussex, 1975). Thus dormancy of seeds is an optional, not an obligate developmental pathway.

If dormancy is optional, a question for developmental biologists is to determine how development is regulated so that only one of two alternative pathways is active and the other is suppressed (Fig. 1b). An efficient way for this to be achieved is to have the signal, or one of its early products, that initiates one developmental pathway also suppress the alternative pathway. In seed development this seems to be the situation. Abscisic acid, a hormone that accumulates to high concentrations during the later stages of seed development (King, 1976; Hsu, 1978) acts to suppress the translation in the embryo of mRNAs that code for germination related enzymes (Dure, 1975), thus suppressing the germination pathway, and it also inhibits RNA synthesis, leading to the metabolic arrest of the dormant embryo (Sussex *et al.*, 1975; Walbot *et al.*, 1975).

If dormancy is a genetically regulated developmental pathway it is of interest to determine what signal initiates activity of the first genes in this pathway, and how this is achieved in terms of cellular biochemistry. In many cases dormancy appears to be initiated in anticipation of a future environmental change. The signal that activates the dormancy pathway cannot, therefore, be the stress itself but an advance signal that is in some way keyed to the stress. This requires a regularity of temporal relationship between the initiating signal and the stress that dormancy is protection against. In bud dormancy in plants, insect diapause, and mammalian hibernation, the signal initiating dormancy has been found to be the seasonal change in day length which occurs predictably year after year and which is closely related to the change in seasons. In plants, where the most critical work has been done on the way in which day length affects development, it has been shown that this is mediated via the pigment phytochrome, the chemical state of which is determined by the duration and quality of light. Changes in the state of phytochrome are followed by changes in gene activity, but how phytochrome brings about these changes is still unknown (Galston and Davies, 1970). Another kind of anticipatory signal that could be keyed to seasonal change, and which could operate as an advance initiating signal for dormancy is the annual endogenous rhythm of the organism. It has been shown that seeds maintained in a constant environment, where they are supposed to be isolated from cues from the external environment, still exhibit a cyclic increase and decrease in percent germination with the cycle length being a year (Baskin and Baskin, 1977).

Some kinds of dormancy are probably not initiated by an advance informational signal from the external environment because in these cases dormancy occurs in environments that are unpredictable. Initiation of dormancy in seeds cannot be a response to the external environment. In annual plants especially, new seeds are formed and become dormant over much of the growing season while the daylength and temperature are changing widely. Similarly, the dormant spores of bacteria and the cysts of amoebae and acellular slime molds are formed in response to a decrease in the availability of water. Here the unpredictability of the environmental stress precludes an anticipatory signal for the initiation of dormancy, and the onset of the stress itself must signal the initiation of the developmental response. The signal that initiates dormancy in all of these cases seems to be a decrease in the concentration of cellular water. In the embryo of bean it has recently been shown that a lowering of the cellular water potential leads to synthesis of abscisic acid and the initiation of dormancy (Hsu and Sussex, 1978). In bacteria, amoebae, and acellular slime molds decreased water availability leads to dormancy, as does raising the osmotic concentration of the medium in which the organisms are growing. Whether it is the decreased water concentration in the cell or the increased concentration of ions that results that initiates gene activity leading to dormancy has not been determined, but a possible line for future investigation is provided by the observation that puffing in the giant chromosomes of *Drosophila* and the initiation of RNA synthesis in the puffs can be induced by increasing the ionic strength of the bathing solution. Thus in this case it appears that specific gene activity can be induced by such alterations (Beermann and Clever, 1964).

As Levins (1969) has pointed out in discussing dormancy as an adaptive strategy, once the environmental stimulus has penetrated the organism its effects spread out along many pathways affecting numerous aspects of structure, function, and behavior. The differences in spread in different organisms accounts for the diversity of organismal changes associated with the dormant state. However, the generalization to be made is that ultimately most of these changes are genetically controlled, and that dormancy then becomes a question of selective gene expression.

REFERENCES

Amen, R. D. (1968). *Botan. Rev.* **34**, 1–31.

Baskin, J. M. and Baskin, C. C. (1977). *Amer. J. Bot.* **64**, 1174–1176.

Beermann, W. and Clever, U. (1964). *Sci. Amer.* **210** (April), 50–58.

Delevoryas, T. (1966). "Plant Diversification." Holt, Rinehart and Winston, New York.

Dure, L. (1975). *Annu. Rev. Plant Physiol.* **26**, 259–287.

Foster, A. S. (1936). *Botan. Rev.* **2**, 349–372.

Galston, A. W. and Davies, P. J. (1970). "Control Mechanisms in Plant Development," Prentice Hall, New Jersey.

Hsu, F. C. (1978). *Plant Physiol.*, in press.

Hsu, F. C. and Sussex, I. M. (1978). *Plant Physiol.*, in press.

Irving, L. (1969). *In* Dormancy and Survival, 23rd Symp. Soc. Exper. Biol., pp. 551–564.

King, R. W. (1976). *Planta* **132**, 43–51.

Levins, R. (1969). *In* Dormancy and Survival, 23rd Symp. Soc. Exper. Biol., pp. 1–10.

Losick, R. and Sonenshein, A. L. (1969). *Nature* **224**, 35–37.

Neff, R. J. and Neff, R. H. (1969). *In* Dormancy and Survival, 23rd Symp. Soc. Exper. Biol., pp. 51–81.

Raghavan, V. (1976). "Experimental Embryogenesis in Vascular Plants." Academic Press, New York.

Robertson, D. (1955). *Genetics* **40**, 745–760.

Sauer, H. W., 1973. *In* Microbial Differentiation, 23rd Symp. Soc. Gen. Microbiol., pp. 373–405.

Sussex, I. M. (1975). *Amer. J. Bot.* **62**, 948–953.

Sussex, I. M., Clutter, M. E., and Walbot, V. (1975). *Plant Physiol.* **56**, 575–578.

Szulmajster, J. (1973). *In* Microbial Differentiation, 23rd Symp. Soc. Gen. Microbiol., pp. 45–83.

Walbot, V., Clutter, M. E. and Sussex, I. M. (1975). *Plant Physiol.* **56**, 570–574.

Index

A

ABA, *see* Abscisic acid
ADP, *see* Adenosine diphosphate
AMP, *see* Adenosine monophosphate
ATP, *see* Adenosine triphosphate
Abscisic acid, *see also* specific compound;
    Cis-abscisic acid; *Trans-trans*-abscisic
    acid
  in *Avena fatua*, 214
  in barley aleurone cells
    regulator molecule, synthesis of, 143
  in cotton cotyledons
    germination, inhibition of, 140–143
  in dormancy
    as inhibitor, 299
  dormancy, in buds
    cytokinin inhibitor, 253, 260
    role in, 244–246, 248–251, 257
  in *Fraxinus*, inhibitor, 146–147
  messenger ribonucleic acid, inhibitor of,
    140–143
  in *Phaseolus vulgaris*, 133–135, 145, 147,
    159
  in plant embryogeny, 132–135, 148,
    161–164
    metabolism, 148–157
    regulator, 117–118, 133–135, 143, 149,
      153–157, 161–164, 200
    water content, relation to, 135, 153–157,
      161–164
  ribonucleic acid, regulator of, 140–143
Actinomycin D
  in *Avena fatua*
    protein synthesis, inhibitor, 209
  in barley aleurone cells
    α-amylase synthesis, effect on, 143–144
  dormancy, in buds, role in, 224

in mammalian diapause, 34–36
in plant embryogeny, 136–138
  inhibitor, 131, 137–140, 143
  ribonucleic acid synthesis, inhibitor of, 140,
    143
[^{3}H]Adenosine
  in cotton cotyledons
    incorporation into ribonucleic acid,
      137–141
[2-,8-^{3}H]Adenosine
  label of adenosine triphosphate, 126
Adenosine diphosphate
  in *Avena fatua*
    in seed metabolism, 208–209, 216
Adenosine monophosphate
  in *Avena fatua*, 216
  messenger ribonucleic acid chain lengths,
    determination of, 138–139, 141
Adenosine triphosphate
  in *Avena fatua*
    in seed metabolism, 208–209, 211, 213,
      215–216
  in *Phaseolus vulgaris*
    levels during development, 124, 126, 131,
      135
    specific activity, 126
Aegilops kotchyi
  developmental arrest in, 160–161
Aestivation, in insects, 51
After-ripening
  in *Avena fatua*, 174, 181, 187–189,
    191–193, 196
    definition, 184
    seed development, role in, 182–186, 194,
      206, 212–213
Aleurone cells
  in *Avena fatua*
    development, role in, 179–182

A
B   8
C   9
D   0
E   1
F   2
G   3
H   4
I   5
J   6